Principles of Genome Analysis and Genomics

D0358045

Sandy B. Primrose

Business and Technology Management
High Wycombe
Buckinghamshire, UK

THIRD EDITION

Blackwell
Publishing

First published 1995
Second edition 1998
Third edition 2003

Library of Congress Cataloging-in-Publication Data

Primrose, S.B.
Principles of genome analysis and genomics/
Sandy B. Primrose, Richard M. Twyman.—3rd ed.
p. cm.
Includes bibliographical references and index.
ISBN 1-40510-120-2 (pbk.: alk. paper)
1. Gene mapping. 2. Nucleotide sequence. 3. Genomics.
I. Twyman, Richard M. II. Title.
QH445.2.P75 2002
572.8′633—dc21   2002070937

A catalogue record for this title is available from the British Library.

Set in $9\frac{1}{2}$/12pt Photina
by Graphicraft Limited, Hong Kong
Printed and bound in Italy
by G. Canale & C.S.P.A., Turin

For further information on
Blackwell Publishing, visit our website:
http://www.blackwellpublishing.com

Contents

Preface

For most of the 20th century, a central problem in genetics was the creation of maps of entire chromosomes. These maps were crucial for the understanding of the structure of genes, their function and their evolution. For a long time these maps were created by genetic means, i.e. as a result of sexual crosses. Starting about 15 years ago, recombinant DNA technology was used to generate molecular or physical maps, defined here as the ordering of distinguishable DNA fragments by their position along the chromosome. Two key features of physical maps are that they can be generated much more quickly than genetic maps, and they usually have a much denser array of markers. The existence of physical maps now is greatly facilitating the analysis of a number of key questions in genetics such as the molecular basis of polygenic disorders and quantitative traits.

Physical mapping embraces a wide range of manipulative and analytical techniques which are detailed in specialist journals using a specialist language. This means that it is difficult for even the experienced biologist entering the field to comprehend the latest developments or even what has been achieved. The first edition of this book was written to provide these new entrants with an overview of the methodologies employed.

The ultimate physical map is a complete genome sequence. Shortly after the first edition of this book was published in 1995, the complete sequence of a bacterial genome was reported for the first time. From a technical point of view this was a particularly noteworthy achievement because the genome sequenced had a size of 1.8 million base pairs, yet the longest individual piece of DNA that can be sequenced is only 600–800 nucleotides. Soon thereafter, the sequences of several mammalian chromosomes were reported as well as the entire genome of the yeast *Saccharomyces cerevisiae*. The key issue was no longer how to sequence a genome but how to handle the sequence data. These changes were reflected in the second edition published in 1998.

As we entered the 21st century, the list of sequenced genomes included over 60 bacteria plus those of yeast, the nematode, the fruit fly, a flowering plant and humans. The size of the human genome (3 billion base pairs) indicates the progress made since the first edition was published. Sequencing of whole genomes is progressing at a rapid rate but the emphasis is now shifting back to biological questions. For example, how do the different components of the genome and the different gene products interact? Answers to questions such as this are being provided using yet another new set of tools in combination with the established mapping and sequencing methodologies. This branch of biology is known as genomics and its importance is such that half of this third edition is devoted to it.

We would like to thank our friends and colleagues who supplied material for this book and took the time to read and make comments on individual chapters, in particular Drs Phillip Gardner, Dylan Sweetman, Ajay Kohli, Gavin Craig, Eva Stoger and Professor Christine Godson. Thanks also to Sue Goddard and her staff in the CAMR library who cheerfully found references and papers at short notice. Richard Twyman would like to thank Kathryn Parry and family (Cain, Evan, Tate and Imogen) for their kindness and hospitality during the summer of 2001 when this project began. He would like to dedicate this book to his parents, Peter and Irene, his children, Emily and Lucy, and to Hannah, Joshua and Dylan.

Abbreviations

2DE	two-dimensional gel electrophoresis
Ac	*Activator*
ADME	adsorption, distribution, metabolism and excretion
AFBAC	affected family-based control
AFLP	amplified fragment length polymorphism
ALL	acute lymphoblastic leukaemia
AML	acute myeloid leukaemia
APL	acute promyelocytic leukaemia
ARS	autonomously replicating sequence
ATRA	all-*trans*-retinoic acid
BAC	bacterial artificial chromosome
BCG	Bacille Calmette–Guérin
bFGF	basic fibroblast growth factor
BIND	Biomolecular Interaction Network Database
BLAST	Basic Local Alignment Search Tool
BLOSUM	Blocks Substitution Matrix
BMP	bone morphogenetic protein
bp	base pair
BRET	bioluminescence resonance energy transfer
CAPS	cleaved amplified polymorphic sequences
CASP	Critical Assessment of Structural Prediction
CATH	Class, Architecture, Topology and Homologous superfamily (database)
CCD	charge couple device
CD	circular dichroism
cDNA	complementary DNA
CEPH	Centre d'Etude du Polymorphisme Humain
CHEF	contour-clamped homogeneous electrical field
CID	collision-induced dissociation
cM	centimorgan
COG	cluster of orthologous groups
cR	centiRay
cRNA	complementary RNA

CSSL	chromosome segment substitution line
ct	chloroplast
DDBJ	DNA Databank of Japan
DIP	Database of Interacting Proteins
DMD	Duchenne muscular dystrophy
DNA	deoxyribonucleic acid
dNTP	deoxynucleoside triphosphate
Ds	*Dissociation*
dsDNA	double-stranded DNA
dsRNA	double-stranded RNA
EGF	epidermal growth factor
ELISA	enzyme-linked immunosorbent sandwich assay
EMBL	European Molecular Biology Laboratory
ENU	ethylnitrosourea
ES	embryonic stem (cells)
ESI	electrospray ionization
EST	expressed sequence tag
EUROFAN	European Functional Analysis Network (consortium)
FACS	fluorescence-activated cell sorting
FEN	flap endonuclease
FIAU	Fialuridine (1-2′-deoxy-2′-fluoro-β-D-arabinofuranosyl-5-iodouracil)
FIGE	field-inversion gel electrophoresis
FISH	fluorescence *in situ* hybridization
FPC	fingerprinted contigs
FRET	fluorescence resonance energy transfer
FSSP	Fold classification based on Structure–Structure alignment of Proteins (database)
GASP	Genome Annotation Assessment Project
G-CSF	granulocyte colony stimulating factor
GeneEMAC	gene external marker-based automatic congruencing
GGTC	German Gene Trap Consortium

GST	gene trap sequence tag	MudPIT	multidimensional protein identification technology
GST	glutathione-*S*-transferase		
HAT	hypoxanthine, aminopterin and thymidine	NGF	nerve growth factor
		NIL	near isogenic line
HDL	high-density lipoprotein	NMR	nuclear magnetic resonance
HERV	human endogenous retrovirus	NOE	nuclear Overhauser effect
HPRT	hypoxanthine phosphoribosyl-transferase	NOESY	NOE spectroscopy
		nt	nucleotide
HTF	*Hpa*II tiny fragment	OFAGE	orthogonal-field-alternation gel electrophoresis
htSNP	haplotype tag single nucleotide polymorphism		
		ORF	open-reading frame
IDA	interaction defective allele	ORFan	orphan open-reading frame
Ihh	Indian hedgehog	P/A	presence/absence polymorphism
IPTG	isopropylthio-β-D-galactopyranoside	PAC	P1-derived artificial chromosome
IST	interaction sequence tag	PAGE	polyacrylaminde gel electrophoresis
IVET	*in vivo* expression technology	PAI	pathogenicity island
kb	kilobase	PAM	percentage of accepted point mutations
LCR	low complexity region		
LD	linkage disequilibrium	PCR	polymerase chain reaction
LINE	long interspersed nuclear element	PDB	Protein Databank (database)
LOD	logarithm$_{10}$ of odds	Pfam	Protein families database of alignments
LTR	long terminal repeat		
m : z	mass : charge ratio	PFGE	pulsed field gel electrophoresis
MAD	multiwavelength anomalous diffraction	PM	'perfect match' oligonucleotide
		poly(A)$^+$	polyadenylated
MAGE	microarray and gene expression	PQL	protein quantity loci
MAGE-ML	microarray and gene expression mark-up language	PRINS	primed *in situ*
		PS	position shift polymorphism
MAGE-OM	microarray and gene expression object model	PSI-BLAST	Position-Specific Iterated BLAST (software)
MALDI	matrix assisted laser desorption ionization	PVDF	polyvinylidine difluoride
		QTL	quantitative trait loci
Mb	megabase	RACE	rapid amplification of cDNA ends
MGED	Microarray Gene Expression Database	RAPD	randomly amplified polymorphic DNA
MIAME	minimum information about a microarray experiment	RARE	RecA-assisted restriction endonuclease
MIP	molecularly imprinted polymer	RC	recombinant congenic (strains)
MIPS	Munich Information Center for Protein Sequences	RCA	rolling circle amplification
		rDNA/RNA	ribosomal DNA/RNA
MM	'mismatch' oligonucleotide	RFLP	restriction fragment length polymorphism
MPSS	massively parallel signature sequencing		
		RIL	recombinant inbred line
mRNA	messenger RNA	R-M	restriction-modification
MS	mass spectrometry	RNA	ribonucleic acid
MS/MS	tandem mass spectroscopy	RNAi	RNA interference
mt	mitochondrial	RNase	ribonuclease
MTM	Maize Targeted Mutagenesis project	RPMLC	reverse phase microcapillary liquid chromatography
Mu	*Mutator*		

RT-PCR	reverse transcriptase polymerase chain reaction	SSR	simple sequence repeat
RTX	repeats in toxins	STC	sequence-tagged connector
SAGE	serial analysis of gene expression	STM	signature-tagged mutagenesis
SCOP	Structural Classification of Proteins (database)	STS	sequence-tagged site
		TAC	transformation-competent artificial chromosome
SDS	sodium dodecylsulphate	TAFE	transversely alternating-field electrophoresis
SELDI	surface-enhanced laser desorption and ionization	TAR	transformation-associated recombination
SGDP	*Saccharomyces* Gene Deletion Project	T-DNA	*Agrobacterium* transfer DNA
Shh	sonic hedgehog	TIGR	The Institute for Genomic Research
SINE	short interspersed nuclear element	TIM	triose phosphate isomerase
SINS	sequenced insertion sites	TOF	time of flight
SNP	single nucleotide polymorphism	tRNA	transfer RNA
SPIN	Surface Properties of protein–protein Interfaces (database)	TUSC	Trait Utility System for Corn
Spm	*Suppressor–mutator*	UPA	universal protein array
SPR	surface plasmon resonance	UTR	untranslated region
SRCD	synchrotron radiation circular dichroism	VDA	variant detector array
		VIGS	virus-induced gene silencing
SSLP	simple sequence length polymorphism	Y2H	yeast two-hybrid
		YAC	yeast artificial chromosome

CHAPTER 1

Setting the scene: the new science of genomics

Introduction

Genetics is the study of the inheritance of traits from one generation to another. As such, it examines the phenotypes of the offspring of sexual crosses. Useful as these data may be, they cannot provide an explanation for the biological basis of a phenotype for that requires biochemical information. In some cases the jump from phenotype to biochemical explanation was relatively simple. Good examples are amino acid auxotrophy and antibiotic resistance in microorganisms and phenylketonuria and sickle cell disease in humans. However, until recently, it was almost impossible to determine the biochemical basis for most of the traits in most organisms.

The first major advance in understanding phenotypes came in the mid-1970s with the development of methods for manipulating genes *in vitro* ('genetic engineering'). This permitted genes to be cloned and sequenced which in turn provided data on the amino acid sequence of the gene product. It also became possible to overexpress the gene product, thereby facilitating its purification and characterization. As the number of characterized gene products has grown, the determination of gene function has become easier, as it is possible to search databases for closely related proteins whose properties are known. Other techniques that have facilitated the analysis of phenotypes are site-directed mutagenesis, where specific base changes or deletions can be made in genes, and gene replacement. All of these techniques, and many others, are described in our companion volume *Principles of Gene Manipulation* (Primrose *et al.* 2001).

Over the past 25 years a vast amount of data has been generated for thousands of different gene products from many different organisms, most of it as a direct result of the ability to manipulate genes. Impressive as this is, gene manipulation on its own cannot meet all the needs of biologists. First, in many instances, a gene needs to be mapped close to a convenient marker before it can be cloned. While this may be easy in an organism such as *Drosophila* where many mutants are available, it is much more difficult in humans or in organisms whose genetics have been poorly studied. Secondly, understanding the phenotype of one or a few genes gives little information about the whole organism and how all its components interact, e.g. its metabolic capabilities or how it controls its development. Thirdly, the analysis of a few genes does not enable us to answer the big questions in biology. For example, how did speech and memory evolve, what changes at the DNA level occurred as the primates evolved, etc.? However, these needs now are being met as a result of efforts to sequence the entire genomes of a number of organisms.

Physical mapping of genomes

In the mid-1980s, scientists began to discuss seriously how the entire human genome might be sequenced. To put these discussions in context, the largest stretch of DNA that can be sequenced in a single pass is 600–800 nucleotides and the largest genome that had been sequenced was the 172 kb Epstein–Barr virus DNA (Baer *et al.* 1984). By comparison, the human genome has a size of 3000 Mb. One school of thought was that completely new sequencing methodology would be required and a number of different technologies were explored but with little success. Early on, it was realized that in order to sequence a large genome it would be necessary to break the genome down into more manageable pieces for sequencing and then join the pieces together again. The problem here was that there were not enough markers on the human genome. It should be noted that humans represent an extreme case of difficulty in creating a genetic map. Not only are directed matings not possible, but the length of the breeding cycle (15–20 years) makes conventional

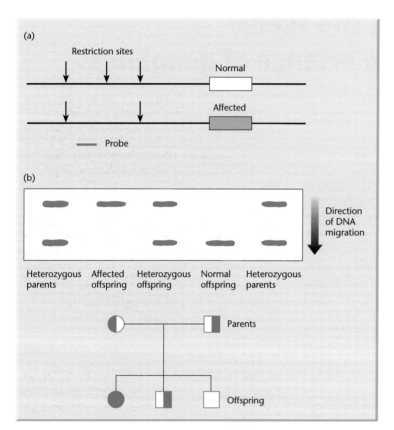

Fig. 1.1 Example of a RFLP and its use for gene mapping. (a) A polymorphic restriction site is present in the DNA close to the gene of interest. In the example shown, the polymorphic site is present in normal individuals but absent in affected individuals. (b) Use of the probe shown in Southern blotting experiments with DNA from parents and progeny for the detection of affected offspring.

analysis impossible. A major breakthrough was the development of methods for using DNA probes to identify polymorphic sequences (Botstein *et al.* 1980). The first such DNA polymorphisms to be detected were differences in the length of DNA fragments after digestion with sequence-specific restriction endonucleases, i.e. restriction fragment length polymorphisms (RFLPs; Fig. 1.1).

To generate an RFLP map the probes must be highly informative. This means that the locus must not only be polymorphic, it must be *very* polymorphic. If enough individuals are studied, any randomly selected probe will eventually discover a polymorphism. However, a polymorphism in which one allele exists in 99.9% of the population and the other in 0.1% is of little utility because it seldom will be informative. Thus, as a general rule, the RFLPs used to construct the genetic map should have two, or perhaps three, alleles with equivalent frequencies.

The first RFLP map of an entire genome (Fig. 1.2) was that described for the human genome by Donis-

Keller *et al.* (1987). They tested 1680 clones from a phage library of human genomic DNA to see whether they detected RFLPs by hybridization to Southern blots of DNA from five unrelated individuals. DNA from each individual was digested with 6–9 restriction enzymes. Over 500 probes were identified that detected variable banding patterns indicative of polymorphism. From this collection, a subset of 180 probes detecting the highest degree of polymorphism was selected for inheritance studies in 21 three-generation human families (Fig. 1.3). Additional probes were generated from chromosome-specific libraries such that ultimately 393 RFLPs were selected. The various loci were arranged into linkage groups representing the 23 human chromosomes by a combination of mathematical linkage analysis and physical location of selected clones. The latter was achieved by hybridizing probes to panels of rodent–human hybrid cells containing varying human chromosomal complements (see p. 38). RFLP maps have not been restricted to the human genome. For example, RFLP maps have

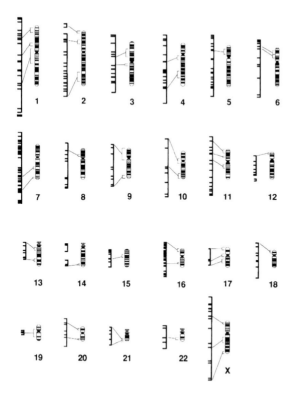

Fig. 1.2 The first RFLP genetic linkage map of the entire human genome. (Reproduced from Donis-Keller *et al.* 1987, with permission from Elsevier Science.)

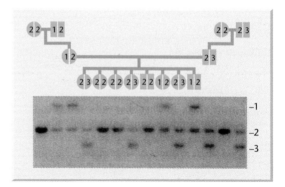

Fig. 1.3 Inheritance of a RFLP in three generations of a family. The RFLP probe used detects a single locus on human chromosome 5. In the family shown, three alleles are detected on Southern blotting after digestion with *TaqI*. For each of the parents it can be inferred which allele was inherited from the grandmother and which from the grandfather. For each child the grandparental origin of the two alleles can then be inferred. (Redrawn from Donis-Keller *et al.* 1987, with permission from Elsevier Science.)

been published for most of the major crops (see for example Moore *et al.* 1995).

The human genome map produced by Donis-Keller *et al.* (1987) was a landmark publication. However, it identified RFLP loci with an average spacing of 10 centimorgans (cM). That is, the loci had a 10% chance of recombining at meiosis. Given that the human genome is 4000 cM in length, the distance between the RFLPs is 10 Mb on average. This is too great to be of use for gene isolation. However, if the methodology of Donis-Keller *et al.* (1987) was used to construct a 1 cM map, then 100 times the effort would be required! This is because 10 times as many probes would be required and 10 times more families studied. The solution has been to use more informative polymorphic markers and other mapping techniques and these are described in detail in Chapter 4. Use of these techniques has led to the generation of a human map with the desired density of markers. More important, these advances in gene mapping were not restricted to the human genome. Rather, the methodology is generic and now has been applied to a wide range of animal and plant genomes.

Sequencing whole genomes

The late 1980s and early 1990s saw much debate about the desirability of sequencing the human genome. This debate often strayed from rationale scientific debate into the realms of politics, personalities and egos. Among the genuine issues raised were questions such as: Is the sequencing of the human genome an intellectually appropriate project for biologists?; Is sequencing the human genome feasible?; What benefits might arise from the project?; Will these benefits justify the cost and are there alternative ways of achieving the same benefits?; Will the project compete with other areas of biology for funding and intellectual resources? Behind the debate was a fear that sequencing the human genome was an end in itself, much like a mountaineer who climbs a new peak just because it is there.

In early 2001 two different groups (International Human Genome Sequencing Consortium 2001; Venter *et al.* 2001) reported the draft sequence of the

Table 1.1 Increases in sizes of genomes sequenced.

Genome sequenced	Year	Genome size	Comment
Bacteriophage φX174	1977	5.38 kb	First genome sequenced
Plasmid pBR322	1979	4.3 kb	First plasmid sequenced
Bacteriophage λ	1982	48.5 kb	
Epstein–Barr virus	1984	172 kb	
Yeast chromosome III	1992	315 kb	First chromosome sequenced
Haemophilus influenzae	1995	1.8 Mb	First genome of cellular organism to be sequenced
Saccharomyces cerevisiae	1996	12 Mb	First eukaryotic genome to be sequenced
Ceanorhabditis elegans	1998	97 Mb	First genome of multicellular organism to be sequenced
Drosophila melanogaster	2000	165 Mb	
Arabidopsis thaliana	2000	125 Mb	First plant genome to be sequenced
Homo sapiens	2001	3000 Mb	First mammalian genome to be sequenced
Rice (*Oryza sativa*)	2002	430 Mb	First crop plant to be sequenced
Pufferfish (*Fugu rubripes*)	2002	400 Mb	Smallest known vertebrate genome
Mouse (*Mus musculis*)	2002/3	2700 Mb	Closest model organism to man

human genome. An analysis of this achievement provides clear answers to the questions raised above. Those opposed to the idea of sequencing the human genome had cited the resources (thousands of scientists and billions of dollars) and time that would be required to accomplish the task. Furthermore, they believed that once the human genome was sequenced there would be a major logistical problem in handling the sequence data. What happened was that the scientific community developed new strategies for sequencing genomes, rather than new methods for sequencing DNA, and complemented these with the development of highly automated methodologies. The net effect was that by the time the human genome had been sequenced, the complete sequence was already known for over 30 bacterial genomes plus that of a yeast (*Saccharomyces cerevisiae*), the fruit fly (*Drosophila melanogaster*), a nematode (*Caenorhabditis elegans*) and a plant (*Arabidopsis thaliana*) (Table 1.1). Furthermore, a whole new science, *bioinformatics*, had been developed to handle and analyse the vast amounts of information being generated by these sequencing projects. Fortuitously, the global development of the Internet occurred at the same time and this enabled scientists around the world to have access to the bioinformatics tools developed in global centres of excellence.

The development of bioinformatics not only facilitated the handling and analysis of sequence data but the development of sequencing strategies as well. For example, when a European consortium set themselves the goal of sequencing the entire genome of the budding yeast *Saccharomyces* (15 Mb), they segmented the task by allocating the sequencing of each chromosome to different groups. That is, they subdivided the genome into more manageable parts. At the time this project was initiated there was no other way of achieving the objective and when the resulting genomic sequence was published (Goffeau *et al.* 1996), it was the result of a unique multicentre collaboration. While the *Saccharomyces* sequencing project was underway, a new genomic sequencing strategy was unveiled: shotgun sequencing. In this approach, large numbers of genomic fragments are sequenced and sophisticated bioinformatics algorithms used to construct the finished sequence. In contrast to the consortium approach used with *Saccharomyces*, a single laboratory set up as a sequencing factory undertook shotgun sequencing.

The first success with shotgun sequencing was the complete sequence of the bacterium *Haemophilus influenzae* (Fleischmann *et al.* 1995) and this was quickly followed with the sequences of *Mycoplasma genitalium* (Fraser *et al.* 1995), *Mycoplasma pneumoniae* (Himmelreich *et al.* 1996) and *Methanococcus jannaschii* (Bult *et al.* 1996). It should be noted that *H. influenzae* was selected for sequencing because so little was known about it: there was no genetic map and not much biochemical data either. By contrast, *S. cerevisiae* was a well-mapped and well-characterized organism. As will be seen in Chapter 5, the relative merits of shotgun sequencing vs. ordered, map-based sequencing still are being debated today. Nevertheless, the fact that a major sequencing laboratory can turn out the entire sequence of a bacterium in 1–2 months shows the power of shotgun sequencing.

Benefits of genome sequencing

Fears that sequencing the human genome would be an end in itself have proved groundless. Because so many different genomes have been sequenced it now is possible to undertake comparative analyses, a topic known as *comparative genomics*. By comparing genomes from distantly related species we can begin to decipher the major stages in evolution. By comparing more closely related species we can begin to uncover more recent events such as genome rearrangement and mutation processes. Currently, the most fertile area of comparative genomics is the analysis of bacterial genomes because so many have been sequenced. Already this analysis is throwing up some interesting questions. For example, over 25% of the genes in any one bacterial genome have no analogues in any other sequenced genome. Is this an artefact resulting from limited sequence data or does it reflect the unique evolutionary events that have shaped the genomes of these organisms? Again, comparative analysis of the genomes of a wide range of thermophiles has revealed numerous interesting features, including strong evidence of extensive horizontal gene transfer. However, what is the genomic basis for thermophily? We still do not know. Comparative genomics also is of value with higher organisms. Selective breeding is not acceptable with humans so we use other mammals as surrogates. Which is the best species to select as a surrogate? Comparative genomics can answer this question.

One of the fascinating aspects of the classic paper of Fleischmann *et al.* (1995) was their analysis of the metabolic capabilities of *H. influenzae* which they deduced from sequence information alone. This analysis (*metabolomics*) has been extended to every other sequenced genome and is providing tremendous insight into the physiology and ecological adaptibility of different organisms. For example, obligate parasitism in bacteria is linked to the absence of genes for certain enzymes involved in central metabolic pathways. Another example is the correlation between genome size and the diversity of ecological niches that can be colonized. The larger the bacterial genome, the greater are the metabolic capabilities of the host organism and this means that the organism can be found in a greater number of habitats.

Analysis of genomes has enabled us to identify most of the genes that are present. However, we still do not know what functions many of these genes perform and how important these genes are to the life of the cell. Nor do we know how the different gene products interact with each other. Because most gene products are proteins, the study of these interactions is known as *proteomics*. It is a discipline which is rooted in experiments rather than computer analysis. For example, the principal way of determining the function of a gene is to delete it and then monitor the fitness of the deletion strain under a variety of selective conditions. This is an enormous task but strategies have been developed for simplifying it. Of course, such a methodology cannot be used with humans, hence the need for comparative genomics.

The biochemistry and genetics of many human diseases have been elucidated but most of these are simple single-gene disorders. Unfortunately, most of the common diseases of humans, and those that put most financial burden on health provision, are polygenic disorders, e.g. hypertension and cancer. Currently we know very little about the *causes* of these diseases and hence we are reduced to treating the *symptoms*. One consequence of this is that many drugs have undesirable side-effects or else only work with certain individuals. Advances in genomics are enabling us to get a better handle on the causes of disease and this will lead to new therapies. Advances in physical mapping of the human genome are enabling us to better predict the effectiveness of a drug and the likely side-effects. Another advance in human genetics is the beginning of an understanding of complex social traits. For example, the first gene controlling speech has just been identified (Lai *et al.* 2001). Polygenic traits are of equal importance in agronomy. Many different characteristics such as weight gain, size, yield, etc. are controlled by many different loci. Again, physical mapping is enabling us to identify the different loci involved and this will facilitate more rational breeding programmes.

Another benefit of genome mapping and sequencing that deserves mention is international scientific collaboration. In magnitude, the goal of sequencing the human genome was equivalent to putting a man on the moon. However, putting a man on the moon was a race between two nations and was driven by global political ambitions as much as by scientific challenge. By contrast, genome sequencing truly has been an international effort requiring laboratories in Europe, North America and Japan to collaborate in a way never seen before. The extent of this collaboration can be seen from an analysis of the affiliations of the authors of the papers on the sequencing of the genomes of *Arabidopsis* (The *Arabidopsis* Genome Initiative 2000) and humans (International Human Genome Sequencing Consortium 2001), for example. The fact that one US company, Celera, has successfully undertaken many sequencing projects in no way diminishes this collaborative effort. Rather, they have constantly challenged the accepted way of doing things and have increased the efficiency with which key tasks have been undertaken.

Three other aspects of genome sequencing and genomics deserve mention. First, in other branches of science such as nuclear physics and space exploration, the concept of 'superfacilities' is well established. With the advent of whole genome sequencing, biology is moving into the superfacility league and a number of sequencing 'factories' have been established. Secondly, high throughput methodologies have become commonplace and this has meant a partnering of biology with automation, instrumentation and data management. Thirdly, many biologists have eschewed chemistry, physics and mathematics but progress in genomics demands that biologists have a much greater understanding of these subjects. For example, methodologies such as mass spectrometry, X-ray crystallography and protein structure modelling are now fundamental to the identification of gene function. The impact that this has on undergraduate recruitment in the sciences remains to be seen.

Outline of the rest of the book

The remainder of the book is divided into three parts. The first part concentrates on the methods developed for mapping and sequencing genomes and on the basics of bioinformatics (Fig. 1.4). The second part deals with genomics; i.e. the analysis of genome

data and the use of map and sequence data to locate genes of interest and to understand phenotypes and fundamental biological phenomena (Fig. 1.5). Thus in the second part of the book, we provide a solution to the problem of understanding the phenotype as outlined earlier in this chapter. Finally, in the last part (Chapter 12) we review some of the applications of the methodologies discussed in the preceding two sections.

The genomes of free-living cellular organisms range in size from less than 1 Mb for some bacteria to millions, or tens of millions, of megabases for some plants. It may even come as a surprise to some to know that a protozoan, never mind a plant, can have a larger genome than that of humans. However, size does not necessarily equal gene content, a phenomenon first elaborated as the C-value paradox (p. 11). Rather, size is often a reflection of genome structure and organization, particularly that of repetitive DNA. This topic is covered in Chapter 2.

The sheer size of the genome of even a simple bacterium is such that to handle it in the laboratory we need to break it down into smaller pieces that are handled as clones. The methods for doing this are covered in Chapter 3. The process of putting the pieces back together again involves mapping. Many different sequence markers are used to do this, as well as some novel mapping methods, and these are described in Chapter 4. DNA sequencing technology is such that only short stretches (~600 bp) can be analysed in a single reaction. Consequently, the genome has to be fragmented and the sequence of each fragment determined and the total sequence reassembled (Chapter 5). Fortunately, the tools and techniques used for mapping also can be applied to genome sequencing. Finally, all the sequence data generated need to be stored and the information that they contain extracted. An introduction to this topic of bioinformatics is provided in Chapter 6.

Sequencing a genome is not an end in itself. Rather, it is just the first stage in a long journey whose goal is a detailed understanding of all the biological functions encoded in that genome and their evolution. To achieve this goal it is necessary to define all the genes in the genome and the functions that they encode. There are a number of different ways of doing this and these are covered in the second part of the book (Chapters 7–11; Fig. 1.5). One such technique is comparative genomics (Chap-

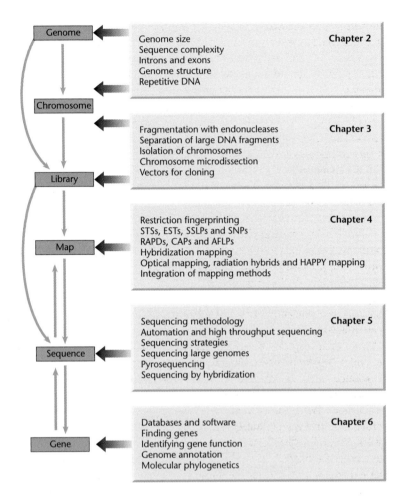

Fig. 1.4 'Road map' outlining the different methodologies used for mapping and sequencing genomes.

ter 7). The premise here is that DNA sequences encoding important cellular functions are likely to be conserved whereas dispensable or non-coding sequences will not. However, comparative genomics only gives a broad overview of the capabilities of different organisms. For a more detailed view one needs to identify each gene in the genome and its function. Such whole genome annotation involves a combination of computer and experimental analysis and is described in Chapters 8–11.

In classical biochemistry one starts by purifying a protein of known function and then determining its structure. In structural genomics, as described in Chapter 8, one does the opposite on the basis that peptide sequences with a similar primary or secondary structure are likely to have similar functions. Chapter 9 is devoted to the burgeoning field of

high throughput expression analysis which is being used with great success to determine the function of anonymous genes. The methodologies used also give qualitative and quantitative information about gene expression as it relates to the biology of the whole organism. For example, it is possible to identify all the genes being transcribed in any cell or tissue at any time and the extent to which these RNAs are translated into proteins. Chapter 10 explores the idea of determining gene function by mutation. Whereas this is carried out on a gene-by-gene basis in classical genetics, in genomics it is performed on a genome-wide scale. Chapter 11 describes the investigation of protein–protein interactions and how these interactions are being mapped and assembled into databases in an attempt to link all proteins in the cell into a functional network.

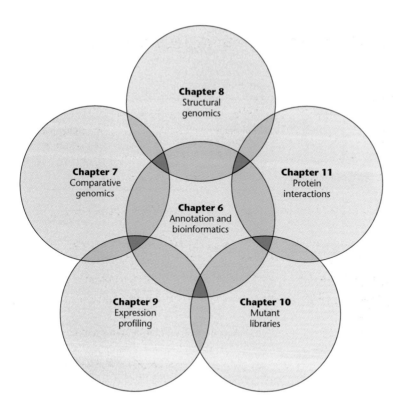

Fig. 1.5 Organization of the second part of the book. Note that Chapter 6, which discusses bioinformatics and the structural annotation of genomes, is the central underpinning theme. This links genome maps and sequences (Chapters 1–5) with comparative genomics (Chapter 7) and functional genomics (Chapters 8–11). Chapters 8–11 consider different aspects of functional genomics but, as shown by the overlapping circles, none is a totally isolated field. This figure conveys how a holistic approach is the best way to mine the genome for functional information.

Terminology

Workers in the field of genomics have coined a whole series of '-omics' terms to describe sub-disciplines of what they do. Confusingly, not everyone uses these terms in the same way. Our use of these terms is defined in Box 1.1.

Keeping up to date

The science of genomics is moving forward at an incredible pace and significant new advances are being reported weekly. This in turn has led to the publication of a plethora of new journals with '-omics' in their titles and which many hard-pressed libraries will be unable to afford. Fortunately, much of this material can now be accessed through the Internet and in the chapters that follow reference is made to relevant websites whenever possible. Any reader not familiar with the PubMed website

(http://www.ncbi.nlm.nih.gov/PubMed/) is strongly advised to spend some time browsing it as it provides very useful access to a wide range of literature. It also has links to the contents pages of many journals.

Suggested reading

Donis-Keller H. *et al.* (1987) A genetic linkage map of the human genome. *Cell* **51**, 319–337. *This is a classic paper and describes the first comprehensive human genetic map to be constructed using DNA-based markers.*

Fleischmann R.D. *et al.* (1995) Whole-genome random sequencing and assembly of *Haemophilus influenzae* Rd. *Science* **269**, 496–512. *This is another classic paper and the wealth of information about the biology of the bacterium that was inferred from the sequence data provided the justification, if one was needed, for whole genome sequencing.*

Primrose S.B., Twyman R.M. & Old R.W. (2001) *Principles of Gene Manipulation* (6th edn.) Blackwell Science, Oxford. *This textbook is widely used around the world and provides a detailed introduction to the many different techniques of gene manipulation that form the basis of the methods for genome analysis.*

Box 1.1 Genomics definitions used throughout this book

Term	Definition
Genomics	The study of the structure and function of the genome
Functional genomics	The high throughput determination of the function of a gene product. Included within this definition is the expression of the gene, the relationship of the sequence and structure of the gene product to other gene products in the same or other organisms, and the molecular interactions of the gene product
Structural genomics	The high throughput determination of structural motifs and complete protein structures and the relationship between these and function
Comparative genomics	The use of sequence similarity and comparative gene order (synteny) to determine gene function and phylogeny
Proteomics	The study of the proteome, i.e. the full complement of proteins made by a cell. The term includes protein–protein and protein–small molecule interactions as well as expression profiling
Transcriptomics	The study of the transcriptome, i.e. all the RNA molecules made by a cell, tissue or organism
Metabolomics	The use of genome sequence analysis to determine the capability of a cell, tissue or organism to synthesize small molecules
Bioinformatics	The branch of biology that deals with *in silico* processing and analysis of DNA, RNA and protein sequence data
Annotation	The derivation of structural or functional information from unprocessed genomic DNA sequence data

Useful websites

http://www3.ncbi.nlm.nih.gov/
This is the website of the National Center for Biotechnology Information. It contains links to many other useful websites. The OMIM pages on this site contain a wealth of information on Mendelian inheritance in humans. This site also is the entry point to PubMed which enables researchers to access abstracts and journal articles on-line.

http://www.sciencemag.org/feature/plus/sfg/resources/
This is the website on functional genomics resources hosted by *Science* magazine. It contains many useful pages and the ones on model organisms are well worth visiting. There also are features pages which are revised regularly.

CHAPTER 2

The organization and structure of genomes

Introduction

There is no such thing as a common genome structure. Rather, there are major differences between the genomes of bacteria, viruses and organelles on the one hand and the nuclear genomes of eukaryotes on the other. Within the eukaryotes there are major differences in the types of sequences found, the amounts of DNA and the number of chromosomes. This wide variability means that the mapping and sequencing strategies involved depend on the individual genome being studied.

Genome size

Because the different cells within a single organism can be of different ploidy, e.g. germ cells are usually haploid and somatic cells diploid, genome sizes always relate to the haploid genome. The size of the haploid genome also is known as the C-value. Measured C-values range from 3.5×10^3 bp for the smallest viruses, e.g. coliphage MS2, to 10^{11} bp for some amphibians and plants (Fig. 2.1). The largest viral genomes are $1-2 \times 10^5$ bp and are just a little smaller than the smallest cellular genomes, those of some mycoplasmas (5×10^5 bp). Simple unicellular eukaryotes have a genome size ($1-2 \times 10^7$ bp) that is not much larger than that of the largest bacterial genomes. Primitive multicellular organisms such as nematodes have a genome size about four times larger. Not surprisingly, an examination of the genome sizes of a wide range of organisms has shown that the *minimum* C-value found in a particular phylum is related to the structural and organizational complexity of the members of that phylum. Thus the minimum genome size is greater in organisms that evolutionarily are more complex (Fig. 2.2).

A particularly interesting aspect of the data shown in Fig. 2.1 is the range of genome sizes found within

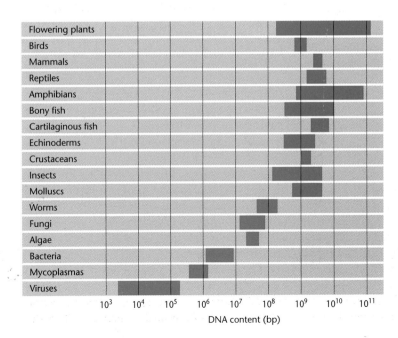

Fig. 2.1 The DNA content of the haploid genome of a range of phyla. The range of values within a phylum is indicated by the shaded area. (Redrawn from Lewin 1994 by permission of Oxford University Press.)

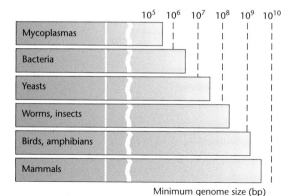

Fig. 2.2 The minimum genome size found in a range of organisms. (Redrawn from Lewin 1994 by permission of Oxford University Press.)

each phylum. Within some phyla, e.g. mammals, there is only a twofold difference between the largest and smallest C-value. Within others, e.g. insects and plants, there is a 10- to 100-fold variation in size. Is there really a 100-fold variation in the number of genes needed to specify different flowering plants? Are some plants really more organizationally complex than humans, as these data imply? Although there is evidence that birds with smaller genomes are better flyers (Hughes & Hughes 1995) and that plants are more responsive to elevated carbon dioxide concentrations (Jasienski & Bazzaz 1995) as their genomes increase in size, this is not sufficient to explain the size differential. The resolution of this apparent C-value paradox was provided by the analysis of sequence complexity by means of reassociation kinetics.

Sequence complexity

When double-stranded DNA in solution is heated, it denatures ('melts') releasing the complementary single strands. If the solution is cooled quickly the DNA remains in a single-stranded state. However, if the solution is cooled slowly reassociation will occur. The conditions for efficient reassociation of DNA were determined originally by Marmur *et al.* (1963) and since then have been extensively studied by others (for a review, see Tijssen 1993). The key parameters are as follows. First, there must be an adequate concentration of cations and below 0.01 M

sodium ion there is effectively no reassociation. Secondly, the temperature of incubation must be high enough to weaken intrastrand secondary structure. In practice, the optimum temperature for reassociation is 25°C below the melting temperature (T_m), that is, the temperature required to dissociate 50% of the duplex. Thirdly, the incubation time and the DNA concentration must be sufficient to permit an adequate number of collisions so that the DNA can reassociate. Finally, the size of the DNA fragments also affects the rate of reassociation and is conveniently controlled if the DNA is sheared to small fragments.

The reassociation of a pair of complementary sequences results from their collision and therefore the rate depends on their concentration. As two strands are involved the process follows second-order kinetics. Thus, if C is the concentration of DNA that is single stranded at time t, then

$$\frac{dC}{dt} = -kC^2$$

where k is the reassociation rate constant. If C_0 is the initial concentration of single-stranded DNA at time $t = 0$, integrating the above equation gives

$$\frac{C}{C_0} = \frac{1}{1 + k \cdot C_0 t}.$$

When the reassociation is half complete, $C/C_0 = 0.5$ and the above equation simplifies to

$$C_0 t_{1/2} = \frac{1}{k}.$$

Thus the greater the $C_0 t_{1/2}$ value, the slower the reaction time at a given DNA concentration. More important, for a given DNA concentration the half-period for reassociation is proportional to the number of different types of fragments (sequences) present and thus to the genome size (Britten & Kohne 1968). This can best be seen from the data in Table 2.1. Because the rate of reassociation depends on the concentration of complementary sequences, the $C_0 t_{1/2}$ for organism B will be 200 times greater than for organism A.

Experimentally it has been shown that the rate of reassociation is indeed dependent on genome size (Fig. 2.3). However, this proportionality is only true in the absence of repeated sequences. When the

Table 2.1 Comparison of sequence copy number for two organisms with different genome sizes.

	Organism A	**Organism B**
Starting DNA concentration (C_0)	10 pg ml^{-1}	10 pg ml^{-1}
Genome size	0.01 pg	2 pg
No. of copies of genome per ml	1000	5
Relative concentration (A vs. B)	200	1

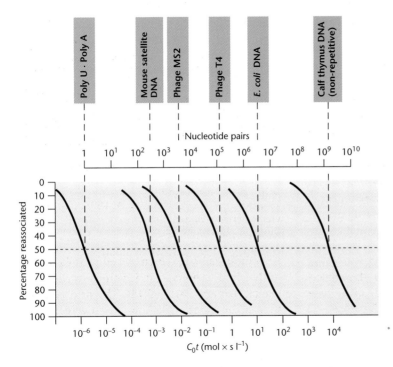

Fig. 2.3 Reassociation of double-stranded nucleic acids from various sources. (Redrawn from Lewin 1994 by permission of Oxford University Press.)

reassociation of calf thymus DNA was first studied, kinetic analysis indicated the presence of two components (Fig. 2.4). About 40% of the DNA had a $C_0t_{1/2}$ of 0.03, whereas the remaining 60% had a $C_0t_{1/2}$ of 3000. Thus the concentration of DNA sequences that reassociate rapidly is 100 000 times, the concentration of those sequences that reassociate slowly. If the slow fraction is made up of unique sequences, each of which occurs only once in the calf genome, then the sequences of the rapid fraction must be repeated 100 000 times, on average. Thus the $C_0t_{1/2}$ value can be used to determine the sequence complexity of a DNA preparation. A comparative analysis of DNA from different sources has shown that repetitive DNA occurs widely in eukaryotes (Davidson & Britten 1973) and that different

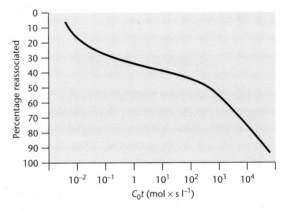

Fig. 2.4 The kinetics of reassociation of calf thymus DNA. Compare the shape of the curve with those shown in Fig. 2.3.

types of repeat are present. In the example shown in Fig. 2.5 a fast-renaturing and an intermediate-renaturing component can be recognized and are present in different copy numbers (500 000 and 350, respectively) relative to the slow component which is unique or non-repetitive DNA. The complexities of each of these components are 340 bp, 6×10^5 bp and 3×10^8 bp, respectively. The proportion of the genome that is occupied by non-repetitive DNA versus repetitive DNA varies in different organisms (Fig. 2.6), thus resolving the C-value paradox. In general, the length of the non-repetitive DNA component tends to increase as we go up the evolutionary tree to a maximum of 2×10^9 bp in mammals. The fact that many plants and animals have a much higher C-value is a reflection of the presence of large amounts of repetitive DNA. Analysis of messenger RNA (mRNA) hybridization to DNA shows that most of it anneals to non-repetitive DNA, i.e. most genes are present in non-repetitive DNA. Thus genetic complexity is proportional to the content of non-repetitive DNA and not to genome size.

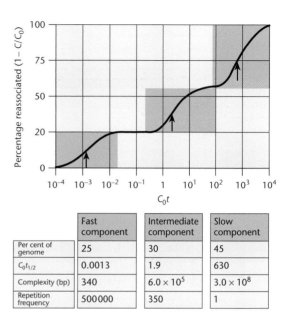

	Fast component	Intermediate component	Slow component
Per cent of genome	25	30	45
$C_0 t_{1/2}$	0.0013	1.9	630
Complexity (bp)	340	6.0×10^5	3.0×10^8
Repetition frequency	500 000	350	1

Fig. 2.5 The reassociation kinetics of a eukaryotic DNA sample showing the presence of two types of repeated DNA. The arrows indicate the $C_0 t_{1/2}$ values for the three components. (Redrawn from Lewin 1994 by permission of Oxford University Press.)

Introns and exons

Introns were initially discovered in the chicken ovalbumin and rabbit and mouse β-globin genes (Breatnach *et al.* 1977; Jeffreys & Flavell 1977).

Both these genes had been cloned by isolating the mRNA from expressing cells and converting it to complementary DNA (cDNA). The next step was to use the cloned cDNA to investigate possible

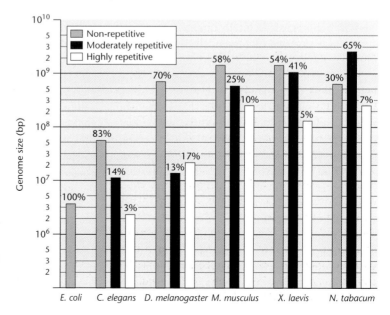

Fig. 2.6 The proportions of different sequence components in representative eukaryotic genomes. (Redrawn from Lewin 1994 by permission of Oxford University Press.)

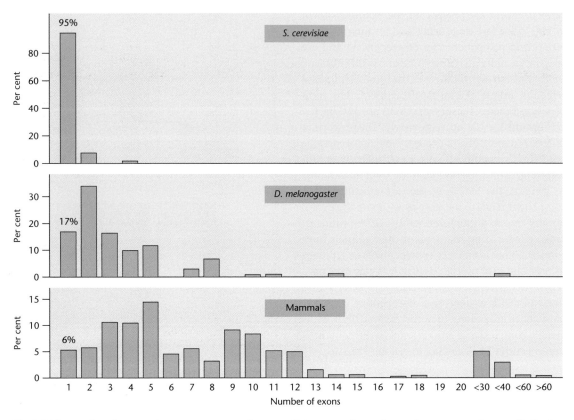

Fig. 2.7 The number of exons in three representative eukaryotes. Uninterrupted genes have only one exon and are totalled in the left-hand column. (Redrawn from Lewin 1994 by permission of Oxford University Press.)

differences in the structure of the gene from expressing and non-expressing cells. Here the Southern blot hybridizations revealed a totally unanticipated situation. It was expected that the analysis of genomic restriction fragments generated by enzymes that did not cut the cDNA would reveal only a single band corresponding to the entire gene. Instead several bands were detected in the hybridized blots. The data could be explained only by assuming the existence of interruptions in the middle of the protein-coding sequences. Furthermore, these insertions appeared to be present in both expressing and non-expressing cells. The gene insertions that are not translated into protein were termed *introns* and the sequences that are translated were called *exons*.

Since the original discovery of introns, a large number of split genes has been identified in a wide variety of organisms. These introns are not restricted to protein-coding genes for they have been found in rRNA and tRNA genes as well. Split genes are rare in

prokaryotes (Edgell *et al.* 2000; Martinez-Abarca & Toro 2000). They also are not particularly common in lower eukaryotes (see below) but the mitochondrial DNA of *Euglena* is an exception with 38% of the genome consisting of intron DNA (Hallick *et al.* 1993).

In *Saccharomyces cerevisiae*, sequencing of the complete genome suggests that there are 235 introns compared with over 6000 open-reading frames and that introns account for less than 1% of the genome (Goffeau *et al.* 1996). Those genes which do have introns usually have only one small one and the longest intron is only 1 kb in size.

However, proceeding up the evolutionary tree, the number of split genes, and the number and size of introns per gene, increases (Fig. 2.7 and Table 2.2). More important, genes that are related by evolution have exons of similar size, i.e. the introns are in the same position. However, the introns may vary in length, giving rise to variation in the length of

Table 2.2 Intron statistics for genes from different species.

Species	Average exon number	Average intron number	Average length (kb)	Average mRNA length (kb)	% Exon per gene
Yeast	1	0	1.6	1.6	100
Nematode	4	3	4.0	3.0	75
Fruit fly	4	3	11.3	2.7	24
Chicken	9	8	13.9	2.4	17
Mammals	7	6	16.6	2.2	13

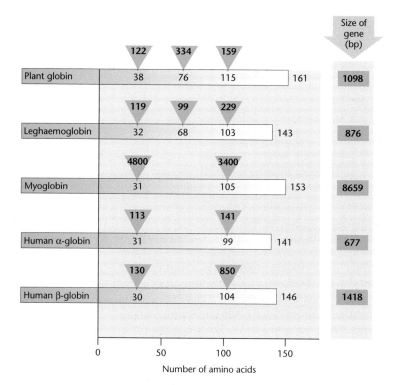

Fig. 2.8 The placement of introns in different members of the globin superfamily. The size of the introns in base pairs is indicated inside the inverted triangles. Note that the size of each polypeptide and the location of the different introns are relatively consistent.

the genes (Fig. 2.8). Note also that introns are much longer than exons, particularly in higher eukaryotes.

If a split gene has been cloned, it is possible to sub-clone either the exon or the intron sequences. If these sub-clones are used as probes in genomic Southern blots, it is possible to determine if these same sequences are present elsewhere in the genome. Often, the exon sequences of one gene are found to be related to sequences in one or more other genes. Some examples of such *gene families* are given in Table 2.3. In some instances the duplicated genes

are clustered, whereas in others they are dispersed. Also, the members may have related, or even identical, functions, although they may be expressed at different times or in different cell types. Thus different globin proteins are found in embryonic and adult red blood cells, while different actins are found in muscle and non-muscle cells.

Functional divergence between members of a multigene family may extend to the loss of gene function by some members. Such *pseudogenes* come in two types. In the first type they retain the usual intron and exon structure but are functionless or

Table 2.3 Some examples of multigene families.

Gene family	Organism	Approximate no. of genes	Clustered (L) or dispersed (D)
Actin	Yeast	1	–
	Slime mould	17	L, D
	Drosophila	6	D
	Chicken	8–10	D
	Human	20–30	D
Tubulin	Yeast	3	D
	Trypanosome	30	L
	Sea urchin	15	L, D
	Mammals	25	D
α-Amylase	Mouse	3	L
	Rat	9	?
	Barley	7	?
β-Globin	Human	6	L
	Lemur	4	L
	Mouse	7	L
	Chicken	4	L

they lack one or more exons. In the second type, found in dispersed gene families, processed pseudogenes are found which lack any sequences corresponding to the introns or promoters of the functional gene members. Multiple copies of an exon also may be found because the same exons occur in several apparently unrelated genes. Exons that are shared by several genes are likely to encode polypeptide regions that endow the disparate proteins with related properties, e.g. adenosine triphosphate (ATP) or DNA binding. Some genes appear to be mosaics that were constructed by patching together copies of individual exons recruited from different genes, a phenomenon known as *exon shuffling* (see pp. 31 and 114).

By contrast with exons, introns are not related to other sequences in the genome, although they contain the majority of dispersed, highly repetitive sequences. Thus, for some genes the exons constitute slightly repetitive sequences embedded in a unique context of introns. It should be noted that introns are not necessarily junk because there now is evidence that some of them encode functional RNA (Moore 1996).

Two intron databases have been constructed (Schisler & Palmer 2000). The Intron DataBase (IDB) contains detailed information about introns

and the other, the Intron Evolution DataBase, provides a statistical analysis of the intron and exon sequences catalogued in the IDB.

Genome structure in viruses and prokaryotes

The genomes of viruses and prokaryotes are very simple structures, although those of viruses show remarkable diversity (for a review see Dimmock *et al.* 2001). Most viruses have a single linear or circular genome but a few, such as reoviruses, bacteriophage ϕ6 and some plant viruses, have segmented RNA genomes. For a long time it was believed that all eubacterial genomes consisted of a single circular chromosome. However, linear chromosomes have been found in *Borrelia* sp., *Streptomyces* sp. and *Rhodococcus fascians* and mapping suggests that *Coxiella burnetii* also has a linear genome. Two chromosomes have been found in a number of bacteria including *Rhodobacter spheroides*, *Brucella melitensis*, *Leptospira interrogans* and *Agrobacterium tumefaciens* (Cole & Saint Girons 1994). In the case of *Agrobacterium*, there is one circular chromosome and one non-homologous linear chromosome (Goodner *et al.* 1999). Linear plasmids have been found in *Borrelia*

Box 2.1 The need for telomeres

The ends of eukaryotic chromosomes are also the ends of linear duplex DNA and are known as *telomeres*. That these must have a special structure has been known for a long time. For example, if breaks in DNA duplexes are not rapidly repaired by ligation they undergo recombination or exonuclease digestion, yet, the ends of chromosomes are stable and chromosomes are not ligated together. Also, DNA replication is initiated in a 5′→3′ direction with the aid of an RNA primer. After removal of this primer there is no way of completing the 5′ end of the molecule (Fig. B2.1). Thus, in the absence of a method for completing the ends of the molecules, chromosomes would become shorter after each cell division.

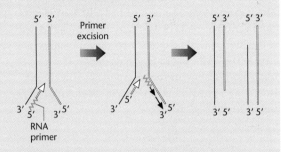

Fig. B2.1 Formation of two daughter molecules with complementary single-stranded 3′ tails after primer excision.

sp. and *Streptomyces* sp. as well as a number of bacteria with circular chromosomes (Hinnebush & Tilley 1993). *Borrelia* has a very complex plasmid content with 12 linear molecules and nine circular molecules (Casjens *et al.* 2000).

Bacterial genomes lack the centromeres found in eukaryotic chromosomes although there may be a partitioning system based on membrane adherence. Duplication of the genomes is initiated at an origin of replication and may proceed unidirectionally. The structure of the origin of replication, the *oriC* locus, has been extensively studied in a range of bacteria and found to consist essentially of the same group of genes in a nearly identical order (Cole & Saint Girons 1994). The *oriC* locus is defined as a region harbouring the *dnaA* (DNA initiation) or *gyrB* (B subunit of DNA gyrase) genes linked to a ribosomal RNA operon.

Many bacterial and viral genomes are circular or can adopt a circular conformation for the purposes of replication. However, those viral and bacterial genomes which retain a linear configuration need a special mechanism to replicate the ends of the chromosome (see Box 2.1). A number of different strategies for replicating the ends of linear molecules have been adopted by viruses (see Dimmock *et al.* 2001) but in bacteria there are two basic mechanisms (Volff & Altenbuchner 2000). In *Borrelia*, the chromosomes have covalently closed hairpin structures

at their termini. Such structures are also found in *Borrelia* plasmids, *Escherichia coli* phage N15, poxviruses and linear mitochondrial DNA molecules in the yeasts *Williopsis* and *Pichia*. Exactly how these hairpin structures facilitate replication of the ends of the molecule is not known. By contrast, in *Streptomyces*, the linear molecules have proteins bound to the 5′ ends of the DNA and such proteins are also found in adenoviruses, and a number of bacteriophages and fungal and plant mitochondrial plasmids. These terminal proteins probably are involved in the completion of replication. In addition, *Streptomyces* linear replicons have palindromic sequences and inverted repeats at their termini.

The bacterial genomes that have been completely sequenced have sizes ranging from 0.6 to 7.6 Mb. The difference in size between the smallest and the largest is not a result of introns for these are rare in prokaryotes (Edgell *et al.* 2000; Martinez-Abarca & Toro 2000). Nor is it a result of repeated DNA. Analysis of the kinetics of reassociation of denatured bacterial DNA did not indicate the presence of repeated DNA in *E. coli* (Britten & Kohne 1968) and only small amounts have been detected in all of the bacterial genomes that have been sequenced. In both *Mycoplasma genitalium* (0.58 Mb) and *E. coli* (4.6 Mb) about 90% of the genome is dedicated to protein-coding genes. Therefore the differences in size reflect the number of genes carried. This begs the question as to the minimal genome size possible for a bacterium. The current view is about 300 genes (Mushegian 1999; see also p. 114). In *Borrelia* the small genome size (1.5 Mb) may be complemented by the high plasmid content (see above) which constitutes 0.67 Mb. These plasmids carry 535 genes, many of which have no counterparts in other organisms, suggesting that they perform specialized functions (Casjens *et al.* 2000).

The organization of organelle genomes

Mitochondria and chloroplasts both possess DNA genomes that code for all of the RNA species and some of the proteins involved in the functions of the organelle. In some lower eukaryotes the mitochondrial (mt) DNA is linear but more usually organelle genomes take the form of a single circular molecule of DNA. Because each organelle contains several copies of the genome and because there are multiple organelles per cell, organelle DNA constitutes a repetitive sequence. Whereas chloroplast (ct) DNA falls in the range 120–200 kb, mtDNA varies enormously in size. In animals it is relatively small, usually less than 20 kb, but in plants it can be as big as 2000 kb.

Organization of the chloroplast genome

The complete sequence of ctDNA has been reported for over a dozen plants including the single-celled protist, *Euglena* (Hallick *et al.* 1993), a liverwort (Ohyama *et al.* 1986), and angiosperms such as *Arabidopsis*, spinach, tobacco and rice (Shinozaki *et al.* 1986; Hiratsuka *et al.* 1989; Sato *et al.* 1999;

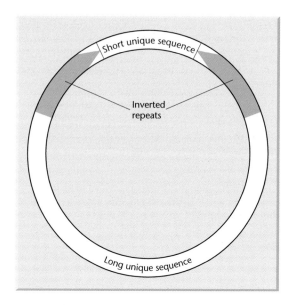

Fig. 2.9 Generalized structure of ctDNA.

Schmitz-Linneweber *et al.* 2001). Overall, there is a remarkable similarity in size and organization (Fig. 2.9 and Table 2.4). The differences in size are accounted for by differences in length of introns and intergenic regions and the number of genes. A general feature of ctDNA is a 10–24 kb sequence that is present in two identical copies as an inverted repeat. The proportion of the genome that is represented by introns can be very high, e.g. in *Euglena* it is 38%.

The chloroplast genome encodes 70–90 proteins, including those involved in photosynthesis, four rRNA genes and 30 or more tRNA genes. Chloroplast mRNAs are translated with the standard genetic code (cf. mitochondrial mRNA). However, editing events cause the primary structures of several

Table 2.4 Key features of chloroplast DNA.

Feature	*Arabidopsis*	Spinach	Maize
Inverted repeats	26 264 bp	25 073 bp	22 748 bp
Short unique sequence	17 780 bp	17 860 bp	12 536 bp
Long unique sequence	84 170 bp	82 719 bp	82 355 bp
Length of total genome	154 478 bp	150 725 bp	140 387 bp
Number of genes	108	108	104
rRNA genes	4	4	4
tRNA genes	37	30	30
Protein-encoding genes	87	74	70

transcripts to deviate from the corresponding genomic sequences by C to U transitions with a strong bias for changes at the second codon position (Maier *et al.* 1995). This editing makes it difficult to convert chloroplast nucleotide sequences into amino acid sequences for the corresponding gene products.

Astasia longa is a colourless heterotrophic flagellate which is closely related to *Euglena gracilis*. It contains a plastid DNA that is 73 kb in length, about half the size of the ctDNA from *Euglena*. Sequencing of this plastid DNA has shown that all chloroplast genes for photosynthesis-related proteins, except that encoding the large subunit of ribulose-1, 5-bisphosphate carboxylase, are missing (Gockel & Hachtel 2000).

Organization of the mitochondrial genome

Mitochondrial DNA (mtDNA) is an essential component of all eukaryotic cells. It ensures consistency of function (cellular respiration and oxidative phosphorylation) despite the great diversity of genome organization. However, many of the genes for mitochondrial proteins are found in the nucleus. Some organisms use the standard genetic code to translate nuclear mRNAs and a different code for their mitochondrial mRNAs.

The mtDNA from animals is about 15–17 kb in size, nearly always circular and very compact. It encodes 37 genes: 13 for proteins, 22 for tRNAs and two for rRNAs. There are no introns and very little intergenic space. It has been found that mtDNA can survive in museum specimens and palaeontological remains and so is proving useful for the study of the genetic relationships of extinct animals (Hofreiter *et al.* 2001).

Plant mtDNA is much larger than that from animals and in angiosperms ranges from 200 kb to 2 Mb (i.e. larger than some bacterial genomes). Plant mtDNAs rival the eukaryotic nucleus in terms of the C-value paradox they present. That is, larger plant mt genomes do not contain more genes than smaller ones but simply have more spacer DNA. Plant mtDNAs do have introns but there is little variation in intron content and size across the range of angiosperms. As an example of the C-value paradox, *Arabidopsis* mtDNA is 367 kb in size whereas human mtDNA is 16.6 kb in size but the former has only 14 more genes (Marienfeld *et al.* 1999). Angiosperm

mtDNAs are larger than their animal counterparts partly because of frequent duplications and partly because of the frequent capture of DNA sequences from the chloroplast and the nucleus. Plant mtDNAs also can lose sequences to the nucleus (Palmer *et al.* 2000).

Fungal mtDNAs resemble plant mtDNAs in that they are larger than those from animals and more heterogeneous in size, e.g. *Saccharomyces* mtDNA is 86 kb in size whereas that from *Mucor* is 34 kb (Foury *et al.* 1998; Papp *et al.* 1999). Fungal mtDNA also contains introns.

Plant mtDNAs differ from those of animals and lower eukaryotes in more than just size. At the sequence level they have an exceptionally low rate of point mutations that is 50–100 times lower than that seen in vertebrate mitochondria. At the structural level, plant mtDNAs have a high rate of genomic rearrangement and duplication.

The organization of nuclear DNA in eukaryotes

Gross anatomy

Each eukaryotic nucleus encloses a fixed number of chromosomes which contain the nuclear DNA. During most of a cell's life, its chromosomes exist in a highly extended linear form. Prior to cell division, however, they condense into much more compact bodies which can be examined microscopically after staining. The duplication of chromosomes occurs chiefly when they are in the extended stage (interphase). One part of the chromosome, however, always duplicates during the contracted metaphase state. This is the *centromere*, a body that controls the movement of the chromosome during mitosis. Its structure is discussed later (p. 30).

The ends of eukaryotic chromosomes are also the ends of linear duplex DNA and are known as *telomeres*. That these must have a special structure has been known for a long time (see Box 2.1). In most eukaryotes the telomere consists of a short repeat of TTAGGG many hundreds of units long but in *Saccharomyces cerevisiae* the repeat unit is TG_{1-3}. These repeats vary considerably in length between species (Table 2.5) but each species maintains a fixed average telomere length in its germline. The enzyme

Table 2.5 Length of the telomere repeat in different eukaryotic species.

Species	Length of telomere repeat
S. cerevisiae (yeast)	300 bp
Mouse	50 kb
Human	10 kb
Arabidopsis	2–5 kb
Cereals	12–15 kb
Tobacco	60–160 kb

telomerase is responsible for maintaining the integrity of telomeres (see Box 2.2).

The fruit fly *Drosophila* differs from most other eukaryotes in having an unusually elaborate method for forming chromosome ends. Instead of telomere repeats it has telomere-specific transposable elements (Pardue *et al.* 1996) that resemble long interspersed nuclear elements (LINEs) (see p. 28).

In many eukaryotes, a variety of treatments will cause chromosomes in dividing cells to appear as a series of light- and dark-staining bands (Fig. 2.10). In G-banding, for example, the chromosomes are subjected to controlled digestion with trypsin before Giemsa staining which reveals alternating positively (dark G-bands) and negatively (R-bands or pale G-bands) staining regions. As many as 2000 light and dark bands can be seen along some mammalian chromosomes. An identical banding pattern (Q-banding) can be seen if the Giemsa stain is replaced with a fluorescent dye such as quinacrine which intercalates between the bases of DNA. A structural basis for metaphase bands has been proposed that is based on the differential size and packing of DNA loops and matrix-attachment sites in G- vs. R-bands

Box 2.2 Telomerase, immortality and cancer

Telomeres are found at the ends of chromosomes and their role is to protect the ends of chromosomes. They also stop chromosomes from fusing to each other. Telomeres consist of repeating units of TTAGGG that can be up to 15 000 bp in length. The very ends of the chromosomes are not blunt-ended but have 3′ single-stranded overhangs of 12 or more nucleotides. The enzyme telomerase, also called telomere terminal transferase, is a ribonucleoprotein enzyme whose RNA component binds to the single-stranded end of the telomere. An associated reverse transcriptase activity is able to maintain the length and structure of telomeres by the mechanism shown in Fig. B2.2. For a detailed review of the mechanism of telomere maintenance the reader should consult the paper by Shore (2001).

Telomerase is found in fetal tissues, adult germ cells and cancer cells. In normal somatic cells the activity of telomerase is very low and each time the cells divide some of the telomere (25–200 bp) is lost. When the telomere becomes too short the chromosome no longer divides and the host cell dies by a process known as apoptosis. Thus, normal somatic cells are mortal and in tissue culture they will undergo 50–60 divisions (Hayflick limit) before they senesce. In contrast to mammals, indeterminately growing multicellular organisms, such as fish and crustacea, maintain unlimited growth potential throughout their entire life and retain telomerase activity (for a review see Krupp *et al.* 2000).

Cancer cells can divide indefinitely in tissue culture and thus are immortal. Telomerase has been found in cancer cells at activities 10- to 20-fold greater than in normal cells. This presence of telomerase confers a selective growth advantage on cancer cells and allows them to grow uncontrollably. Telomerase is an ideal target for chemotherapy because this enzyme is active in most tumours but inactive in most normal cells.

If recombinant DNA technology is used to express telomerase in human somatic cells maintained in culture, senescence is avoided and the cells become immortal. This immortalization usually is accompanied by an increased expression of the *c-myc* oncogene to the levels seen in many cancer cell lines.

Chromosomal rearrangements involving telomeres are emerging as an important cause of human

continued

Box 2.2 *continued*

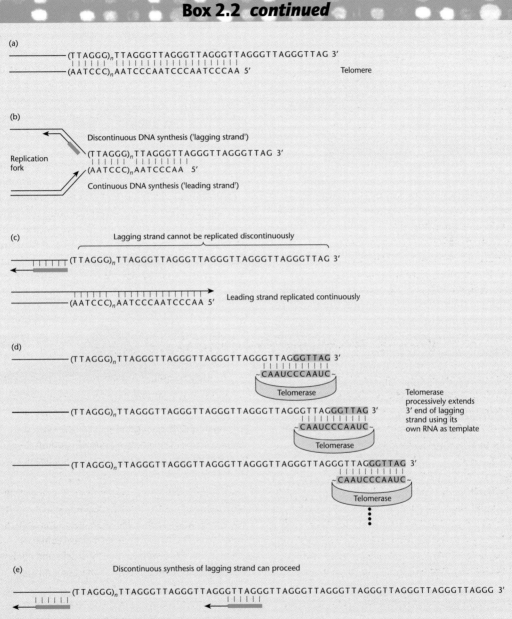

Fig. B2.2 Diagram of telomere replication and the role of telomerase. (a) The hexameric repeat sequence found at telomere ends in human chromosomes. (b) and (c) A replication fork attempting to replicate the telomere. The Okazaki fragment (green arrow) allows replication of all but the most terminal portion of the lagging strand. (d) and (e) Telomerase, carrying its own RNA template (5'-CUAACCCUAAC-3'), extends the lagging strand at the end of the chromosome and allows replication. (Redrawn with permission from Nussbaum *et al.* 2001. W.B. Saunders Publishing, 2001, a division of Harcourt.)

genetic diseases (Knight & Flint 2000). Telomere-specific clones have been used in combination with fluorescence *in situ* hybridization (FISH) to detect abnormalities not found by conventional cytogenetic analysis.

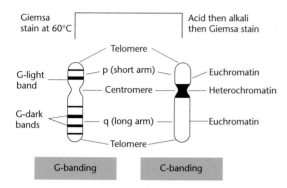

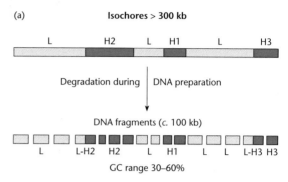

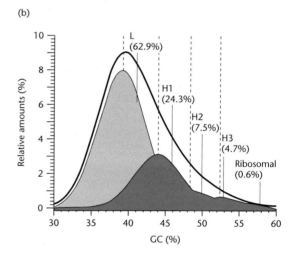

Fig. 2.10 Banding patterns revealed on chromosomes by different staining methods. Note that intercalating fluorescent dyes produce the same pattern as Giemsa stain at 60°C. In the C-banding technique some heterochromatin may be detected at the telomeres.

Fig. 2.11 (a) Scheme of the isochore organization of the human genome. This genome, which is typical of the genome of most mammals, is a mosaic of large (⩾ 300 kb, on average) DNA segments, the isochores, which are compositionally homogeneous (above a size of 3 kb) and can be partitioned into a number of families. Isochores are degraded during routine DNA preparations to fragments of approx. 100 kb in size. The GC-range of the isochores from the human genome is 30–60% (b). (b) The CsCl profile of human DNA is resolved into its major DNA components, namely the families of DNA fragments derived from isochore families L (i.e. L1 + L2), H1, H2, H3. Modal GC levels of isochore families are indicated on the abscissa (broken vertical lines). The relative amounts of major DNA components are indicated. Satellite DNAs are not represented. (Reprinted from Zoubak *et al.* 1996 by permission of Elsevier Science.)

(Saitoh & Laemmli 1994). Bands are classified according to their relative location on the short arm (p) or the long arm (q) of specific chromosomes; e.g. 12q1 means band 1 on the long arm of chromosome 12. If the chromosome DNA is treated with acid and then alkali prior to Giemsa staining, then only the centromeric region stains and this is referred to as *heterochromatin*. The unstained parts of the chromosome are called *euchromatin*.

Because Giemsa stain shows preferential binding to DNA rich in AT base pairs, the dark G-bands should be AT-rich and the light G-bands GC-rich. This has been confirmed by fractionation of DNA by equilibrium centrifugation. When vertebrate DNA is isolated by the usual methods it gets fragmented into pieces approximately 100 kb in size. These DNA preparations can be fractionated according to their GC content in CsCl density gradients (Fig. 2.11) suggesting that the DNA is composed of mosaics of long DNA segments (isochores) differing in composition (Zoubak *et al.* 1996). In the H2 and H3 isochores, gene concentration is very high (one gene per 5–15 kb) whereas in the other two isochores the gene density is very low (one gene per 50–150 kb) (Fig. 2.12). The H2 and H3 isochores have been located in the light G-bands (Saccone *et al.* 1999), whereas the L and H1 isochores are found in the G-dark bands. Analysis of the complete human genome sequence indicates that isochores are not as homogenous as originally thought but the concept still is of value.

CpG islands represent a different form of compositional heterogeneity. They were originally identified as short regions of mammalian DNA which contained many sites for the restriction endonuclease *Hpa*II and for this reason originally were called *Hpa*II Tiny Fragment (HTF) islands. This island DNA is found in short regions of 1–2 kb which together

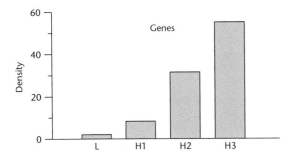

Fig. 2.12 Density of gene sequences in isochore families. Relative numbers of sequences over relative amounts of isochore families are presented in the histograms. (Reprinted from Zoubak *et al.* 1996 by permission of Elsevier Science.)

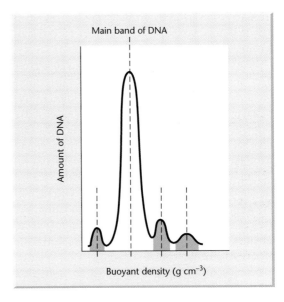

Fig. 2.13 Detection of three satellite DNA bands (dark shading) on equilibrium density gradient centrifugation of total DNA.

account for approximately 2% of the mammalian genome. It has distinctive properties when compared with the DNA in the rest of the genome. It is unmethylated, GC-rich and does not show any suppression of the dinucleotide CpG. By contrast, bulk genomic DNA has a GC content which is much lower, is methylated at CpG and the CpG dinucleotide is present at a much lower frequency than would be predicted from base composition. CpG islands have been found at the 5′ ends of all housekeeping genes and of a large proportion of genes with a tissue-restricted pattern of expression (Craig & Bickmore 1994).

Origins of replication

All of the cell's DNA must be replicated once, and once only, during the cell cycle. In yeast (*S. cerevisiae*) the origins of DNA replication have been identified as autonomously replicating sequence (ARS) elements. Incorporation of the latter on plasmids permits their replication in yeast as yeast artificial chromosomes (YACs) (see p. 40). The sequence requirements for an ARS element are known (Marahrens & Stillman 1992). So far, origins of DNA replication have not been unambiguously identified in other eukaryotes.

Repeated sequences

It will be recalled from the section on reassociation kinetics of DNA that repeated sequences are a common feature of eukaryotic DNA. How are these repeated sequences organized in the genome? Are they tandemly repeated or are they dispersed? The answer is that both organizational patterns are found. In general, highly repetitive DNA is organized around centromeres and telomeres in the form of tandem repeats, whereas moderately repetitive DNA is dispersed throughout the chromosome.

Tandemly repeated sequences

Repeated sequences were first discovered 30 years ago during studies on the behaviour of DNA in centrifugal fields. When DNA is centrifuged to equilibrium in solutions of CsCl, it forms a band at the position corresponding to its own buoyant density. This in turn depends on its percentage G + C content:

$$\rho(\text{density}) = 1.660 + 0.00098\,(\%\,GC)\,\text{g cm}^{-3}.$$

When eukaryotic DNA is centrifuged in this way the bulk of the DNA forms a single, rather broad band centred on the buoyant density which corresponds to the average G : C content of the genome. Frequently one or more minor or *satellite* bands are seen (Fig. 2.13). The behaviour of satellite DNA on density gradient centrifugation frequently is anomalous. When the base composition of a satellite is determined by chemical means it often is different to that predicted from its buoyant density. One reason is that it is methylated which changes its buoyant density.

Once isolated, satellite DNA can be radioactively labelled *in vitro* and used as a probe to determine where on the chromosome it will hybridize. In this technique, known as *in situ hybridization*, the chromosomal DNA is denatured by treating cells that have been squashed on a cover slip. The localization of the sites of hybridization is determined by autoradiography. Using this technique, most of the labelled satellite DNA is found to hybridize to the heterochromatin present around the centromeres and telomeres. Because RNA that is homologous to satellites is found only rarely the heterochromatic DNA most probably is non-coding.

When satellite DNA is subjected to restriction endonuclease digestion only one or a few distinct low-molecular-weight bands are observed following electrophoresis. These distinct bands are a tell-tale sign of tandemly repeated sequences. The reason (Fig. 2.14) is that if a site for a particular restriction endonuclease occurs in each repeat of a repetitious tandem array, then the array is digested to unit-sized fragments by that enzyme. After elution of the DNA band from the gel it can be used for sequence analysis either directly or after cloning. However, the sequence obtained is a consensus sequence and not necessarily the sequence of any particular repeat unit because sequence divergence can and does occur very readily. Note that if such sequence divergence occurs within a restriction endonuclease cleavage site in the repeated units then digestion with the enzyme produces multimers of the repeat unit ('higher order repeats') (Fig. 2.14).

The amount of satellite DNA and its sequence varies widely between species and can be highly polymorphic within a species. Thus 1–3% of the genome of the rat (*Rattus norvegicus*) is centromeric satellite DNA, whereas it is 8% in the mouse (*Mus musculus*) and greater than 23% in the cow (*Bovis domesticus*). The length of the satellite repeat unit varies from the $d(AT)_n : d(TA)_n$ structure found in the land crab (*Gecarcinus lateralis*) to the very complicated structure seen in the domestic cow (Fig. 2.15). Even within a single genus each species can have a distinctive set of satellite sequences. In general, little is known about the detailed structure of repeated DNA arrays but one family of repeated DNA that has

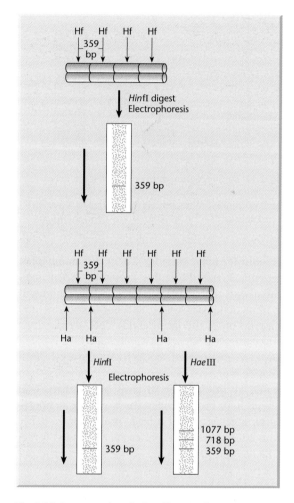

Fig. 2.14 Digestion of purified satellite DNA by restriction endonucleases. The basic repeat unit is 359 bp long and contains one endonuclease *Hin*fI (Hf) site. Digestion with *Hin*fI converts most of the satellite DNA to a set of 359 bp long fragments. These are abundant enough to be seen as a band against the smear of other genomic fragments after gel electrophoresis and staining with ethidium bromide. Digestion of the DNA with endonuclease *Hae*III (Ha) yields a ladder of fragments that are multiples of 359 bp in length. (Redrawn with permission from Singer & Berg 1990.)

been analysed extensively is ribosomal DNA (rDNA) (Williams & Robbins 1992). An example of rDNA repeat structure is shown in Fig. 2.16.

Not all tandem repetitions are restricted to heterochromatin: some are found dispersed throughout the genome, often in the spacer region between genes. One such group is the *minisatellite* sequences whose repeat unit is 14–500 bp in length.

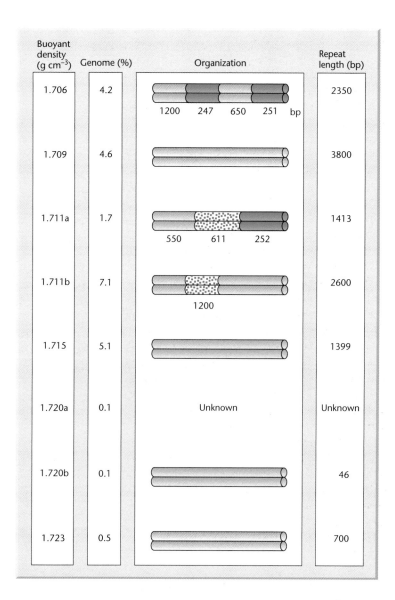

Buoyant density (g cm^{-3})	Genome (%)	Organization	Repeat length (bp)
1.706	4.2	1200 247 650 251 bp	2350
1.709	4.6		3800
1.711a	1.7	550 611 252	1413
1.711b	7.1	1200	2600
1.715	5.1		1399
1.720a	0.1	Unknown	Unknown
1.720b	0.1		46
1.723	0.5		700

Fig. 2.15 The structure of the different types of satellite DNA found in the domestic cow. Homologous portions of the different satellites are indicated by similar colouring. The designations a and b indicate two versions with the same buoyant density. (Redrawn with permission from Singer & Berg 1990.)

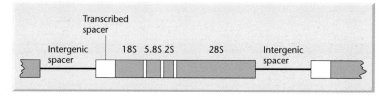

Fig. 2.16 Detailed architecture of a rDNA repeat unit. Transcribed regions are shown in boxes. Coding sequences are shown in green and spacer regions in white.

Minisatellites demonstrate intraspecies polymorphism and this been used for DNA fingerprinting of organisms. When cloned probes containing a minisatellite sequence are annealed with DNA blots containing restriction endonuclease digests of DNA, multiple bands hybridize. The pattern of bands varies from one individual of a species to another but is the same when DNA from several tissues of a single individual is examined. The bands are inherited in a Mendelian fashion and it is possible to identify those

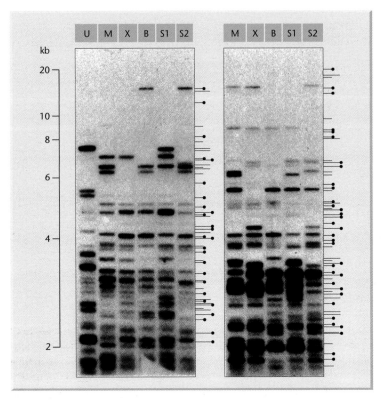

Fig. 2.17 Use of minisatellite sequences to detect polymorphisms in human DNA for forensic purposes. The profiles shown are for blood DNA derived from a family where the familial relationship of one member (X) was questioned. Fragments present in the mother's (M) DNA are indicated by a short horizontal line (to the right of each profile); paternal fragments absent from M but present in at least one of the undisputed siblings (B, S1, S2) are marked with a long line. Maternal and paternal fragments transmitted to X are shown with a dot.

bands inherited from each parent (Fig. 2.17). For this reason the technique has forensic applications (for reviews see Monckton & Jeffreys 1993; Alford & Caskey 1994).

Microsatellite DNA families, or simple sequence repeats (SSRs), are small arrays of tandem repeats of 1–13 bp that are interspersed throughout the genome (for a review see Toth *et al.* 2000). Microsatellites comprise 3% of the human genome with an average density of one SSR per 2 kb of sequence. Of these SSRs, dinucleotide repeats are the most common (0.5% of genome) closely followed by runs of $(dA \cdot dT)_n$ at 0.3% of the genome. Trinucleotide repeats are much rarer. Within the dinucleotide repeats there is a heavy bias towards dAC·dTG and dAT·dTA repeats with dGC·dCG being extremely rare. The significance of these repeats in normal genes is not known but they can be the locus for a number of inherited disorders when they undergo unstable expansion. For example, in fragile X syndrome patients can exhibit hundreds or even thousands of the CGG triplet at a particular site, whereas unaffected individuals only have about 30 repeats.

So far, over a dozen examples of disease resulting from trinucleotide expansion have been described (Sutherland & Richards 1995; Warren 1996; Mitas 1997; Bowater & Wells 2000). Similar trinucleotide repeats have been discovered in bacteria (Hancock 1996) and yeast (Dujon 1996; MarAlba *et al.* 1999; Richard *et al.* 1999) following complete genome sequencing. As in the human case, perfect trinucleotide repeats in yeast are subject to polymorphic size variation while imperfect ones are not. Some pathogenic bacteria use length variation in simple repeats to change the antigens on their surfaces so that they can evade host immune attack.

Tandem repetition of DNA sequences also occurs within coding regions. For example, linked groups of identical or near-identical genes sometimes are repeated in tandem. These are the gene families described earlier (p. 15). However, tandem repetition also occurs within a single gene; for example, the *Drosophila* 'glue' protein gene contains 19 direct tandem repeats of a sequence 21 base pairs long that encodes seven amino acids. The repeats are not perfect but show divergence from a consensus sequence. Another example is the gene for α2(1) collagen found in chicken, mouse and humans. The gene comprises 52 exons with introns varying in length from 80 to 2000 bp. However, all the exon sequences are multiples of 9 bp and most of them are 54 or 108 bp long. This accounts for the observed primary sequence of collagen which has glycine in every third position and a very high content of proline and lysine.

Dispersed repeated sequences

Moderately repetitive DNA is characterized by being dispersed throughout the genome and this was discovered by an extension of the early work on reannealing kinetics. It will be recalled from p. 11 that reannealing of DNA is dependent on DNA fragment size and that it is common practice to shear the DNA before hybridization begins. The reason for this is that it was observed that most high-molecular-weight fragments would reassociate with each other as a result of the interaction of repetitive DNA sequences. Very large particles, termed networks, were formed. This observation indicated that many fragments as long as 10 000 nucleotides contained more than one repetitive sequence element. When the fragment size was reduced, a smaller number of fragments reassociated at low $C_0t_{1/2}$ values. When the percentage reassociation is determined for different lengths of DNA at C_0t values where single-copy DNA cannot anneal, it is possible to determine the length of the interspersed sequences (Fig. 2.18). From the example shown it can be seen that in

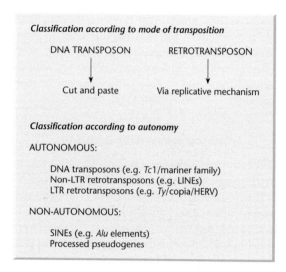

Fig. 2.19 Classification of transposable elements found in eukaryotes.

human DNA 2.2 kb of relatively rare sequences separate the repeated sequences in 60% of the genome. In *Xenopus* about 800 bp of unique sequence separates repeats in about 65% of the genome. By contrast, in *Drosophila* about 12 kbp separate the repeated sequences. Kinetic analysis of this type has revealed that the predominant interspersion pattern of insects and fungi is long range (i.e. like *Drosophila*), whereas in most other plants and animals it is short range. However, molecular cloning has revealed that many organisms have both short- and long-range interspersion patterns superimposed upon one another.

The short and long interspersed repeated sequences described above are examples of transposable elements (Fig. 2.19). These are two broad classes of transposable element: DNA transposons and retrotransposons. Most DNA transposons move as pieces of DNA, cutting and pasting themselves into new genomic locations. In contrast, the genomic expansion of retrotransposons occurs by a replicative mechanism. Retrotransposons are duplicated through an RNA intermediate. The original transposon remains where it was but its RNA transcript is reverse transcribed into DNA which then integrates into a new genomic location.

Transposable elements can be classified in a different way depending on the degree of mechanistic self-sufficiency. Autonomous transposable elements,

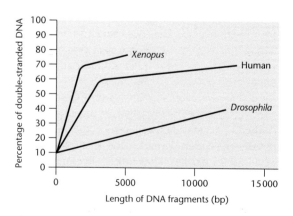

Fig. 2.18 Interspersion analysis of DNA reassociation kinetics (see text for details).

such as LINEs, are thought to encode essentially all of the machinery that they require to move. Non-autonomous transposable elements, such as short interspersed nuclear elements (SINEs) are entirely dependent on other transposable elements for their mobility.

Autonomous retrotransposons

The retroviruses are the paradigm for autonomous retrotransposons because they encode reverse transcriptase and/or integrase activities. The retrotransposons differ from the retroviruses themselves in not passing through an independent infectious form but otherwise resemble them in the mechanism used for transposition. The autonomous retrotransposons are characterized by the presence of two open-reading frames (ORFs) on one strand, one of which encodes reverse transcriptase (Fig. 2.20). There are two classes of autonomous retrotransposon: those with long terminal repeats (e.g. *Ty1–Ty5* elements of yeast, the *copia* and *gypsy* elements of *Drosophila*) and those without long terminal repeats (e.g. LINEs).

The *Ty1/copia* and *Ty3/gypsy* class retroelements have been found in all animal, plant and fungal species that have been examined but their organization and ubiquity show extensive variation. For example, the *Ty1/copia* group of retroposons are found in all plants including single-cell algae, bryophytes, gymnosperms and angiosperms. They usually are present in high copy numbers, from hundreds to millions, and as highly heterogeneous populations. This is in marked contrast to insect and fungal systems where these retroposons are present in much lower numbers (tens to hundreds) and as more homogeneous populations. There is no relationship between the total copy number and the host genome size and the copy number can vary widely between closely related species within a genus (Kumar 1996).

In *S. cerevisiae*, *Ty1–Ty5* retrotransposons have a strong bias for sites in the genome into which they integrate (Goffeau *et al.* 1996). Over 90% of the *Ty1–Ty4* elements are located within 750 bp upstream of genes transcribed by RNA polymerase III, particularly tRNA genes. The *Ty5* elements are located at the telomeres or regions that have telomeric

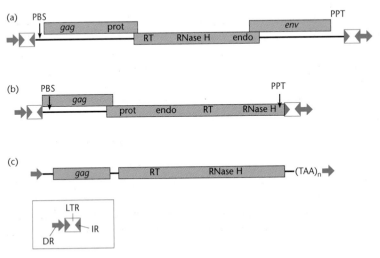

Fig. 2.20 Major types of retroelements. Overall organization of (a) a retrovirus, the avian leukosis virus; (b) a retrotransposon, the yeast *Ty1* element; and (c) a non-LTR (long terminal repeat) retrotransposon, the *Drosophila* I factor. Open-reading frames are depicted by green boxes. The *gag* gene encodes structural proteins of the virion core, including a nucleic-acid-binding protein; the *env* gene encodes a structural envelope protein, necessary for cell-to-cell movement; prot, protease involved in cleavage of primary translation products; RT, reverse transcriptase; RNase H, ribonuclease; endo, endonuclease necessary for integration in the host genome; LTR, long terminal repeats containing signals for initiation and termination of transcription, and bordered by short inverted repeats (IRs) typically terminating in 5′–TG . . . CA–3′; PBS, primer binding site, complementary to the 3′ end of a host tRNA, and used for synthesis of the first (−) DNA strand; PPT, polypurine tract used for synthesis of the second (+) DNA strand; DR, short direct repeats of the host target DNA, created upon insertion. (Reprinted from Grandbastien 1992 by permission of Elsevier Science.)

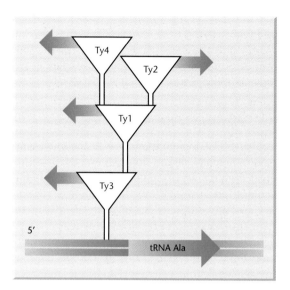

Fig. 2.21 Example from *Saccharomyces cerevisiae* of retrotransposons within retrotransposons which typically are found upstream of tRNA genes. (Reprinted from Voytas 1996 by permission of the American Association for the Advancement of Science.)

500 000 LINEs but most are functionally inactive because of truncations, rearrangements and non-sense mutations. Only 3000–4000 are full length and about 1–2% are able to transpose (Sassaman *et al.* 1997). By contrast, the mouse genome has only about 3000 LINEs (DeBerardinis *et al.* 1998).

DNA transposons

DNA transposon sequences constitute roughly 2.5% of the human genome and are members of the MER 1 or MER 2 families of medium reiterated frequency repeat family. *Tc1* from *Caenorhabditis elegans* and *mariner* from *Drosophila mauritania* are members of the *Tc1/mariner* superfamily of DNA transposons (Plasterk *et al.* 1999). These transposons have a single ORF encoding the transposase.

Non-autonomous retrotransposons

The *Alu* family of SINEs are found in Old World primates and are named for an *Alu* I restriction endonuclease site typical of the sequence. *Alu* elements are 280 nucleotide sequences that lack introns (Fig. 2.22) and there are at least 1 million copies of them in the human genome. They are transcribed but are not translated because they lack an open-reading frame (Schmid 1998). Processed pseudogenes are similar to *Alu* elements and are non-functional genes that have been moved from their native location through an RNA intermediate.

chromatin. Regions targeted by yeast retroposons are typically devoid of ORFs and reiterative integration can generate blocks of elements within elements (Fig. 2.21). These element landing pads provide a safe haven for elements to integrate without causing deleterious mutations.

In humans, the closest elements to long terminal repeat (LTR) retrotransposons are endogenous retroviruses (HERVs) that make up roughly 7% of the genome (Smit 1999). Most HERV sequences have accumulated several nonsense mutations and no longer encode functional retroviral proteins.

LINEs pervade mammalian genomes and are preferentially located in the dark G-bands of metaphase chromosomes. The human genome contains about

Distribution of transposable elements

The distribution of transposable elements in the human genome is quite different to that in the fruit fly, nematode and *Arabidopsis* genomes. First, the euchromatic portion of the human genome has a much higher density of transposable element copies.

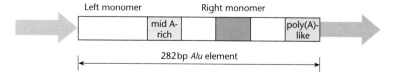

Fig. 2.22 Structure of a typical 282-bp *Alu* element flanked by direct repeats (green arrows). The *Alu* element itself consists of an inexact duplication of two monomer units separated by a mid A-rich region. The right monomer contains an additional sequence of approximately 30 bp (dark green) that is absent from the left monomer. The consensus sequence usually is followed by an A-rich region resembling a poly(A) tail.

Second, the human genome is filled with copies of ancient transposons whereas the transposons in other organisms, particularly the fruit fly, are of more recent origin. This probably reflects the efficiency of housecleaning through genomic deletion. Thirdly, two repeat families (LINE1 and *Alu*) account for 60% of all interspersed repeat sequences in the human genome but the other genomes studied contain many different transposon families. Similarly DNA transposons represent only 6% of all interspersed repeats in humans but 25, 49 and 87% in the fruit fly, *Arabidopsis* and the nematode, respectively.

Within an organism there can be tremendous variation in the distribution of repeat sequences. Some regions of the human genome are extraordinarily dense in repeats whereas others are nearly devoid of repeats. For example, a 525 kb region on chromosome Xp11 has an overall transposable element density of 89% and includes a 200 kb segment with a density of 98%. By contrast, the density in the four homeobox clusters is less than 2%.

Segmental duplications

Such duplications involve the transfer of 1–200 kb blocks of genomic sequence to one or more locations in the genome. In human DNA, the pericentromeric and subtelomeric regions of chromosomes are filled with segmental duplications but they are much less common in the *Saccharomyces*, nematode and fruit fly genomes. In *Arabidopsis* there are 24 segmental duplications that account for 58% of the genome but only one is in the centromeric region.

Segmental duplications can be divided into two categories: interchromosomal and intrachromosomal duplications. The former are defined as segments that are duplicated among non-homologous chromosomes. For example, a 9.5 kb genomic segment from the human adrenoleukodystrophy locus from Xq28 has been duplicated to regions near the centromeres of chromosomes 2, 10, 16 and 22. Intrachromosomal duplications occur within a particular chromosome or chromosomal arm. This category includes several duplicated segments, also known as low copy repeat sequences, which mediate recurrent chromosomal structural rearrangements associated with genetic disease in humans. For example, on chromosome 17 there are three copies

of a 200 kb repeat separated by 5 Mb of sequence and two copies of a 24 kb repeat separated by 1.5 Mb of sequence. These sequences are so similar that they can undergo recombination (for a review of this topic the reader should consult Eichler 2001).

Plasticity of the eukaryotic genome

Genomic DNA is often thought of as the stable template of heredity, largely dormant and unchanging, apart from the occasional point mutation. Nothing could be further from the truth. DNA is dynamic and is constantly subjected to rearrangements, insertions and deletions as a result of the activity of transposable elements. For example, LINE insertions have been found to be responsible for certain cases of haemophilia, thalassaemia, Duchenne muscular dystrophy and chronic granulomatous disease. A particularly striking example is a LINE insertion in the *APC* gene of adenocarcinoma cells from a colon cancer patient, but not in the surrounding normal tissue (Miki *et al.* 1992). In addition to duplicating themselves, LINEs can carry with them genomic flanking sequences that are downstream from their 3′-untranslated regions and could have a role in exon shuffling (see Box 2.3). SINEs may be hot-spots for recombination and a number of examples have been identified in human genetic disorders (the reader interested in this topic should consult the review of Luning Prak & Kazazian 2000).

Centromere structure

During mitosis, the key chromosomal element responsible for directing operations is the centromere and its associated kinetochore complex. Most organisms have monocentric chromosomes. That is, the centromere is located at a single point on the chromosome. A few species, such as the sedge *Luzula* and the nematode *C. elegans*, have holocentric chromosomes in which the microtubules attach throughout the length of the chromosome.

The structure of centromeric DNA varies widely in different organisms. In *S. cerevisiae* the centromere is precisely defined and only 125 bp are sufficient to mediate spindle attachment. By contrast, in the fission yeast *Schizosaccharomyces pombe* the centromere is 40–120 kb in size and consists of

Box 2.3 Exon shuffling mediated by a LINE

Moran *et al.* (1999) were able to demonstrate in cultured cells the transduction of a gene by a long interspersed nuclear element (LINE). A neomycin (*neo*) reporter gene was placed downstream from a LINE sequence and both the promoter and the initiation codon of the *neo* gene were replaced with a 3′ splice site. The only way that this modified *neo* gene could be expressed after transduction was if it was inserted into an actively transcribed gene and spliced onto the transcript derived from that gene. Such events were readily detected and characterized as authentic LINE insertions in several genes.

The above result suggests a simple mechanism by which exons can be shuffled in the human genome (see Fig. B2.3). Transcription of a LINE element within a gene *X* fails to terminate at its own weak polyadenylation (poly(A)) signal but terminates instead at the poly(A) signal of the gene. The reverse transcriptase and endonuclease encoded by this transcript bind to the poly(A) tail and insert a cDNA copy into gene *Y*. This step results in the transduction of gene *X* into gene *Y* thereby creating a new gene.

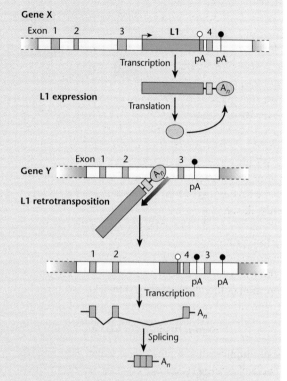

Fig. B2.3 Possible mechanism for LINE involvement in exon shuffling, L1, LINE element; pA, polyadenylation signal; A_n, poly(A) tail. Reverse transcriptase and endonuclease are shown in grey. (Reprinted from Eickbush 1999 by permission of the American Association for the Advancement of Science.)

moderately repetitive DNA. At least 12 kb is required to provide segregation function but such minimal centromeres are subject to epigenetic regulation. In *Drosophila*, *Arabidopsis*, mammals and the filamentous fungus *Neurospora*, centromere activity has been mapped to highly repetitive regions containing satellite DNA and other repetitive DNA (Fig. 2.23; Craig *et al.* 1999). In contrast with *S. pombe*, the centromere of human chromosomes has a size of 3 Mb (Schueler *et al.* 2001). A key component of mammalian centromeres is alphoid DNA which consists of a 171 bp motif repeated in a tandem head-to-tail fashion and then is organized into higher order repeat arrays.

Summary of structural elements of eukaryotic chromosomes

The different strutural elements of eukaryotic chromosomes that have been discussed in the previous sections are summarized in Fig. 2.24. A key feature of eukaryotic DNA is that as genome size increases so too does the amount of repeat sequences and hence the density of coding sequences decreases. This is illustrated by data on gene density derived from the different DNA sequencing projects (Table 2.6).

Although advantage can sometimes be taken of repeat sequences, generally speaking they tend to cause problems. For example, when DNA containing

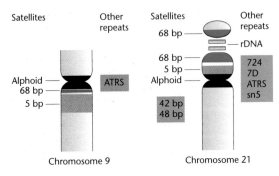

Fig. 2.23 Structures of the centromeres of human chromosomes 9 and 21, showing tandemly repeated satellite DNA and other centromeric repeated sequences. The sequences shown in boxes are known to be present but their precise locations have not yet been mapped. Satellite DNA is named by the size of the repeat unit except for alphoid DNA which has a 171 bp repeat. ATRS, A + T rich sequence. (Reprinted from Tyler-Smith & Willard 1993 by permission of Elsevier Science.)

Table 2.6 Average gene density in different organisms.

Organism	Gene density (genes/Mb)
Bacteria	800–1100
Saccharomyces cerevisiae	446
Plasmodium (malarial parasite)	221
Caenorhabditis elegans (nematode)	196
Arabidopsis thaliana	175
Fugu (puffer fish)	150
Ciona (sea squirt)	97
Drosophila	71
Human	20

repeated sequences is used as a hybridization probe it will anneal to many different regions of the genome. During cloning of genome fragments recombination can occur between repeats leading to 'scrambling' of DNA sequences. During the actual DNA sequencing reactions, slippage can occur, particularly with microsatellites. Finally, during data assembly, incorrect positioning of genome fragments or sequences can occur because repeat units are incorrectly recognized as being unique. These

issues are discussed in more detail in subsequent chapters.

One final aspect of genome structure deserves mention here. DNA methylation is a covalent modification of cytosine, hereditable by somatic cells after cell division, and is generally associated with transcriptional silencing. It occurs at CpG dinucleotides and approximately 70% of such sites are modified, the exceptions being CpG islands (p. 22) which are CpG-rich. The pattern of DNA methylation carries a substantial information content over and above that inherent in the underlying DNA sequence (Primrose *et al.* 2001).

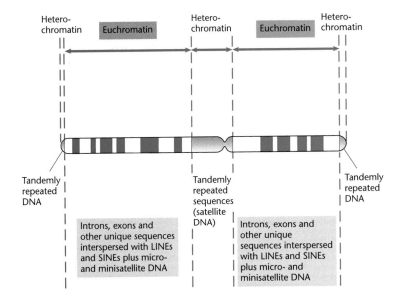

Fig. 2.24 The location of repeated sequences within a typical chromosome.

Suggested reading

Bernardi G. (2000) Isochores and the evolutionary genomics of vertebrates. *Gene* **241**, 3–17. *This excellent review pulls together historical data on the physical analysis of genomes and relates the compositional heterogeneity that was found to gene organization and chromosome banding.*

Eichler E.E. (2001) Recent duplication, domain accretion and the dynamic mutation of the human genome. *Trends in Genetics* **17**, 661–669. *An excellent review showing how duplicative transposition has shaped the human genome.*

Heslop-Harrison J.S. (2000) Comparative genome organization in plants: from sequence and markers to chromatin and chromosomes. *Plant Cell* **12**, 617–635. *Relates the material discussed in this chapter to chromatin structure and nuclear architecture from a purely plant perspective.*

Luning Prak E.T. & Kazazian H.H. (2000) Mobile elements and the human genome. *Nature Reviews Genetics* **1**, 134–144. *This is an excellent review on how DNA transposons and retrotransposons can contribute to genome plasticity.*

Pidoux A.L. & Allshire R.C. (2000) Centromeres: getting a grip of chromosomes. *Current Opinion in Cell Biology* **12**, 308–319. *This review is for the reader who wishes a better understanding of the interaction of the centromere with the other parts of the mitotic apparatus.*

Schueler M.G. *et al.* (2001) Genomic and genetic definition of a functional human centromere. *Science* **294**, 109–115. *This paper describes the first really successful attempt at determining the structure of a centromere from a higher eukaryote.*

Shore D. (2001) Telomeric chromatin: replicating and wrapping up chromosome ends. *Current Opinion in Genetics and Development* **11**, 189–198. *This review presents a biological perspective on telomeres and focuses on the replication of telomeres, the protection of chromosome ends from DNA damage, and the establishment and maintenance of telomeric heterochromatin.*

Wood V. *et al.* (2002) The genome sequence of *Schizosaccharomyces pombe*. *Nature* **415**, 871–80. *This key paper was published after this edition went to press but is important because it details the differences in genome organization between* S. cerevisiae *and* S. pombe. *For example,* S. pombe *has 20 times more introns than* S. cerevisiae.

Useful websites

http://www.cbs.dtu.dk/databases/DOGS/
This webpage provides links to a number of databases of genome sizes. Each database is very detailed and covers a broad spectrum of organisms.

http://isis.bit.uq.edu.au/
This website hosts the intron sequence and information database and is searchable by many different criteria.

http://www.cs.rhbnc.ac.uk/home/jhancock/simple/
This website provides an analysis of the mean levels of repetition for different genomes.

http://www3.edi.ac.uk/Research/Mitbase/mitbase.pl
This site provides a comprehensive database on mitochondria from vertebrates, invertebrates, plants and protists and has links to other websites on mitochondria.

http://charon.girinst.org/
This is the site of the Genetic Information Research Institute and is the entry point for Repbase, a database of prototypic sequences representing repetitive DNA from different eukaryotic species.

CHAPTER 3

Subdividing the genome

Introduction

As outlined in Chapter 1, the first step in sequencing a genome is to divide the individual chromosomes in an ordered manner into smaller and smaller pieces that ultimately can be sequenced. That is, one begins by creating a genomic library. However, at some stage the different clones have to be ordered into a physical map corresponding to that found in the intact organism. The magnitude of this task depends on the average size of the cloned insert in the library: the larger the insert, the fewer clones that have to be ordered.

The number of clones required can be calculated easily. If the average size of the cloned insert is 20 kb and the genome has a size of 2.8×10^6 kb, e.g. the human genome, then the size of the genome relative to the size of the cloned insert, designated n, is $2.8 \times 10^6/20 = 1.4 \times 10^5$. The number of independent recombinants required in the library must be greater than n. This is because sampling variation will lead to the inclusion of some sequences several times and the exclusion of other sequences in a library of just n recombinants. Clarke & Carbon (1976) have derived a formula that relates the probability (P) of including any DNA sequence in a random library of N independent recombinants:

$$N = \frac{\ln (1 - P)}{\ln (1 - 1/n)}.$$

Therefore, to achieve a 95% probability ($P = 0.95$) of including any particular sequence in a random human genomic library of 20 kb fragment size:

$$N = \frac{\ln (1 - 0.95)}{\ln (1 - 1/1.4 \times 10^5)} = 4.2 \times 10^5.$$

If the probability is to be increased to 99%, then N becomes 6.5×10^5. Put a different way, a threefold coverage gives a 95% probability of including any sequence and a fivefold coverage a 99% probability.

These calculations assume equal representation of sequences, but this is not true in practice.

Fragmentation of DNA with restriction enzymes

Type II restriction endonucleases have target sites which are 4–8 bp in length. If all bases are equally frequent in a DNA molecule then we would expect a tetranucleotide to occur on average every 4^4 (i.e. 256) nucleotide pairs in a long random DNA sequence. Similarly, a hexanucleotide would occur every 4^6 (i.e. 4096) bp and an octanucleotide every 4^8 (i.e. 65 536) bp. Table 3.1 shows the *expected* number of fragments that would be produced when different genomes are digested completely with different restriction endonucleases. In practice, the actual number of fragments is quite different because the distribution of nucleotides is non-random and most organisms do not have an equal number of the four bases; e.g. human DNA has overall only 40% G + C and the frequency of the dinucleotide CpG is only 20% of that expected. In addition, many cytosine residues are methylated and this can prevent restriction endonuclease digestion. Thus the enzyme *Not*I, which recognizes the sequence GCGGCCGC, cuts human DNA into fragments of average size 1000–1500 kb rather than the 65 kb expected. Similarly, the *Escherichia coli* genome is cut by *Not*I into only 20 fragments, not the 72 expected from Table 3.1 (Smith *et al.* 1987). Again, the *Schizosaccharomyces pombe* genome, which is slightly smaller than that of *Saccharomyces cerevisiae*, is cut into only 14 fragments (Fan *et al.* 1989). In addition to *Not*I there are a number of other restriction endonucleases with 8-bp recognition sequences (Table 3.2). In genomes rich in G + C the tetranucleotide CTAG is particularly rare (McClelland *et al.* 1987), such that the enzymes *Spe*I (ACTAGT), *Xba*I (TCTAGA), *Nhe*I (GCTAGC) and *Bln*I (CCTAGG) can produce a more

Table 3.1 Expected number of DNA fragments produced from different genomes by restriction endonucleases with tetra-, hexa- and octanucleotide recognition sequences. The expected number assumes that the DNA has a 50% G + C content and there is a totally random distribution of the four bases along any one strand of DNA.

Organism	Haploid genome size (n)	Expected number of fragments		
		4-Cutter ($n/256$)	6-Cutter ($n/4096$)	8-Cutter ($n/65\ 536$)
Escherichia coli	4.70×10^6 bp	18 359	1147	72
Saccharomyces cerevisiae	1.35×10^7 bp	52 734	3296	206
Drosophila melanogaster	1.80×10^8 bp	703 125	43 945	2746
Homo sapiens	2.80×10^9 bp	1093 750	683 593	42 274

Table 3.2 Restriction endonucleases with 8-bp recognition sequences.

Endonuclease	Recognition sequence
*Not*I	GCGGCCGC
*Sfi*I	GGCCnnnnGGCC
*Swa*I	ATTTAAAT
*Pac*I	TTAATTAA
*Pme*I	GTTTAAAC
*Sgr*AI	CACCGGTG
*Sse*8387I	CCTGCAGG
*Srf*I	GCCCGGGC
*Sgf*I	GCGATCGC
*Fse*I	GGCCGGCC
*Asc*I	GGCGCGCC

limited number of DNA fragments than might at first be expected.

Gelfand and Koonin (1997) have analysed those bacterial genomes that have been completely sequenced and have found that short palindromic sequences, like restriction endonuclease recognition sites, are deficient at a statistically significant level. They suggest that in the course of evolution bacterial DNA has been exposed to a wide spectrum of restriction enzymes, probably as a result of latent transfer mediated by mobile genetic elements.

The average size of DNA fragment produced by digestion with restriction enzymes with 4- and 6-base recognition sequences is too small to be of much use for preparing gene libraries except in special circumstances (see below). Even enzymes with 8-base recognition sequences may not be of particular value because, although the average size of fragment should be 65.5 kb, in the case of *S. pombe*, in practice the fragments range in size from 4.5 kb to 3.5 Mb (Fan *et al.* 1989). If more uniform-sized fragments are required it is usual to partially digest the target DNA with an enzyme with a 4-base recognition sequence. The partial digest then can be fractionated (see next section) to separate out fragments of the desired size. Because the DNA is randomly fragmented there will be no exclusion of any sequence. Furthermore, clones will overlap one another (Fig. 3.1) and this is particularly important when trying to order different clones into a map (see Chapter 4).

Some introns encode endonucleases that are site-specific and have 18–30 bp recognition sequences (Dujon *et al.* 1989). These endonucleases can be used to produce a very limited number of fragments, some or all of which are produced by cleavage within related genes. For example, the intron-encoded endonuclease I-*Cen*I from the chloroplast large rRNA gene of *Chlamydomonas eugametos* cuts the chromosomes of *E. coli* and *Salmonella* seven times, once within each of the seven *rrn* operons (Liu *et al.* 1993). The I-*Sce*I intron encoded endonuclease from *S. cerevisiae* has an 18 bp recognition site and the VDE endonuclease from the same organism has a 30 bp recognition site. Assuming a random organization of sequences, the frequency of occurrence of an 18 bp site is one per 7×10^{10} bp (4^{18}). Thus there are probably few or no endogenous sites even within very complex genomes (Jasin 1996).

New sites for rare-cutting endonucleases can be introduced into genomes to facilitate gene mapping.

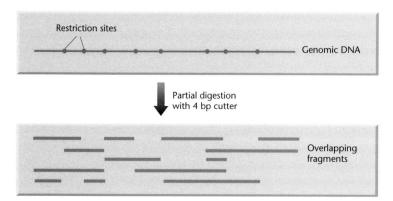

Fig. 3.1 The generation of overlapping fragments of genomic DNA following partial digestion with a restriction enzyme with a tetranucleotide recognition sequence.

For example, in bacteria a number of temperate phages and transposons contain single *Not*I sites and their insertion into the chromosome will create one or more new sites depending on the number of integration events (Smith *et al.* 1987; Le Bourgeois *et al.* 1992). Similarly, an integrative vector was used to generate sites in yeast for the intron-encoded endonuclease I-*Sce*I (Thierry & Dujon 1992). A variation of this technique can be used to reduce the number of cleavage sites. A novel *E. coli lac* operator containing a site for *Hae*II (Pu GCGC Py) was introduced into the genomes of *E. coli* and yeast. Addition of the *lac* repressor protein protects this site from methylation by the M.*Hha* methyltransferase which recognizes the sequence GCGC found in all *Hae*II recognition sites. Consequently, a single site remains that is susceptible to cleavage by *Hae*II, the so-called *Achilles heel* site (Koob & Szbalski 1990).

The RecA-assisted restriction endonuclease (RARE) technique for reducing the number of cleavage sites involves the RecA-mediated formation of a triplex structure between an oligonucleotide and a chosen locus on the chromosome containing the site of a given restriction enzyme (Koob & Szbalski 1990; Ferrin & Camerini-Otero 1991). After methylation of the genome by the corresponding methyltransferase and removal of the RecA protein, only the protected site is cleaved by the endonuclease. Two such targeted cuts permit the excision of a unique DNA fragment from the genome (Szybalski 1997) and this methodology has been used to facilitate optical mapping (see p. 64). Although these methods for reducing the number of restriction sites potentially are very powerful, they are technically difficult and are not widely used.

Separating large fragments of DNA

As a DNA molecule is digested with a restriction endonuclease it is cut into smaller and smaller pieces. As such, the number of smaller molecules will always exceed the number of larger molecules. Small fragments ligate more efficiently and clones with small inserts transform with a higher efficiency. If there is no size selection then in a random cloning exercise the recombinants will have a preponderance of small inserts. Some vectors have an automatic size selection, e.g. the λ phage (5–25 kb) and cosmid (35–45 kb) vectors, but most do not. Thus it is common practice after DNA digestion to isolate DNA fragments of the desired size.

The most widely used method of separating large DNA molecules is by electrophoresis in agarose. An agarose gel is a complex network of polymeric molecules whose average pore size is dependent on the buffer composition and the type and composition of agarose used. Upon electrophoresis, DNA molecules display elastic behaviour by stretching in the direction of the applied field and then contracting into balls. The larger the pore size of the gel, the greater the ball of DNA that can pass through and hence the larger the molecules that can be separated, i.e. the gel acts as a sieve. Once the globular volume of the DNA molecule exceeds the pore size, the DNA molecule can only pass through by reptation (i.e. a snake-like movement). This leads to size-independent mobility which occurs with conventional agarose gel electrophoresis above 20–30 kb. To achieve reproducible separations of DNA fragments larger than this it is necessary to use pulsed electrical fields.

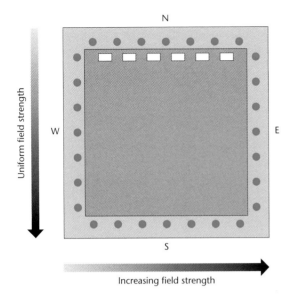

Fig. 3.2 Instrumentation for pulsed field gel electrophoresis. The platinum electrodes are shown as green dots and the gel slots are positioned along one edge of the gel matrix which is shown in grey. The letters N, S, E and W refer to the electrical field orientation. (Adapted from Schwartz & Cantor 1984.)

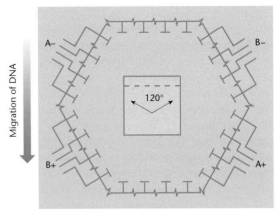

Fig. 3.3 Schematic representation of contour-clamped homogeneous electrical field (CHEF) pulsed field gel electrophoresis.

Pulsed field gel electrophoresis (PFGE) was developed by Schwartz & Cantor (1984). The technique employs alternately pulsed perpendicularily oriented electrical fields (Fig. 3.2). The duration of the applied electrical pulses is varied from 1 to 90 s permitting separation of DNAs with sizes from 30 kb to 2 Mb. The mechanism by which this occurs is believed to be as follows. When a large DNA molecule enters a gel in response to an electrical field it must elongate parallel to the field. This field then is cut off and a new field applied at right angles to the long axis of the DNA. The molecule is unable to move until it re-orientates in a new direction and the time required for this will depend on molecular weight. Repeating the cycle results in each DNA molecule having a characteristic net mobility along the diagonal of the gel. The use of non-uniform electrical fields is critical in achieving high resolution. The reason for this is that the leading edge of a DNA band is exposed to a weaker electrical field than the trailing edge and thus the band is subject to constant compression. The usefulness of this method has been demonstrated by the separation of the different fragments of DNA produced by *Not*I digestion of *E. coli* and *S.*

pombe DNA (Smith *et al.* 1987; Fan *et al.* 1989) and the separation of intact *S. cerevisiae* chromosomes (Schwartz & Cantor 1984).

A major disadvantage of PFGE, as originally described, is that the DNA samples do not run in straight lines, making analysis by Southern blotting, etc., difficult. This problem has been overcome by the development of improved methods for alternating electrical fields including orthogonal-field-alteration gel electrophoresis (OFAGE), field-inversion gel electrophoresis (FIGE), transversely alternating-field electrophoresis (TAFE) and contour-clamped homogeneous electrical field electrophoresis (CHEF) (Cantor *et al.* 1988). CHEF, which was first described by Chu *et al.* (1986), is the most commonly used method for genomes of all types. It uses a hexagonal array of fixed electrodes (Fig. 3.3) and this creates a homogeneous electrical field resulting in enhanced resolution of DNA fragments. The DNA tracks run in straight lines and are comparable across the gel, thus making calibration with markers and the interpretation of Southern blots simpler than with the original PFGE. With these developments the reproducible size resolution has been increased to 10 Mb which is greater in size than most prokaryotic genomes.

Isolation of chromosomes

The task of lining up the different clones that make up a gene library can be simplified if the individual chromosomes are separated prior to digestion. PFGE permits the separation of yeast chromosomes (Fan *et al.* 1989), the largest of which is 3–5 Mb. However, the chromosomes of higher eukaryotes are much larger than this. For example, *Drosophila* chromosome 2 is 67 Mb in length and human chromosomes vary in size from 50 Mb (chromosome 21) to 263 Mb (chromosome 1). Electrophoresis cannot separate molecules this large. Instead, fluorescence-activated cell sorting (FACS), also known as flow karyotyping, has to be used (Davies *et al.* 1981).

In FACS, chromosome preparations are stained with a DNA binding dye which can fluoresce in a laser beam. The amount of fluorescence exhibited by a given chromosome is proportional to the amount of dye bound. This, in turn, is proportional to the amount of the DNA and hence the size of the chromosome. Chromosomes can therefore be fractionated by size in a FACS machine (Fig. 3.4). A stream of droplets containing single stained chromosomes is passed through a laser beam at the rate of 2000 chromosomes per second and the fluorescence from each is monitored by a photomultiplier. When the fluorescence intensity indicates that the chromosome illuminated by the laser is the one desired, the charging collar puts an electrical charge on the droplet. When droplets containing the desired chromosome pass between charged deflection plates, they are deflected into a collection vessel. Uncharged droplets lacking the desired chromosome pass into a waste collection vessel. In a variation of the above technique, two fluorescent dyes are used: one binds preferentially to AT-rich DNA and the other binds preferentially to GC-rich DNA. The stained chromosomes pass through a point on which a pair of laser beams are focused, one beam to excite the fluorescence of each dye. Each chromosome type has characteristic numbers of AT and GC base pairs so chromosomes can be identified by a combination of the total fluorescence and the ratio of the intensities of the fluorescence emissions from the two dyes.

The kind of separation that can be achieved with human chromosomes is shown in Fig. 3.5. Note that some of the chromosomes cannot be separated in the FACS because they are of similar size and AT : GC ratio. However, separation can be achieved if *somatic cell hybrids* are used as the source of DNA (D'Eustachio & Ruddle 1983). These hybrids con-

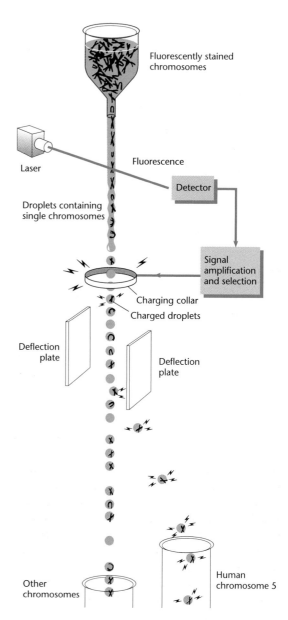

Fig. 3.4 Schematic representation of single chromosome separation by fluorescence-activated cell sorting (FACS; see text for details). (Redrawn with permission from Dogget 1992, courtesy of University Science Books.)

tain a full complement of chromosomes of one species but only one or a limited number of chromosomes from the second. They are formed by fusing the cells of the two different species and applying conditions that select against the two donor cells. One of the parents may be sensitive to a particular

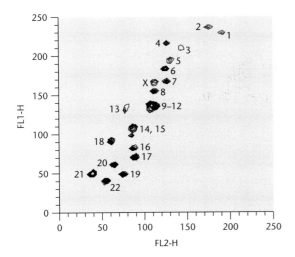

Fig. 3.5 Separation of different human chromosomes by FACS using two fluorescent dyes, Hoechst 33258 (FL1-H) and chromomycin A3 (FL2-H). (Kindly supplied by Dr D. Davies, Imperial Cancer Research Fund.)

drug and the other might be a mutant requiring special conditions for growth, e.g. thymidine kinase negative cells do not grow in hypoxanthine, aminopterin and thymidine (HAT) medium. In the presence of the drug and HAT medium, only hybrid cells containing a functional thymidine kinase gene can grow. If hybrid cells are grown under non-selective conditions after the initial selection, chromosomes from one of the parents tend to be lost more or less at random. In the case of human–rodent fusions, which are the most common, the human chromosomes are preferentially lost. Eventually only one or a few chromosomes from one parent remain. In this way rodent cell lines containing one or two human chromosomes have been constructed and these can be used to isolate the individual human chromosomes by FACS.

Lee *et al.* (1994) have shown that chromosome sorting from hybrid cells can be simplified if rodent cells are replaced with cells from the Indian Muntjac deer. Cells of the Muntjac deer have only a very small number of giant chromosomes: two autosomes plus X and Y. Thus donor chromosomes in hybrid cells are easily separated from host chromosomes. Not only were highly purified human chromosomes obtained by flow cytometry but they were obtained at over 90% purity by rate zonal centrifugation in sucrose gradients.

Chromosome microdissection

Particular subchromosomal regions, such as a chromosomal band (see Fig. 2.24), can be obtained by microdissection of metaphase chromosomes. A micromanipulator with very fine needles is used to cut out the desired band from individual chromosomes. When sufficient material has been collected the DNA is cloned. The technique originally was developed for use with *Drosophila* and mouse DNA and at least 100 chromosomes were needed for dissection. Ludecke *et al.* (1989) have simplified this technique. Only a few chromosomes are dissected and after cloning into a vector this DNA is amplified by means of polymerase chain reaction (PCR).

Vectors for cloning DNA

Once the genome has been fragmented it is essential to propagate each fragment to enable it to be mapped and ultimately sequenced. Although many different types of vector have been developed (for review see Primrose *et al.* 2001) only a few are of use for large-scale genome sequencing projects. The early work on construction of gene libraries made use of bacteriophage λ vectors. With such vectors the largest size of DNA that can be accommodated is about 25 kb. If *in vitro* packaging is used, then a large number of independent recombinants can be selected.

In place of the phage λ vectors, cosmid vectors may be chosen. These also have the high efficiency afforded by packaging *in vitro* and have an even higher capacity than any phage λ vector. However, there are two drawbacks in practice. First, most workers find that screening libraries of phage λ recombinants by plaque hybridization gives cleaner results than screening libraries of bacteria containing cosmid recombinants by colony hybridization. Plaques usually give less of a background hybridization than do colonies. Secondly, it may be desirable to retain and store an amplified genomic library. With phage, the initial recombinant DNA population is packaged and plated out. It can be screened at this stage. Alternatively, the plates containing recombinant plaques can be washed to give an *amplified* library of recombinant phage. The amplified library can then be stored almost indefinitely: phage have a long shelf-life. The amplification is so great that samples of this amplified library could be plated out and screened with different probes on hundreds of occasions. With bacterial colonies containing cosmids it is also possible to store an amplified library (Hanahan & Meselson 1980), but bacterial populations cannot be stored as readily as phage populations. There is often an unacceptable loss of viability when the bacteria are stored.

A word of caution is necessary when considering the use of any amplified library. There is the possibility of *distortion*. Not all recombinants in a population will propagate equally well, e.g. variations in target DNA size or sequence may affect replication of a recombinant phage, plasmid or cosmid. Therefore, when a library is put through an amplification step particular recombinants may be increased in frequency, decreased in frequency or lost altogether. Factors which affect sub-clone representation include the nature and complexity of repeat sequences, length of the repeat region and insert orientation (Chissoe *et al.* 1997). Notable differences in sub-clone representation also can occur between related vectors. Development of modern vectors and cloning strategies has simplified library construction to the point where many workers now prefer to create a new library for each screening, rather than risk using a previously amplified one.

Genomic DNA libraries in phage λ vectors are expected to contain most of the sequences of the genome from which they have been derived. However, deletions can occur, particularly with DNA containing repeated sequences. As noted in Chapter 2, repeated sequences are widespread in the DNA from higher eukaryotes. The deletion of these repeated sequences is not prevented by the use of recombination-deficient strains.

Yeast artificial chromosomes

The upper size limit of 35–45 kb for cloning in a cosmid means that it would take 4500 clones to cover the *D. melanogaster* genome and 70 000 clones to cover the human genome. What is required is a vector that permits a larger insert size and yeast artificial chromosomes (YACs) make this possible. It will be recalled from Chapter 2 that the minimum structural elements for a linear chromosome are an origin of replication (*ars*), telomeres and a centromere. Murray and Szostak (1983) combined an *ars* and a centromere from yeast with telomeres from *Tetrahymena* to generate a linear molecule that behaved as a chromosome in yeast. When the requirements for normal replication and segregation were studied it was found that the length of the YAC was important. When the YAC was less than 20 kb in size, centromere function was impaired. However, much larger YACs segregated normally. Burke *et al.* (1987) made use of this fact in developing a vector (Fig. 3.6) for cloning large DNA molecules. They showed that YACs could be used to generate whole libraries from the genomes of higher organisms with insert sizes at least 10-fold larger than those that can be accommodated by bacteriophage λ and cosmid vectors.

A major drawback with YACs is that although they are capable of replicating large fragments of DNA, manipulating such large DNA fragments in the liquid phase prior to transformation, and keeping them intact, is very difficult. Thus many of the early YAC libraries had average insert sizes of only 50–100 kb. By removing small DNA fragments by PFGE fractionation prior to cloning, Anand *et al.* (1990) were able to increase the average insert size to 350 kb. By including polyamines to prevent DNA degradation, Larin *et al.* (1991) were able to construct YAC libraries from mouse and human DNA with average insert sizes of 700 and 620 kb, respectively. Later Bellanné-Chantelot *et al.* (1992) constructed a human library with an average insert size of 810 kb and with some inserts as large as 1800 kb.

There are a number of operational problems associated with the use of YACs (Kouprina *et al.* 1994; Monaco & Larin 1994). The first of these is that it is estimated that 10–60% of clones in existing libraries represent chimaeric DNA sequences: i.e. sequences from different regions of the genome cloned into a single YAC. Chimaeras may arise by co-ligation of DNA inserts *in vitro* prior to yeast transformation, or by recombination between two DNA molecules that were introduced into the same yeast cell. It is possible to detect chimaeras by *in situ* hybridization of the YAC to metaphase chromosomes: hybridization to two or more chromosomes or to geographically disparate regions of the same chromosome is indicative of a chimaera.

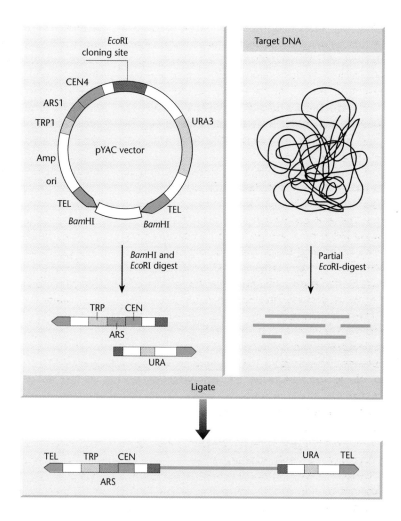

Fig. 3.6 Construction of a yeast artificial chromosome (YAC) containing large pieces of cloned DNA. Key regions of the pYAC vector are as follows: TEL, yeast telomeres; ARS1, autonomously replicating sequence; CEN4, centromere from chromosome 4; URA3 and TRP1, yeast marker genes; Amp, ampicillin-resistance determinant of pBR322; ori, origin of replication of pBR322.

A second problem with YACs is that many clones are unstable and tend to delete internal regions from their inserts. Using a model system, Kouprina *et al.* (1994) were able to show that deletions can be generated both during the transformation process and during mitotic growth of transformants and that the size of the deletions varied from 20 to 260 kb. Ling *et al.* (1993) showed that the frequency of deletion formation could be reduced by use of a strain rendered recombination-deficient as a result of a *rad52* mutation. However, such strains grow more slowly and transform less efficiently than RAD[+] strains and therefore are not ideal hosts for YAC library construction. Le & Dobson (1997) have shown that the *rad54-3* allele significantly stabilizes YAC clones containing human satellite DNA sequences. Strains carrying this allele can undergo meiosis and have

growth and transformation rates comparable with wild-type strains. Heale *et al.* (1994) have shovn that chimaera formation results from the yeast's mitotic recombination system which is stimulated by the sphaeroplasting step of the standard YAC transformation system. Transformation of intact yeast cells is much less recombinogenic. An additional limitation on the use of YACs is the high rate of loss of some YACs during mitotic growth.

The third major problem with YAC clones is that the 15 Mb yeast host chromosome background cannot be separated from the YACs by simple methods, nor is the yield of DNA very high. Unlike plasmid vectors in bacteria, YACs have a structure very similar to natural yeast chromosomes. Thus, purifying YAC from the yeast chromosomes usually requires separation by PFGE. Alternatively, the entire yeast genome is subcloned in bacteriophage or cosmid vectors followed by identification of those clones carrying the original YAC insert.

P1-derived and bacterial artificial chromosomes as alternatives to yeast artificial chromosomes

Just as the problems with YACs were beginning to surface, two new classes of vector were developed

that were based on the *E. coli* transducing phage P1 and the *E. coli* sex factor F. Fortuitously, these new vector systems, termed P1-derived artificial chromosomes (PACs) and bacterial artificial chromosomes (BACs), respectively, are particularly suitable for handling large genomic fragments because insert rearrangements are minimal. The initial P1 vectors were developed by Sternberg *et al.* (Sternberg 1990; Pierce *et al.* 1992) and have a capacity for DNA fragments as large as 100 kb. Thus the capacity is about twice that of cosmid clones but less than that of YAC

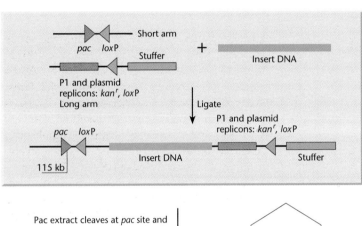

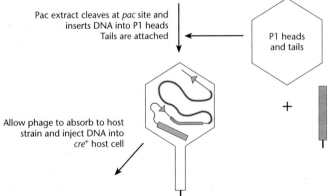

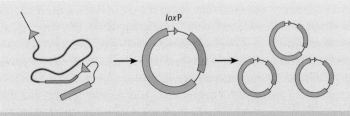

Cre recombinase protein circularizes injected DNA at the *lox*P sites. DNA replicates using plasmid replicon. Plasmid copy number is increased by induction of P1 lytic replicon.

Fig. 3.7 The phage P1 vector system. The P1 vector Ad10 (Sternberg 1990) is digested to generate short and long vector arms. These are dephosphorylated to prevent self-ligation. Size-selected insert DNA (85–100 kb) is ligated with vector arms, ready for a two-stage processing by packaging extracts. First, the recombinant DNA is cleaved at the *pac* site by pacase in the packaging extract. Then the pacase works in concert with head–tail extract to insert DNA into phage heads, *pac* site first, cleaving off a headful of DNA at 115 kb. Heads and tails then unite. The resulting phage particle can inject recombinant DNA into host *E. coli*. The host is *cre*+. The *cre* recombinase acts on *lox*P sites to produce a circular plasmid. The plasmid is maintained at low copy number, but can be amplified by inducing the P1 lytic operon.

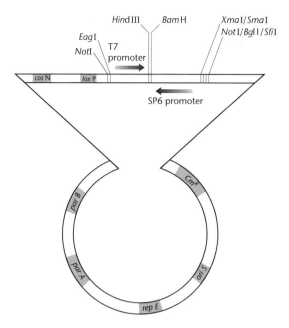

Fig. 3.8 Structure of a BAC vector derived from a mini-F plasmid. The *oriS* and *repE* genes mediate the unidirectional replication of the F factor, while *parA* and *parB* maintain the copy number at a level of one or two per genome. Cm^R is a chloramphenicol-resistance marker. *cosN* and *loxP* are the cleavage sites for λ terminase and P1 *cre* protein, respectively. *Hind*III and *Bam*HI are unique cleavage sites for inserting foreign DNA. (Adapted from Shizuya *et al.* 1992.)

Table 3.3 Maximum DNA insert possible with different cloning vectors.

Vector	Host	Insert size
λ phage	*E. coli*	5–25 kb
λ cosmids	*E. coli*	35–45 kb
P1 phage	*E. coli*	70–100 kb
PACs	*E. coli*	100–300 kb
BACs	*E. coli*	≤300 kb
YACs	*S. cerevisiae*	200–2000 kb

clones. These vectors contain a packaging site (*pac*) that is necessary for *in vitro* packaging of recombinant molecules into phage particles. They also contain two *loxP* sites. These *loxP* sites are recognized by the phage recombinase, the product of the phage *cre* gene, and this leads to circularization of the packaged DNA after it has been injected into an *E. coli* host expressing the recombinase (Fig. 3.7). Clones are maintained in *E. coli* as low copy number plasmids by selection for a vector kanamycin-resistance marker but a high copy number can be induced by exploitation of the P1 lytic replicon.

A vector based on the single-copy sex factor F initially was developed by Shizuya *et al.* (1992) and called a BAC (Fig. 3.8). It carries the λ*cos*N and P1 *loxP* sites, two cloning sites (*Hind*III and *Bam*HI) and several G + C restriction enzyme sites (e.g. *Sfi*I, *Not*I, etc.) for potential excision of the inserts. T7 and SP6 promoters for generating RNA probes also flank the cloning site. BACs can be transformed into *E. coli* very efficiently, thus avoiding the packaging extracts

that are required with the P1 system, and stably maintain genomic fragments in excess of 300 kb. Later, Ioannou *et al.* (1994) developed a PAC by combining features of both the P1 and F-factor systems. The key difference between the two types of vectors is that BACs are always maintained at very low copy numbers whereas a high copy number can be induced in PACs. Like BACs, PAC vectors can also stably maintain inserts in the 100–300 kb range.

Although the inserts in BACs and PACS usually are much smaller than those found in YACs (Tables 3.3 and 3.4), BACs and PACs have become the standard cloning vectors for projects on genomic sequencing and analysis. This is partly because of the ease with which they can be handled and the ability to isolate large quantities of pure clone DNA. However, the most important reason is the greatly increased stability of BAC and PAC clones compared with YAC clones. For example, using these vectors it has proved possible to stably clone highly repetitive alphoid DNA (Harrington *et al.* 1997; Ebersole *et al.* 2000; Mann & Huxley 2000). Because BAC and PAC libraries are being used in genomic sequencing projects, a low level of chimaerism is essential and when Osoegawa *et al.* (2000) made a detailed study of a BAC-based murine library they found chimaeras to be below 1%. A similar study on 169 BAC clones with large human DNA inserts revealed no chimaeras and minor rearrangements in only 11% of them (Osoegawa *et al.* 2001).

The original BAC vector (pBAC108L) was not ideal because it lacked a selectable marker for recombinants. Hence, recombinant clones had to be identified by colony hybridization which is not well-suited for library development. Two new vectors

Table 3.4 Size of different genomes relative to the average DNA insert that is obtained with different cloning vectors.

Organism	Haploid genome size	Size of genome relative to size of cloned DNA (n)			
		Cosmid (40 kb)	P1 (85 kb)	BAC/PAC (250 kb)	YAC (1000 kb)
E. coli	4.70×10^6 bp	118	55	19	5
S. cerevisiae	1.35×10^7 bp	338	159	54	14
D. melanogaster	1.80×10^8 bp	4500	2118	720	180
H. sapiens	2.80×10^9 bp	70 000	32 941	11 200	2800

were developed, pBELOBAC11 and pECBAC1 (Kim et al. 1996; Frijter et al. 1997), in which the *lacZ* gene was introduced as a scorable marker. The only difference between these two vectors is that the former has two sites for *Eco*RI, one in the *lac* insert and the other in the chloramphenicol resistance gene, whereas the latter vector has had the second *Eco*RI site removed. Another improved BAC (pBACe3.6) has been developed by Frengen et al. (1999) in which a *sac*B gene containing multiple cloning sites is used as the selectable marker. E. coli strains carrying *sac*B synthesize a levansucrase which generates toxic metabolites in the presence of high concentrations of sucrose unless the gene is interrupted by a DNA insert. Two additional features of PBACe3.6 deserve comment. First, it contains a single 18 bp recognition site for the PI-*Sce*I nuclease. This feature permits linearization of the recombinant DNA, irrespective of the characteristics of the insert sequence, and is an essential requirement for

the transformation of mammalian cells with BAC and PAC clones. The second noteworthy feature of pBACe3.6 is the presence of an *att*Tn7 sequence which permits retrofitting (see below) by transposon Tn7-mediated insertion of desirable sequence elements.

Retrofitting

Once a set of BAC of PAC clones has been generated, there are many different uses to which they can be put. It is almost axiomatic that, regardless of the application, the clones will require further modification. Such modifications are carried out by retrofitting. Although there are a number of ways of doing retrofitting they share certain features. One method (Chatterjee & Coren 1997; Chatterjee et al. 1999) involves the use of transposons carrying a *lox*P site. Figure 3.9 shows an example of the use of this

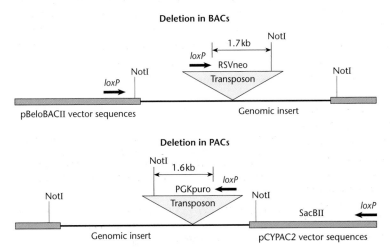

Fig. 3.9 Transposon-mediated retrofitting and deletion formation in a BAC (pBELOBAC 11) and a PAC (pCYPAC2). RSVneo is a neomycin-resistance gene under the control of a respiratory syncytial virus (RSV) promoter and *PGKpuro* is a puromycin-resistance gene under the control of a mammalian PGK promoter. (Redrawn with permission from Chatterjee et al. 1999.)

system to introduce markers selectable in mammalian cells. In one case the transposon carried *lox*P and a gene encoding neomycin resistance driven by a promoter from respiratory syncytial virus. In another, the transposon carried *lox*P and a puromycin resistance gene driven by a mammalian PGK promoter. When transposition is mediated *in vivo*, the transposons are most likely to insert into the genomic DNA because this is so much larger than the vector DNA. Because clones carrying the transposons have two *lox*P sites they are subject to deletion by the *cre* recombinase. If a series of different transposants is selected then a whole series of different deletion variants of the insert can be generated. Furthermore, by linearizing the DNA and then transforming it into mammalian cells it is possible to test the functionality of these deletion derivatives.

Kaname & Huxley (2001) have used a modification of the above procedure to retrofit BACs and PACs. In their procedure one starts with simple cloning vectors carrying a single *lox*P site and inserts any DNA that is to be retrofitted. The vector plus insert then is transformed into a cell carrying the BAC or PAC whereupon the *cre* recombinase will generate a hybrid vector by homologous recombination. The advantage of this method over that described previously is that the genomic insert is left untouched. Another way of retrofitting BACs and PACs while leaving the genomic insert untouched is to make use of the attTn7 site in the vector pBACe3.6 (described above). Any DNA to be retrofitted onto the BAC is first inserted in the transposon Tn7 (Frengen *et al.* 2000). When this transposon is introduced into a cell carrying pBACe3.6 there is a high probability that it will insert at the *att*Tn7 site.

Genomic libraries constructed in BACs and PACs have been extensively utilized in the generation of physical maps (see Chapter 4) and as substrates for genome sequencing (see Chapter 5). However, for functional analysis, cloned DNA needs to be introduced back into the same kind of cells from which it was derived. This can be done by cloning it into a suitable vector or by retrofitting the BAC or PAC as described above. The problem with this approach is that each individual genomic clone has to be manipulated independently. To eliminate this problem a number of groups have developed shuttle BACs and PACs. For example, Liu *et al.* (1999a) have developed a transformation-competent artificial chromosome

(TAC) that can maintain large DNA fragments in *E. coli*, *Agrobacterium tumefaciens* and *Arabidopsis thaliana*. In a similar development, Frengen *et al.* (2000) and Coren & Sternberg (2001) have developed BACs and PACs that can be shuttled from *E. coli* to mammalian cells where they are stably maintained.

Choice of vector

The maximum size of insert that the different vectors will accommodate is shown in Table 3.3, and Table 3.4 shows the size of different genomes relative to the size of insert with the various vectors. However, as noted above, the size of insert is not the only feature of importance. The absence of chimaeras and deletions is even more important. In practice, some 50% of YACs show structural instability of inserts or are chimaeras in which two or more DNA fragments have become incorporated into one clone. These defective YACs are unsuitable for use as mapping and sequencing reagents and a great deal of effort is required to identify them. Cosmid inserts sometimes contain the same abberations and the greatest problem with them arises when the DNA being cloned contains tandem arrays of repeated sequences. The problem is particularly acute when the tandem array is several times larger than the allowable size of a cosmid insert. Potential advantages of the BAC and PAC systems over YACs include lower levels of chimaerism, ease of library generation and ease of manipulation and isolation of insert DNA. BAC clones seem to represent human DNA far more faithfully than their YAC or cosmid counterparts and appear to be excellent substrates for shotgun sequence analysis resulting in accurate contiguous sequence date (Venter *et al.* 1996).

Suggested reading

Brune W., Messerle M. & Koszinowski U.H. (2000) Forward with BACs: new tools for herpesvirus genomics. *Trends in Genetics* **16**, 254–259. *A short review that illustrates how complete viral genomes can be cloned in BACs and details the advantages of doing this.*

Osoegawa K. *et al.* (2000) Bacterial artificial chromosome libraries for mouse sequencing and functional analysis. *Genome Research* **10**, 116–128.

Zeng C. *et al.* (2001) Large-insert BAC/YAC libraries for selective re-isolation of genomic regions by homologous recombination in yeast. *Genomics* **77**, 27–34.
These two papers are good examples of the way BACs and PACs are used to clone large fragments of DNA.

Useful website

http://rebase.neb.com
This is the reference database for information on all known restriction endonucleases.

CHAPTER 4

Assembling a physical map of the genome

Introduction

After a genome has been fragmented and the fragments cloned to generate a genomic library, it is necessary to assemble the cloned fragments in the same linear order as found in the chromosomes from which they were derived. Positioning cloned DNA fragments is analogous to completing the outside edge of a jigsaw puzzle but, rather than looking for interlocking pieces, detectable overlaps between clones are sought. Two or more clones that can be shown to overlap make up a *contig*. Ideally, one would like to identify a whole series of fragments that can be unambiguously shown to form a series of contigs that correspond exactly to the chromosome structure of the parent organism. Physical markers, which essentially are unique DNA sequences, are used to identify the overlaps and as a consequence these markers get 'mapped' on the genome thereby creating a physical map.

Originally, physical maps were seen as tools to facilitate the sequencing of complete genomes. However, it now is possible to shotgun sequence an entire genome without the existence of a clone-based map but for genomes whose size exceeds 100 Mb there still are great benefits to having a map (see p. 82). Today, physical maps have assumed an importance that far exceeds their utility in facilitating whole genome sequencing. Perhaps the easiest way of understanding their importance is by analogy. Suppose we decide to drive the 3500 miles from Los Angeles to New York. If the only map we have says that we will have to drive through Albuquerque and St Louis, and that these are so many miles from Los Angeles, then we have very limited information. Such a map is analogous to a genetic map of a chromosome: the mapped genes are not close enough to be of much utility. On the other hand, suppose that every mile between Los Angeles and New York there is a milepost giving the exact distance to both places. In this situation we will know exactly where we are

at any time but the information is of no other interest. This is analogous to knowing the exact base sequence of a chromosome or genome. What we really want for our journey is a map that has different features of interest (towns, rivers, national parks, museums) evenly spread along our route that will enable us to relate observations to locations and position to geographical features. A physical map of a chromosome is exactly the same in that different kinds of markers are identified by physical means and mapped along the chromosome. Genetic markers can be located with reference to the physical markers and vice versa. Although the exact DNA sequence of the chromosome enables the exact location of physical features to be determined, the physical map is more useful for day to day use.

Three general methodologies for mapping have been developed: restriction enzyme fingerprinting, marker sequences and hybridization assays. In each case the objective is to create a landmark map of the genome under study with markers dispersed at regular intervals throughout the map. In practice, composite maps are created in which different kinds of markers are located on the same map. Although most physical maps have been created by locating markers on cloned genomic fragments and then building contigs, a number of other methods for mapping markers have been developed. These include optical mapping, radiation hybrid mapping and HAPPY mapping.

Restriction enzyme fingerprinting

The principle of restriction enzyme fingerprinting was originally developed for the nematode *Caenorhabditis elegans* (Coulson *et al.* 1986) and yeast (Olson *et al.* 1986). In its original format the starting material was a genomic library made in cosmids. Each cosmid clone was digested with a restriction endonuclease with a hexanucleotide recognition

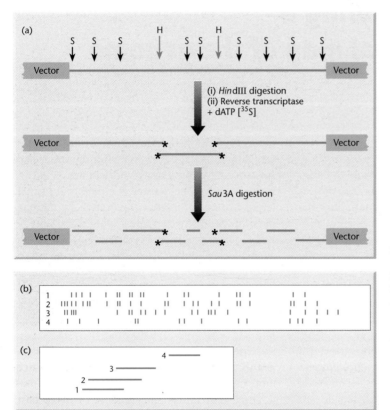

Fig. 4.1 The principle of restriction-fragment fingerprinting. (a) The generation of labelled restriction fragments (see text for details). (b) Pattern generated from four different clones. Note the considerable band sharing between clones 1, 2 and 3 indicating that they are contiguous whereas clone 4 is not contiguous and has few bands in common with the other three. (c) The contig map produced from data shown in (b). (Adapted and redrawn with permission from Coulson *et al.* 1986.)

sequence and which leaves staggered ends, e.g. *Hind*III. The ends of the fragments were labelled by end-filling with reverse transcriptase in the presence of a radioactive nucleoside triphosphate. The *Hind*III was destroyed by heating and the fragments cleaved again with a restriction endonuclease with a tetra-nucleotide recognition sequence, e.g. *Sau*3A. The fragments were separated on a high resolution gel and detected by autoradiography, the output being a clone fingerprint (Fig. 4.1). Note that the fingerprint is not an order of restriction sites; rather, it is a series of clusters of bands based on the probability of overlap of clones.

Although restriction enzyme fingerprinting was applied to genomes such as those from *Drosophila* (Siden-Kiamos *et al.* 1990), *Arabidopsis* (Hauge *et al.* 1991) and the human genome (Bellane-Chantelot *et al.* 1992; Trask *et al.* 1992), the method generally fell out of favour except for microbial genomes (for reviews see Cole & Saint Girons 1994; Fonstein & Haselkorn 1995). The reasons for this were twofold. First, cosmids with an average insert size of 40 kb

have relatively few sites for any single restriction endonuclease. This means that to generate the necessary numbers of fragments to allow statistically significant matching of overlaps it is necessary to use double digests and radioactive labelling methods. This methodology is not reproducible, nor is it amenable to large-scale mapping efforts (Marra *et al.* 1997; Little 2001). Secondly, the problems of detecting overlaps from fingerprint patterns grow exponentially as the size of the genome being mapped increases.

Two developments have led to a resurgence in restriction enzyme fingerprinting. The first of these was the development of a high throughput method of fingerprinting using large-insert clones (Marra *et al.* 1997). In this method bacterial artificial chromosome (BAC) and P1-derived artificial chromosome (PAC) clones replace the cosmids used by Coulson *et al.* (1986). Because these clones have insert sizes of 150 kb they yield many easily detectable fragments on digestion with a single restriction enzyme (Fig. 4.2). By measuring the relative mobilities of the

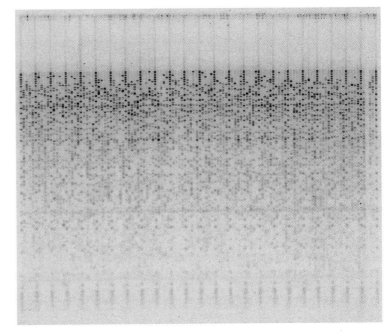

Fig. 4.2 A typical agarose-mapping gel showing human P1-derived artificial chromosomes (PACs) digested with *Hind*III. Clones are present in triplicate to verify stability during propagation and to check for the possibility of cross-contamination. DNA size standards are present in every fifth lane. (Photograph kindly supplied by Dr M. Marra.)

fragments it is possible to develop a fingerprint of each clone and to identify other clones that share a large proportion of fragments with the same relative mobilities. In this way it is possible to infer the overlap of clones and construct contigs. Two other features of the method are worth noting. First, because only single digestion is involved, non-radioactive detection methods can be used. However, because of the small quantities of DNA obtainable with low copy number vectors, such as BACs and PACs, high sensitivity detection is required. This is achieved by using fluorescent dyes and highly sensitive fluorescence imagers. Secondly, the method enables the size of each restriction fragment and each clone to be estimated. This is particularly useful when designing sequencing strategies.

The second development that led to the resurgence of restriction fragment mapping was the creation of powerful software called fingerprinted contigs (FPC) to carry out the massive task of comparing the fingerprints of different clones and determining which ones overlap (Soderlund *et al.* 2000). The FPC software is used in conjunction with the methodology of Marra *et al.* (1997). To determine if two clones overlap, the number of shared bands is counted where two bands are considered 'shared' if they have the same size within a tolerance. The

probability that the number of shared bands is a coincidence is computed and, if this score is below a user specified cutoff, the clones are considered to overlap. In a test of this software the largest contig constructed consisted of 9534 clones. Such a contig could not be constructed manually. An additional benefit of the FPC software is that it can accommodate the use of physical markers such as sequence-tagged sites (STSs), etc. as described later.

Any reader who wishes to see the benefits of the methodologies of Marra *et al.* (1997) and Soderlund *et al.* (2000) should compare the papers on the restriction mapping of the *E. coli* genome (Kohara *et al.* 1987; described in previous editions of this book) and that of *Bradyrhizobium japonicum* (Tomkins *et al.* 2001).

Marker sequences

Sequence-tagged sites and sequence-tagged connectors

The concept of sequence-tagged sites (STSs) was developed by Olson *et al.* (1986) in an attempt to systematize landmarking of the human genome. Basically, an STS is a short region of DNA about

```
  5  5'-GAATTCCTGA  CCTCAGGTGA  TCTGCCCGCC  TCGGCCTCCC  AAAGTGCTGG
 51     GATTTACAGG  CATGAGGCAC  CACACCTGGC  CAGTTGCTTA  GCTCTCTAAG
101     TCTTATTTGC  TTTACTTACA  AAATGGAGAT  ACAACCTTAT  AGAACATTCG
151     ACATATACTA  GGTTTCCATG  AACAGCAGCC  AGATCTCAAC  TATATAGGGA
201     CCAGTGAGAA  ACCAATGTCA  GGTAGCTGAT  GATGGGCAAa  GGgATGGGgA
251     CTGATATGCC  cNNNNNGACG  ATTCGAGTGA  CAAGCTACTA  TGTACCTCAC
301     CTTTtCATCT  tGATCTTCAC  CACCCATGGg  TAGGTGTCAC  TGAAaTT-3'
               3'-CTAGAAGTG  GTGGGTACCC  AT-5'—————Primer B
```

Primer A
Primer B

Melting
temperature

Primer A 5'-GTT TCC ATG AAC AGC AGT CAG-3'   69.4°C
Primer B 5'-TAC CCA TGG GTG GTG AAG AGC-3'   68.7°C

Fig. 4.3 Example of a sequence-tagged site (STS). The STS developed from the sequence shown is 171 bases long. It starts at base 162 and runs through base 332. Primer A is 21 bases long and lies on the sequenced strand. Primer B is also 21 bases long and is complementary to the shaded sequence towards the 3' end of the sequenced strand. Note that the melting temperatures of the two primers are almost equal. (Reproduced with permission from Dogget 1992, courtesy of University Science Books.)

200–300 bases long whose exact sequence is found nowhere else in the genome. Two or more clones containing the same STS must overlap and the overlap must include the STS.

Any clone that can be sequenced may be used as an STS provided it contains a unique sequence. A better method to develop STS markers is to create a chromosome-specific library in phage M13. Random M13 clones are selected and 200–400 bases sequenced. The sequence data generated are compared with all known repeated sequences to help identify regions likely to be unique. Two PCR primer sequences are selected from the unique regions which are separated by 100–300 bp and whose melting temperatures are similar (Fig. 4.3). Once identified, the primers are synthesized and used to PCR amplify genomic DNA from the target organism and the amplification products analysed by agarose gel electrophoresis. A functional STS marker will amplify a single target region of the genome and produce a single band on an electrophoretic gel at a position corresponding to the size of the target region (Fig. 4.4). Alternatively, an STS marker can be used as a hybridization probe.

Operationally, an STS is specified by the sequence of the two primers that make its production possible. Thus it is defined once and for all regardless of whether the region under study is cloned in a phage, a cosmid, a BAC, PAC or yeast artificial chromosome (YAC). Moreover, the STS will remain valid if the corresponding area of the genome is re-cloned sometime in the future in a new and, as yet, unknown vector. The STS is fully portable once the two sequences of the primer are known and these can be obtained from databanks (see Chapter 5).

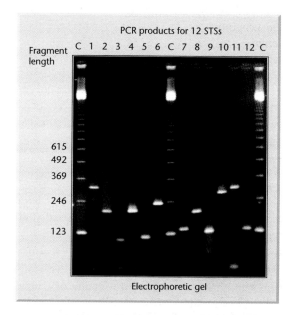

PCR products for 12 STSs

Fragment length   C 1 2 3 4 5 6 C 7 8 9 10 11 12 C

615
492
369
246
123

Electrophoretic gel

Fig. 4.4 Confirmation that an STS is a unique sequence on the genome. Note that the 12 STSs from chromosome 16 shown above appear as single bands after amplification and hybridization to a chromosome 16 genomic library. (Reproduced with permission from Dogget 1992, courtesy of University Science Books.)

Sequence-tagged connectors (STCs) essentially are long STSs and were first proposed by Venter *et al.* (1996) as an aid to sequencing the human genome. The basic idea is that a BAC library is constructed with an average insert size of 150 kb and a 15-fold coverage of the genome. In the case of the human genome this would require 300 000 clones. Both ends of each BAC insert then are sequenced for 500 bases from the point of insert. In the case of the human genome this would generate 600 000 sequences scattered every 5 kb across the genome. Zhao *et al.* (2000) undertook such a task and sequenced both ends of 186 000 BAC clones. These sequences can act as connectors because they will allow any one BAC to be connected to about 30 others (150 kb insert 'divided' by 5 kb represents 30 BACs). In this way a physical map can be constructed. Ideally, this technique is used in conjunction with restriction enzyme fingerprinting as described on p. 47.

Expressed sequence tags

In organisms with large amounts of repetitive DNA the generation of an appropriate sequence, and confirmation that it is an STS, can be time consuming. Adams *et al.* (1991) suggested an alternative approach. The principle of their method is based on the observation that spliced mRNA contains sequences that are largely free of repetitive DNA. Thus partial cDNA sequences, termed expressed sequence tags (ESTs), can serve the same purpose as the random genomic STSs but have the added advantage of pointing directly to an expressed gene. In a test of this concept, partial DNA sequencing was conducted on 600 randomly selected human cDNA clones to generate ESTs. Of the sequences generated, 337 represented new genes, including 48 with similarity to genes from other organisms, and 36 matched previously sequenced human nuclear genes. Forty-six ESTs were mapped to chromosomes.

In practice, there are a number of operational considerations associated with the use of ESTs. First, they need to be very short to ensure that the two ends of the sequence are contiguous in the genome, i.e. are not separated by an intron. Secondly, large genes may be represented by multiple ESTs which may correspond to different portions of a transcript or various alternatively spliced transcripts. For example, one of the major databases holds over 1300 different EST sequences for a single gene product, serum albumin. While this may or may not be a problem in constructing a physical map, it is problematical in the construction of a genetic map.

If it is desirable to select a single representative sequence from each unique gene, then this is accomplished by focusing on 3′ untranslated regions (3′ UTRs) of mRNAs. This can be achieved using oligo(dT) primers if the mRNA has a poly(A) tail. Two advantages of using the 3′ UTRs are that they rarely contain introns and they usually display less sequence conservation than do coding regions (Makalowski *et al.* 1996). The former feature leads to PCR product sizes that are small enough to amplify. The latter feature makes it easier to discriminate among gene family members that are very similar in their coding regions.

Polymorphic sequence-tagged sites (simple sequence length polymorphisms)

So far it has been suggested that an STS yields the same product size from any DNA sample. However, STSs can also be developed for unique regions along the genome that vary in length from one individual to another. This variation in length most often occurs because of the presence of microsatellites. In humans these usually take the form of CA (or GT) repeats with the dinucleotide being repeated 5–50 times. These sequences are very attractive because they are highly polymorphic, i.e. they will occur as $(CA)_{17}$ in one person, $(CA)_{15}$ in another, and so on. These repeat units are flanked by unique sequences (Fig. 4.5) which can act as primers for the generation of an STS. By definition, such STSs are polymorphic and can be traced through families along with other DNA markers. The frequency of such polymorphic STSs in different eukaryotic genomes is shown in Table 4.1 and a more detailed analysis has been presented by Toth *et al.* (2000).

Many polymorphic STSs have been identified either in DNA sequence databases or following the screening and sequencing of small-insert genomic libraries. Cregan *et al.* (1999) developed a method for identifying them by subcloning a limited number of BAC clones and Qi *et al.* (2001) have described a method for doing so without subcloning.

Single nucleotide polymorphisms

Single nucleotide polymorphisms (SNPs, pronounced 'snips') are single base pair positions in genomic DNA at which different sequence alternatives (alleles) exist in a population (Fig. 4.6). In highly outbred populations, such as humans, polymorphisms are considered to be SNPs only if the least abundant allele has a frequency of 1% or more. This is to distinguish

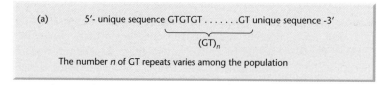

(a)   5'- unique sequence GTGTGT . . . . . . .GT unique sequence -3'

(GT)$_n$

The number n of GT repeats varies among the population

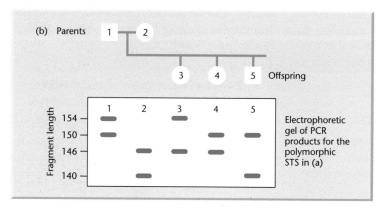

Fig. 4.5 The use of a polymorphic STS in inheritance studies. (a) Structure of a polymorphic STS. (b) Schematic representation of a polymorphic STS in a five-member family. The two alleles carried by the father are different from those carried by the mother. The children inherit one allele of the STS from each parent. (Redrawn with permission from Dogget 1992, courtesy of University Science Books.)

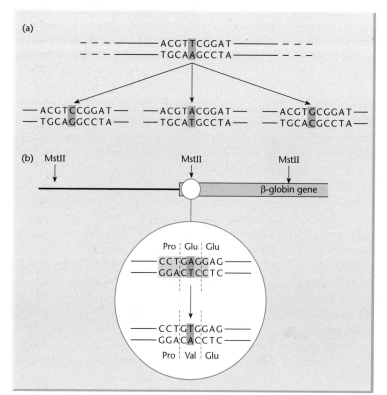

Fig. 4.6 Examples of single nucleotide polymorphisms. (a) Three possible SNP variants of a T/A base pair. (b) A single base change in the β-globin gene that destroys a restriction site for endonuclease *Mst*II (CC/TNAGG) thereby generating a restriction fragment length polymorphism (RFLP). This base change results in amino acid residue 6 being changed from glutamate to valine and is the cause of sickle cell disease.

Table 4.1 Total lengths (base pairs per megabase of DNA) of simple sequence repeats in different groups of organisms. (From Toth *et al*. 2000.)

| Taxonomic group | Genomic region | Length of repeated motif (bp) | | | | | | |
		1	2	3	4	5	6	Total
Primates	All	3429	1643	477	1368	898	341	8156
	Intergenic regions	3880	1709	517	1464	991	385	8946
	Introns	4137	1506	424	1428	988	392	8875
	Exons	49	10	1126	29	57	244	1515
Human chromosome 22	All	5141	1511	604	1906	1097	419	10 678
Rodentia	All	1839	5461	1196	2942	1417	1034	13 889
	Intergenic regions	2192	5928	1230	2823	1577	740	14 490
	Introns	2182	5837	1123	3009	1399	922	14 472
	Exons	62	70	1557	63	116	620	2488
Mammalia	All	1397	2312	532	915	774	693	6623
	Intergenic regions	1954	4666	531	1529	1115	1155	10 950
	Introns	1967	2202	395	792	685	637	6678
	Exons	69	88	876	19	18	356	1426
Vertebrata	All	1418	2449	1069	1279	709	220	7144
	Intergenic regions	2193	3363	1127	1766	1201	320	9970
	Introns	1476	3193	861	1502	585	142	7759
	Exons	49	0	823	0	26	75	973
Arthropoda	All	985	1403	956	439	732	875	5390
	Intergenic regions	1462	2259	1128	621	1110	1090	7670
	Introns	950	1627	728	461	735	917	5418
	Exons	12	34	1566	0	21	591	2224
C. elegans	All	428	556	337	144	225	449	2139
	Intergenic regions	573	822	414	198	310	574	2891
	Introns	512	549	228	169	283	556	2297
	Exons	43	54	308	18	38	116	577
Embryophyta	All	1245	1067	880	184	491	272	4139
	Intergenic regions	2012	1715	869	303	781	334	6014
	Introns	1380	1322	576	260	547	207	4292
	Exons	18	50	1119	2	29	303	1521
S. cerevisiae	All	1075	580	646	93	204	406	3004
	Intergenic regions	3140	1875	512	273	494	532	6826
	Introns	3012	1437	516	162	509	288	5924
	Exons	36	19	706	7	52	330	1150
Fungi	All	905	272	485	194	395	426	2677
	Intergenic regions	2080	555	550	421	925	548	5079
	Introns	2075	1013	951	458	659	661	5817
	Exons	9	4	381	2	35	219	650

SNPs from very rare mutations. In practice, the term SNP is typically used more loosely than required by the above definition. For example, single base variants in cDNAs (cSNPs) are usually classed as SNPs because most of these will reflect underlying genomic DNA variants although they could result from RNA editing. Single basepair insertions or deletions (indels) also are considered to be SNPs by some workers. A special sub-set of SNPs is one where the base change alters the sensitivity of a sequence to cleavage by a restriction endonuclease. These are known as restriction fragment length polymorphisms (RFLPs, see p. 2) or 'snip-SNPs'. Another important sub-set is one in which the two alleles can be distinguished by the presence or absence of a particular phenotype. Good examples here are emphysema in humans, caused by a C > T change in the gene for α_1-antitrypsin, and sickle cell anaemia. In the latter case, an A > T change results in the replacement of a glutamine residue in β-globin with a valine residue but also destroys an *Mst*II site thereby generating an RFLP.

SNPs probably are the most important sequence markers for physical mapping of genomes. The reason for this is that they have the potential to provide the greatest density of markers. For example, their frequency is estimated to be 1 per 1000 basepairs which, in the case of the human genome, means 3 million in total. Indeed, the Human SNP Consortium plans to map 300 000 of them (Masood 1999). Given their importance, many different methods for discovering SNPs have been developed (for review see Landegren *et al.* 1998) but only a few will be discussed here. One method of discovering SNPs is simply to mine the sequence data stored in the major databases (see p. 97). For example, Irizarry *et al.* (2000) analysed more than 542 million basepairs of EST and mRNA sequences and identified 48 196 candidate SNPs. A small number of these candidate SNPs, which also were RFLPs, was selected for further study and a high proportion of them were confirmed as RFLPs. To a large extent, the success of this method was because of the use of statistical methods to detect true polymorphisms as opposed to sequencing or other technical errors.

A variation on the above method is that adopted for detecting SNPs in the nematode *Caenorhabditis elegans*. The genome of the reference Bristol N2 strain had already been sequenced completely. Consequently, Wicks *et al.* (2001) sequenced 5.4 Mb of DNA from 11 000 clones prepared with genomic DNA from a different isolate of *C. elegans* and then undertook sequence comparisons. They identified 6222 potential polymorphisms of which 4670 were single basepair substitutions, including 3457 RFLPs. An analysis of these polymorphisms is shown in Fig. 4.7. Many of these candidate SNPs have since

(a)

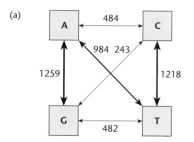

(b)

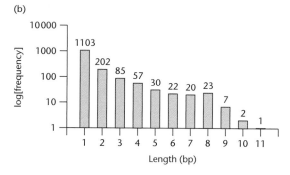

Fig. 4.7 Characteristics of 6222 polymorphisms detected in the genome of an isolate of *Caenorhabditis elegans*. (a) Base changes observed in 4670 SNPs detected amongst the 6222 polymorphisms. The numbers beside the arrows show the numbers of each type of base change. (b) The frequency distribution of the size of all insertion and deletion events (indels). (Reprinted from Wicks *et al.* 2001 by permission of Nature Publishing Group, New York.)

been confirmed by more detailed experimental work.

Because they wanted to understand the role of SNPs in susceptibility of humans to common diseases, Cargill *et al.* (1999) took a complementary approach to that of Wicks *et al.* (2001). They started by selecting over 100 genes relevant to cardiovascular diseases, endocrinology and neuropsychiatry and screened 114 individuals from different ethnic backgrounds for SNPs in these genes. They PCR-amplified relevant sequences, either from coding regions or adjacent non-coding regions (UTRs and introns), and screened them by two different methods. The first method involved hybridization to high-density DNA probe arrays containing oligonucleotides specific for the sequences under study. Such arrays are known as variant detector arrays (VDAs) or DNA microchips and are described in more detail on p. 87. Suffice it to say here that variant sequences typically give rise to altered hybridization patterns. The second method involved subjecting

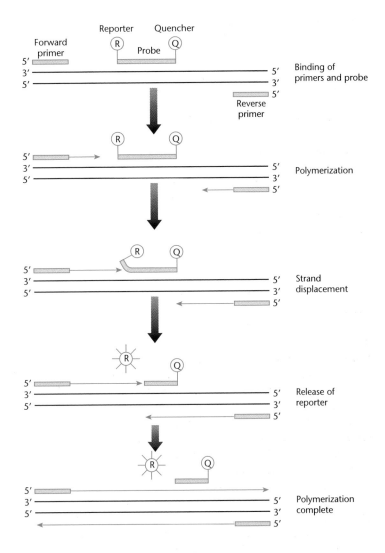

Fig. 4.8 The TaqMan assay (see text for details). (Redrawn with permission from Primrose *et al.* 2001.)

PCR products to denaturing HPLC at a critical temperature where heterozygous individuals typically give rise to heteroduplex products with altered denaturation and migration properties.

The methods described above are used to *discover* new SNPs for placement on physical maps. However, if the objective is to associate particular SNPs with particular phenotypes (e.g. disease states, agronomic traits, etc.) then high-throughput methods for genotyping are required. One such method has been described by Ranade *et al.* (2001) and makes use of the TaqMan assay (Fig. 4.8) of Livak *et al.* (1995). In this method, a 100 bp region flanking the polymorphism is amplified in the presence of two probes, each specific for one or other allele. The

probes have a reporter fluor at the 5′ end but do not fluoresce when free in solution because they have a quencher at the 3′ end. During PCR, the Taq polymerase encounters a probe specifically base paired with its target and liberates the fluor, thereby increasing net fluorescence. The presence of two probes, each labelled with a different fluor, allows one to detect both alleles in a single tube without any post-PCR processing.

SNPs occur at a frequency of approximately 1 per 1000 bp. Therefore, in a 50 000 bp sequence one would expect 50 SNPs and theoretically these could come in 2^{50} different variations. The reality is somewhat different, at least in humans, where only four or five different patterns of 50 SNPs (haplotypes) are seen in 80–90% of the population (Helmuth 2001). Although the reasons why such haplotypes are conserved is not fully understood, their existence facilitates the association of SNPs with disease states.

Problems with the PCR in the detection of STSs and SNPs

There are a number of problems associated with the use of the PCR in the detection of STSs and SNPs. The first of these is that the method is labour intensive, although automation can help. The second is the cost of synthesizing large numbers of different primers. The third problem is that STSs are defined by the PCR conditions that generate it. Many thermal cyclers, particularly early models, are not accurately calibrated and/or do not display uniform temperatures and this makes it difficult to exactly reproduce reaction conditions in different laboratories. For these reasons, a number of groups have developed alternative methods of polymorphism detection. For example, Dong *et al.* (2001) have developed a generic PCR amplification protocol for use in SNP detection with DNA microarrays. In this method, genomic DNA is digested with *Eco*RI and

the digestion products resolved by agarose gel electrophoresis. Fragments of length 250–350 bp are selected and ligated to a single adaptor molecule. PCR amplification then is undertaken by using a primer that hybridizes to the adaptor.

Lyamichev *et al.* (1999, 2000) have developed an invasive cleavage method for SNP detection that does not require the use of the PCR at all. The method makes use of flap endonucleases (FENs) which recognize and cleave a structure formed when two overlapping oligonucleotide probes hybridize to a target DNA strand (Fig. 4.9). Two types of probe are used. The first are downstream signal probes that have a three-nucleotide non-complementary arm at the 5' end plus a fluorescent label. The second type are invasive probes that are complementary to the upstream region of the target and are designed to overlap the site of hybridization of the signal probe by 0, 1, 3, 5 or 8 nucleotides. As can be seen from Fig. 4.9, cleavage of the fluorescent arm only occurs when the two probes overlap. Thus the FEN requires at least one overlapping nucleotide between the signal and invasive probes to recognize and cleave the displaced 5' end of the signal probe. However, mismatch of the signal probe one base upstream of

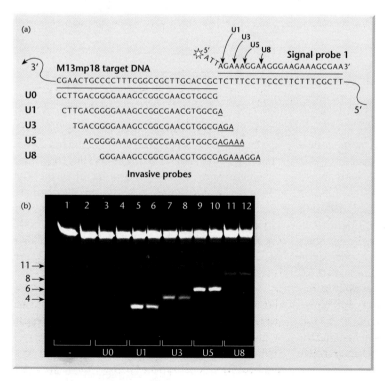

Fig. 4.9 The principle of the invasive cleavage assay. (a) Sequence of the target region of M13-single stranded DNA, and the structure of the signal probe and five invasive probes, The underlined nucleotides at the 3' end of the invasive probes indicate the extent of the overlap with the signal probe. The labelled arrows above the signal probe show the cleavage points induced by each invasive probe. Fluorescent labels are indicated by stars. (b) Gel analysis of cleavage products of signal probe produced by digestion with *Pfu* FEN in duplicate reactions. Note that the cleavage products represent the sum of the three-base flap plus the invaded bases. (Reprinted from Lyamichev *et al.* 1999 by permission of Nature Publishing Group, New York.)

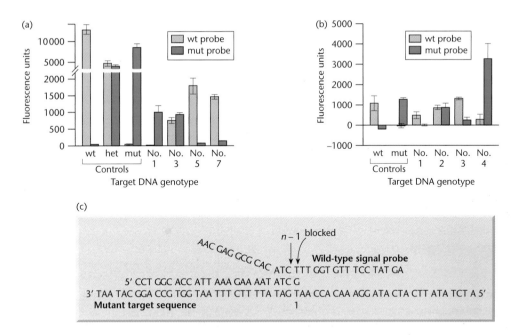

Fig. 4.10 The detection of SNPs by invasive cleavage. Control samples were single-stranded synthetic DNA templates of the indicated genotype. The light green bars indicate the signal generated from probes complementary to wild-type target sequences; the dark green bars show signal generated with probes complementary to the mutations. Error bars indicate the standard deviations obtained from triplicate measurements on each target. (a) Assay for the presence of the wild-type sequence or Leiden mutation in the factor V gene. The scale for the fluorescence signal is discontinuous to permit comparison of strong and weak signals within one graph. (b) Assay for the presence of the wild-type sequence or the ΔF508 mutation in the CFTR gene. (c) Schematic representation of the complex formed on the ΔF508 mutant target with the wild-type invasive and signal probes. Arrows indicate the site blocked for cleavage by the mismatch between the probe and the target strand, and the n-1 position at which adventitious cleavage could occur. (Reprinted from Lyamichev *et al.* 1999 by permission of Nature Publishing Group, New York.)

the cleavage site creates a structure equivalent to that formed by a non-overlapping invasive probe in that no cleavage occurs. This specificity is the basis of an alternative method of SNP detection (see below).

To test the utility of the invasive cleavage method, invasive and signal probes were designed that would bind either the wild-type sequence of the human factor V gene or the Leiden mutant. These two alleles differ by a single nucleotide (C > T) at position 1 of the sequence shown in Fig. 4.10. When the wild-type signal probe is hybridized to the mutant sequence, or the mutant signal probe is hybridized to the wild-type sequence, no FEN cleavage occurs. Conversely, hybridization of the mutant signal probe to the mutant sequence, or wild-type signal probe to wild-type sequence, results in FEN cleavage. The method also can be used to detect 1 bp indels. More recently, Wilkins Stevens *et al.* (2001) have converted this assay into a solid phase format.

Randomly amplified polymorphic DNA and cleaved amplified polymorphic sequences

Polymorphic DNA can be detected by amplification in the absence of the target DNA sequence information used to generate STSs. Williams *et al.* (1990) have described a simple process, distinct from the PCR process, which is based on the amplification of genomic DNA with *single* primers of arbitrary nucleotide sequence. The nucleotide sequence of each primer was chosen within the constraints that the primer was nine or 10 nucleotides in length, between 50 and 80% G + C in composition and contained no palindromes. Not all the sequences amplified in this way are polymorphic but those that are (randomly amplified polymorphic DNA, RAPD) are easily identified. RAPDs are widely used by plant molecular biologists (Reiter *et al.* 1992; Tingey & Del Tufo 1993) to construct maps because they provide

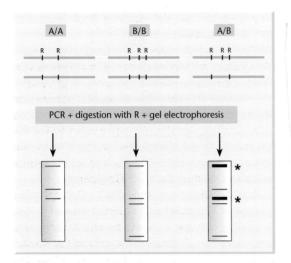

Fig. 4.11 Generation and visualization of CAPS markers. Unique-sequence primers are used to amplify a mapped DNA sequence for two different homozygous strains (A/A and B/B) and from the heterozygote A/B. The amplified fragments from strains A/A and B/B contain two and three recognition sites, respectively, for endonuclease R. In the case of the heterozygote A/B, two different PCR products will be obtained, one of which is cleaved twice and the other three times. After fractionation by agarose gel electrophoresis the PCR products from the three strains give readily distinguishable patterns. The asterisks indicate bands that will appear as doublets. (Redrawn with permission from Konieczny & Ausubel 1993.)

very large numbers of markers and are very easy to detect by agarose gel electrophoresis. However, they have two disadvantages. First, the amplification of a specific sequence is sensitive to PCR conditions, including template concentration, and hence it can be difficult to correlate results obtained by different research groups. For this reason, RAPDs may be converted to STSs after isolation (Kurata *et al.* 1994). A second limitation of the RAPD method is that usually it cannot distinguish heterozygotes from one of the two homozygous genotypes. Nevertheless, Postlethwait *et al.* (1994) have used RAPDs to develop a genetic linkage map of the zebrafish (*Danio rerio*).

A different method for detecting polymorphisms, which is not subject to the problems exhibited by RAPDs, has been described by Konieczny and Ausubel (1993). In this method, STSs are derived from genes which have already been mapped and sequenced. Where possible the primers used are cho-

sen such that the PCR products include introns to maximize the possibility of finding polymorphisms. The primary PCR products are subjected to digestion with a panel of restriction endonucleases until a polymorphism is detected. Such markers are called cleaved amplified polymorphic sequences (CAPS). The way in which CAPS are detected is shown in Fig. 4.11. Note that whereas RFLPs are well suited to mapping newly cloned DNA sequences, they are not convenient to use for mapping genes, such as plant genes, which are first identified by mutation. CAPS are much more useful in this respect.

Amplified fragment length polymorphism

Amplified fragment length polymorphism (AFLP) is a diagnostic fingerprinting technique that detects genomic restriction fragments and in that respect resembles the RFLP technique (Vos *et al.* 1995). The major difference is that PCR amplification rather than Southern blotting is used for detection of restriction fragments. The resemblance to the RFLP technique was the basis for choosing the name AFLP. However, the name AFLP should not be used as an acronym because the technique detects *presence* or *absence* of restriction fragments and *not* length differences. The AFLP approach is particularly powerful because it requires no previous sequence characterization of the target genome. For this reason it has been widely adopted by plant geneticists. It also has been used with bacterial and viral genomes (Vos *et al.* 1995). It has not proved useful in mapping animal genomes because it is dependent on the presence of high rates of substitutional variation in the DNA; RFLPs are much more common in plant genomes compared to animal genomes.

The AFLP technique is based on the amplification of subsets of genomic restriction fragments using PCR (Fig. 4.12). To prepare an AFLP template, genomic DNA is isolated and digested simultaneously with two restriction endonucleases, *Eco*RI and *Mse*I. The former has a 6 bp recognition site and the latter a 4 bp recognition site. When used together these enzymes generate small DNA fragments that will amplify well and are in the optimal size range (< 1 kb) for separation on denaturing polyacrylamide gels. Following heat inactivation of the restriction enzymes the genomic DNA fragments are ligated to *Eco*RI and *Mse*I adapters to generate tem-

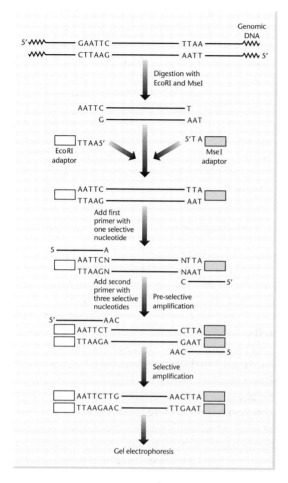

Fig. 4.12 Principle of the amplified fragment length polymorphism (AFLP) method (see text for details).

ated on a gel and the resulting DNA fingerprint detected by autoradiography.

The AFLP technique will generate fingerprints of any DNA regardless of the origin or complexity. The number of amplified fragments is controlled by the cleavage frequency of the rare cutter enzyme and the number of selective bases. In addition, the number of amplified bands may be controlled by the nature of the selective bases. Selective extension with rare di- or trinucleotides will result in a reduction of the number of amplified fragments.

The AFLP technique is not simply a fingerprinting technique; rather, it is an enabling technology that can bridge the gap between genetic and physical maps. Most AFLP fragments correspond to unique positions on the genome and hence can be exploited as landmarks. In higher plants AFLPs may be the most effective way to generate high-density maps. The AFLP markers also can be used to detect corresponding genomic clones. Finally, the technique can be used for fingerprinting of cloned DNA segments. By using no or few selective nucleotides, restriction fragment fingerprints will be produced which subsequently can be used to line up individual clones and make contigs.

Hybridization assays

Joining contigs by walking

When ordering genomic clones into contigs it is not unusual to find that there are many more contigs than chromosomes. That is, many of the contigs should join together but the linking clones are missing or have not been detected. Another observation made during contig assembly is that many clones do not form part of any contig, i.e. they are *singletons*. Hybridization is a very useful means of placing singletons and finding linking clones and the method is known as *chromosome walking* (Bender *et al.* 1983) which was originally developed for the isolation of gene sequences whose function is unknown but whose genetic location is known. The principle of this method is shown in Fig. 4.13. For the purposes of map generation a single cloned fragment is selected. This is used as a probe to detect other clones in the library with which it will hybridize and which represent clones overlapping with it. The overlap

plate DNA for amplification. These common adapter sequences flanking variable genomic DNA sequences serve as primer binding sites on the restriction fragments. Using this strategy it is possible to amplify many DNA fragments without having prior sequence knowledge.

The PCR is performed in two consecutive reactions. In the first pre-amplification reaction, genomic fragments are amplified with AFLP primers each having one selective nucleotide (see Fig. 4.12). The PCR products of the pre-amplification reaction are diluted and used as a template for the selective amplification using two new AFLP primers which have two or three selective nucleotides. In addition, the *Eco*RI selective primer is radiolabelled. After the selective amplification the PCR products are separ-

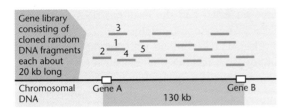

Fig. 4.13 Chromosome walking. It is desired to clone DNA sequences of gene B, which has been identified genetically but for which no probe is available. Sequences of a nearby gene A are available in fragment 1. Alternatively, a sequence close to gene B could be identified by *in situ* hybridization to *Drosophila* polytene chromosomes. In a large random genomic DNA library many overlapping cloned fragments are present. Clone 1 sequences can be used as a probe to identify overlapping clones 2, 3 and 4. Clone 4 can, in turn, be used as a probe to identify clone 5, and so on. It is, therefore, possible to walk along the chromosome until gene B is reached.

can be to the right or the left. This single walking step is repeated many times and can occur in both directions along the chromosome. A potential problem with chromosome walking is created by the existence of repeated sequences. If the clone used as the probe contains a sequence repeated elsewhere in the genome, it will hybridize to non-contiguous fragments. For this reason the probe used for stepping from one genomic clone to the next must be a unique sequence clone, or a sub-clone which has been shown to contain only a unique sequence. Clearly, if chromosome walking is to be employed it makes sense to use very large fragments of DNA as this minimizes the number of steps.

Hybridization mapping

This method starts with a genomic library as before. Five kinds of probes, representing known repetitive sequences (centromeric, telomeric, 17S and 5S ribosomal and the long terminal repeat (LTR) of retrotransposons), are hybridized to the library to identify those clones that contain only unique DNA. A number of clones carrying unique DNA are selected at random and used as hybridization probes to detect overlapping clones (see Fig. 4.14). From those clones which do not give a positive hybridization signal another set is selected at random for use as probes in the next round of experiments. This process is continued until all clones show positive hybridization at least once. In practice, some clones containing

repetitive DNA have to be used to join contigs. Nevertheless, a key feature of this method is that clones are randomly chosen as probes based on only one criterion – that the clone has not yet given a positive hybridization signal. By this means, large numbers of redundant hybridizations are avoided.

Two refinements of the above process simplify the construction of contigs. First, probes can be prepared from either of the ends of the cloned DNA by using a vector with inward-facing T3 and T7 promoters located at the cloning site (Fig. 4.15). This simplifies contig generation compared with STSs because in the latter case it is hoped that the STS will lie in a region of overlap. However, if the ends of the inserts were sequenced, rather than generating probes, then these sequences could be used as STSs. The use of insert ends as probes eliminates the problem of false positives (i.e. the presence of a hybridization signal between two clones that do not overlap), which can arise because of cross-hybridization between repetitive elements. This is done by demanding that all pairs of cosmid overlaps be reciprocal if both cosmids in the pair are used as probes (Zhang *et al.* 1994). That is, if cosmid *x* hybridizes to cosmid *y* then *y* must hybridize to *x*. If reciprocity is not achieved, that particular overlap is eliminated from the data set. Secondly, if a cosmid genomic library and a YAC genomic library are prepared from the same organism, the YAC library can be used to pre-sort the cosmid clones (Fig. 4.16). This mapping procedure can be refined still further: first by using as hybridization probes, short random-sequence oligonucleotides (Fig. 4.16); then, additional information can be obtained by using as probes sequences representing intron–exon boundaries, or zinc fingers or other structural motifs.

The utility of hybridization mapping has been shown by the construction of a complete map of the fission yeast *Schizosaccharomyces pombe* (Maier *et al.* 1992; Hoheisel *et al.* 1993; Mizukami *et al.* 1993) and the plant *Arabidopsis thaliana* (Mozo *et al.* 1999).

In situ hybridization

There are a number of different kinds of genome maps, e.g. cytogenetic, linkage, physical, etc. The classic cytogenetic map gives visual reality to other maps and to the chromosome itself. Because it does not rely on the cloning of DNA fragments it avoids

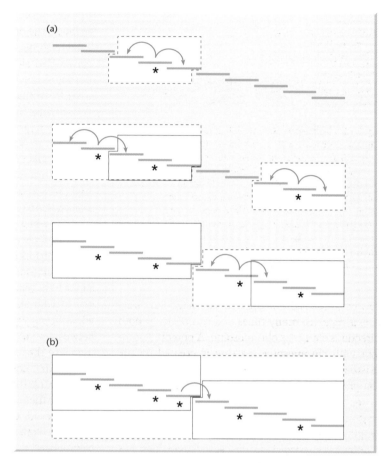

Fig. 4.14 The principle of hybridization mapping. (a) Clones for use as probes are randomly picked (*) from a given set of cosmids whose map order is not known. Hybridization identifies overlapping clones (arrows). From clones that do not give a positive signal in any earlier hybridization assay (unboxed areas), probes for the next round of experiments are chosen until all the clones show positive hybridization at least once. (b) Gaps in the map caused by the lack of probes for certain overlap regions are closed by using terminal contig clones. (Reprinted from Hoheisel 1994 by permission of Elsevier Science.)

the pitfalls that this procedure can introduce, particularly with YACs (see p. 40). Genetic linkage mapping allows the localization of inherited markers relative to each other. As with cytogenetic maps, linkage maps examine chromosomes as they are in cells. Although the methodology used to construct cytogenetic and linkage maps can lead to errors, they nevertheless are used as gold standards against which the physical maps are judged. The importance of *in situ* hybridization is that it enables this comparison to be made. Providing hybridization of repeated sequences is suppressed and provided no DNA chimaeras are present, a cloned fragment or restriction fragment should anneal to a single location on the cytogenetic map. Furthermore, the physical map order should match that found by *in situ* hybridization. Where genetic markers have been located on the cytogenetic map by *in situ* hybridization they also can be positioned on the physical map.

Originally, *in situ* hybridization of unique sequences utilized radioactively labelled probes and it was a technique which required a great deal of technical dexterity. Today, the methods used are all derivatives of the fluorescence *in situ* hybridization (FISH) method developed by Pinkel *et al.* (1986). In this technique, the DNA probe is labelled by addition of a reporter molecule. The probe is hybridized to a preparation of metaphase chromosomes which has been air-dried on a microscope slide and in which the DNA has been denatured with formamide. Following hybridization, and washing to remove excess probe, the chromosome preparation is incubated in a solution containing a fluorescently labelled affinity molecule which binds to the reporter on the hybridized probe. The preparation is then examined with a fluorescence microscope. If large DNA probes are used, they will contain many repetitive sequences which will bind indiscriminately to the target. This

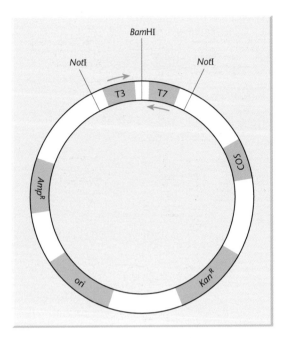

Fig. 4.15 A cosmid vector used to generate probes specific for the ends of cloned inserts. The vector contains bacteriophage T3 and T7 promoters flanking a unique *Bam*HI cloning site. Note also the *Not*I sites to facilitate restriction mapping and excision of the insert DNA. ori, origin of replication; *Amp*^{R} and *Kan*^{R}, genes conferring resistance to ampicillin and kanamycin respectively; *COS*, cohesive sites essential for *in vitro* packaging in phage λ particles. Arrows show the direction of transcription from the T3 and T7 promoters.

that it cannot be used to generate reliable measurements of the distances between signals nor can it be used to localize unknown sequences on a chromosome. This is because the stretching of individual chromosomes is highly variable.

In interphase nuclei the chromatin is less condensed than in metaphase chromosomes and hence provides a good target for high resolution FISH (Trask *et al.* 1989; Yokota *et al.* 1995). The resolution of FISH can be further improved by loosening the organization of the interphase chromatin using high salt, alkali or detergent treatment of the cell preparations (Parra & Windle 1993). These techniques, which are commonly referred to as fibre-FISH, provide a resolution that permits the detection of a probe to a single fibre. Theoretically, the resolution of fibre-FISH is the same as the resolving power of the light microscope, i.e. 0.34 μm. This is equivalent to 1 kb and has been achieved by Florijn *et al.* (1995).

Cytogenetic maps

As noted in the previous section, *in situ* hybridization can be used to visually map a particular genomic clone to a particular chromosomal location. Such cytogenetic maps have been available for a long time, particularly in *Drosophila* and humans. In both cases these maps were constructed piecemeal using probes isolated in many different laboratories. As large, BAC-based clones covering the entire human genome have become available it has become possible to generate a detailed human cytogenetic map (Cheung *et al.* 2001). To do this, 8877 genomic clones were selected that could be identified by one or more STSs, or by STCs (end sequences) or by its entire sequence. Each clone was mapped to a chromosomal band by FISH and on average there was one clone per megabase of DNA. A significant proportion of these clones map to more than one location because they hybridize to duplicated genes or to low-copy repeated sequences. Similar cytogenetic maps are being constructed for other species.

non-specific binding can be eliminated by competitive suppression hybridization. Before the main hybridization the probe is mixed with an aqueous solution of unlabelled total genomic DNA. This saturates the repetitive elements in the probe so that they no longer interfere with the *in situ* hybridization of the unique sequences.

Conventional FISH has a resolving power of ~1 Mb. If a higher resolution is necessary the less condensed chromosomes need to be used as the target. Highly elongated metaphase chromosomes have been prepared by mechanically stretching them by cytocentrifugation. This results in chromosomes that are 5–20 times their normal length. Laan *et al.* (1995) have shown that these stretched chromosomes are excellent for fast and reliable ordering of clones that are separated by at least 200 kb. They also can be used to establish the centromere–telomere orientation of a clone. The disadvantages of the method are

Padlock probes

A new variation on multicolour FISH is the use of padlock probes (Nilsson *et al.* 1997). Padlock probes are oligonucleotides that can be ligated to form a

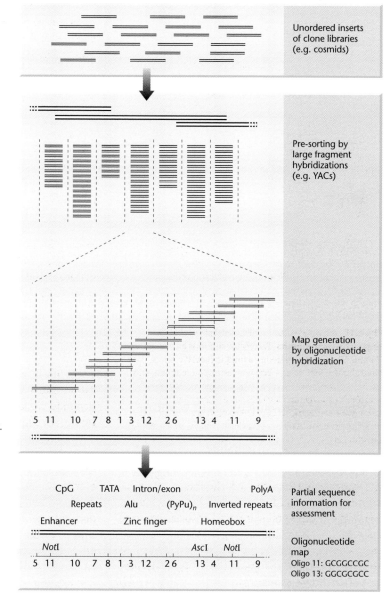

Unordered inserts
of clone libraries
(e.g. cosmids)

Pre-sorting by
large fragment
hybridizations
(e.g. YACs)

Map generation
by oligonucleotide
hybridization

5 11   10   7 8 1 3 12   2 6   13 4   11   9

CpG   TATA   Intron/exon   PolyA   Partial sequence
    Repeats   Alu   (PyPu)$_n$ Inverted repeats   information for
 Enhancer   Zinc finger   Homeobox   assessment

*Not*I   *Asc*I   *Not*I   Oligonucleotide
                                                                map
5 11   10   7 8 1 3 12   2 6   13 4   11   9
                                                                Oligo 11: GCGGCCGC
                                                                Oligo 13: GGCGCGCC

Fig. 4.16 Principle of oligomer
mapping. A high-resolution library (e.g.
cosmids) is subdivided by hybridization
of low-resolution DNA fragments (e.g.
yeast artificial chromosome (YAC)
clones). Fingerprinting data for
establishing the order of clones are
produced by hybridization with short
oligonucleotides. Besides providing
mapping data, this method yields an
oligonucleotide map and partial
sequence information concurrently.
(Reprinted from Hoheisel 1994 by
permission of Elsevier Science.)

circle if they bind to a sequence of exact complementarity. The lateral arms of the oligonucleotide twist
around the DNA target forming a double helix and
their termini are designed to juxtapose so that they
may be ligated enzymatically. Nilsson *et al.* (1997)
used two different oligonucleotide probes, each corresponding to a different sequence variant of a centromeric alpha-satellite repeat, and differing by only
a single base pair. The closure of the two alternative

padlocked probes occurred only when there was perfect sequence recognition (Fig. 4.17). By labelling
the two probes with different fluorescent dyes it
was possible to monitor the two sequence variants
simultaneously. Antson *et al.* (2000) have extended
the use of padlock probes to the detection of SNPs in
single-copy sequences although methods for signal
amplification were required (for review see Baner
et al. 2001).

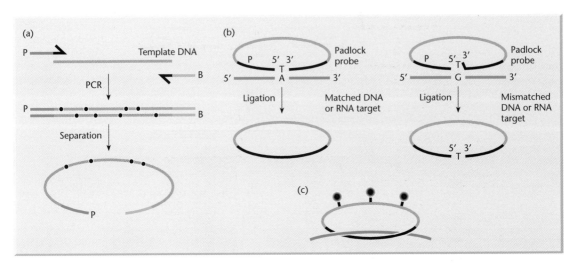

Fig. 4.17 The use of padlock probes to detect SNPs. (a) Synthesis of padlock probes by PCR. The 5′ ends of the PCR primers (dark green and light green) define the two target complementary sequences of the padlock probe to be constructed. One primer has a 5′ phosphate (P) to permit ligation whereas the other has a 5′ biotin (B) which is used to remove this strand by capture on a solid support. Black dots denote labelled nucleotides incorporated during PCR. (b) Hybridization of a padlock probe (black and grey) to a target DNA sequence (green). The padlock probe can be converted to a circle by a ligase (left) but ligation is inhibited if the ends of the probe are mismatched to their target and the probe remains linear (right). (c) Direct detection of a reacted probe catenated to a target sequence. (Reprinted from Baner *et al.* 2001 by permission of Elsevier Science.)

Physical mapping without cloning

Optical mapping

Ideally, one would like to combine the sizing power of electrophoresis with the intrinsic capability of FISH. Schwartz *et al.* (1993) have developed the technique of optical mapping which approaches this ideal by imaging single DNA molecules during restriction enzyme digestion. In practice, a fluid flow is used to stretch out fluorescently stained DNA molecules dissolved in molten agarose and fix them in place during gelation. A restriction enzyme is added to the molten agarose–DNA mixture. Cutting is triggered by the diffusion of magnesium ions into the gelled mixture which has been mounted on a microscope slide. Fluorescence microscopy is used to record at regular intervals the cleavage sites. These are visualized by the appearance of growing gaps in imaged molecules and bright condensed pools of DNA on the fragment end flanking the cut site. Wang *et al.* (1995a) extended the technique to incorporate the use of RecA-assisted restriction endonuclease (RARE) cleavage (p. 36) and Yokota

et al. (1997) developed better methods of straightening the DNA molecules prior to digestion.

Since the original description of optical mapping many different methodological improvements have been introduced (Aston *et al.* 1999), two of which deserve mention here. First, in the original method the mass of each restriction fragment was determined from fluorescence intensity and apparent length. These fragment masses were reported as a fraction of the total clone size and later converted to kilobases by independent measure of clone masses (i.e. cloning vector sequence). Additionally, maps derived from ensembles of identical sequences were averaged to construct final maps. Lai *et al.* (1999) simplified the sizing of fragments by mixing bacteriophage λ DNA with the test sample. After digestion these provide internal size standards (Fig. 4.18). They can also be used to monitor the efficiency of enzyme digestion. Secondly, the method has been automated and map construction algorithms have been developed thereby increasing the power of the method (Giacolone *et al.* 2000).

The accuracy of optical mapping has been determined by comparing the optical maps of *E. coli* and

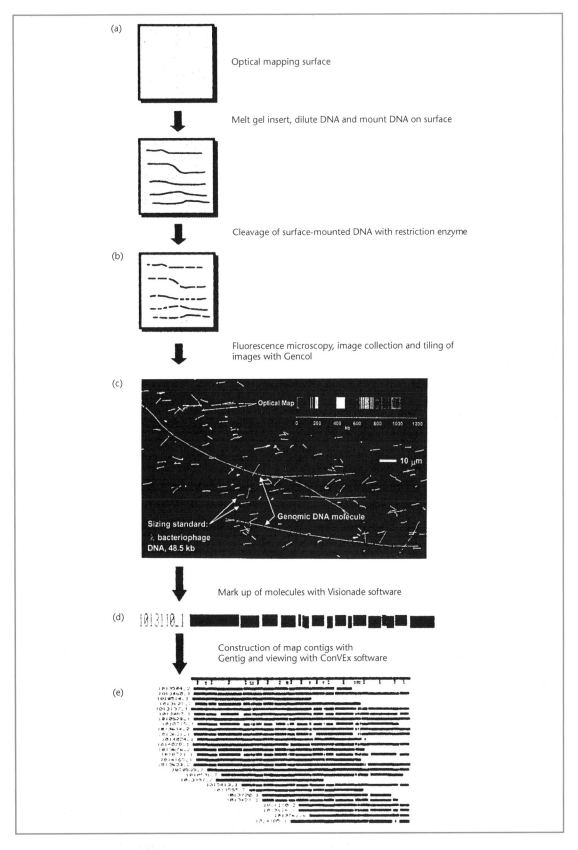

Fig. 4.18 Schematic representation of optical mapping. (Reprinted from Z. Lai *et al.* 1999 by permission of Nature Publishing Group, New York.)

Deinococcus radiodurans with the restriction maps generated by analysis of the complete DNA sequences of these genomes. In both cases the error rate was less than 1% (Lin *et al.* 1999). The power of optical mapping has been demonstrated by its use in the development of maps for the genome of the malarial parasite (*Plasmodium falciparum*) and the *DAZ* locus of the human genome (Lai *et al.* 1999; Giacolone *et al.* 2000). The original plan for the sequencing of the 24.6 Mb malarial genome was to separate the 14 chromosomes by pulsed-field gel electrophoresis (PFGE) (see p. 37) and then construct and map chromosome-specific libraries. There are two problems with this approach. First, chromosomes 5–9 are unseparable by PFGE and migrate as a blob. Secondly, the malarial genome is AT-rich and this presents problems for reliable library construction. By using optical mapping, ordered restriction maps for the enzymes *Bam*HI and *Nhe*I were derived from unfractionated *P. falciparum* DNA and assembled into 14 contigs corresponding to the chromosomes. A schematic of the method used is shown in Fig. 4.18.

The construction of accurate physical maps and the generation of accurate sequence data are very difficult when clones have a high content of repetitive DNA. A good example is the *DAZ* locus which maps to the q arm of the Y chromosome. Prior to optical mapping, our knowledge of the structure of the *DAZ* locus was sketchy but by optical mapping of 16 different BAC clones that hybridize to the locus the structure became apparent (Giacolone *et al.* 2000). Rather than consisting of two copies of the *DAZ* gene, as originally thought, the locus consists of four copies of the gene, each with its own distinctive arrangement of repetitive elements (Fig. 4.19).

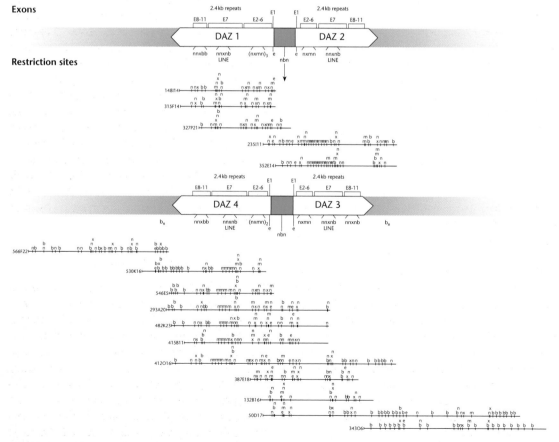

Fig. 4.19 Map of the *DAZ* region of the human Y chromosome generated by optical mapping. Sites for restriction enzymes are represented as follows: x, *Xho*I; b, *Bam*HI; n, *Nhe*I; e, *Eag*I; m, *Mlu*I. (Redrawn with permission from Giacolone *et al.* 2000.)

Radiation hybrid mapping

This method makes use of somatic cell hybrids (see p. 38). A high dose of X-rays is used to break the human chromosome of interest into fragments and these fragments are recovered in rodent cells. The rodent–human hybrid clones are isolated and examined for the presence or absence of specific human DNA markers. The farther apart two markers are on the chromosome, the more likely a given dose of X-rays will break the chromosome between them, placing the markers on two separate chromosomal fragments. By estimating the frequency of breakage, and thus the distance, between markers it is possible to determine their order in a manner analogous to conventional meiotic mapping.

Radiation hybrid mapping has a number of advantages over conventional genetic (meiotic) mapping. First, chromosome breakage is random and there are no hot-spots, interference or gender-specific differences as seen with recombination. Secondly, a much higher resolution can be achieved, e.g. 100–500 kb in radiation mapping as opposed to 1–3 Mb in genetic mapping, and the resolution can be varied by varying the radiation dose. Finally, it is not necessary to use polymorphic markers; monomorphic markers such as STSs can be used as well. The basic theory of radiation hybrid mapping is set out in Box 4.1.

Radiation hybrid mapping was first developed by Goss and Harris (1975). In their experiments, human peripheral blood lymphocytes were irradiated and then fused to hypoxanthine phosphoribosyl transferase (HPRT)-deficient hamster cells. Growth in hypoxanthine, aminopterin and thymidine (HAT) medium resulted in the isolation of a set of clones, each carrying a different X chromosome fragment that included the selected HPRT marker. Goss and Harris (1977) were able to establish the order of three markers on the long arm of the X chromosome and to demonstrate retention of non-selected chromosome fragments. They also derived mathematical approaches for constructing genetic maps on the basis of co-retention frequencies. However, the power of the technology could not be exploited at the time because insufficient genetic markers were available.

Renewed interest in radiation hybrid mapping was prompted by the work of Cox *et al.* (1990) who modified the original approach. They used as a donor cell a rodent–human somatic cell hybrid that contained a single copy of human chromosome 21 and very little other human DNA. This cell line was exposed to 80.0 Gy of X-ray which resulted in an average of five human chromosome-21 pieces per cell. Because broken chromosomal ends are rapidly healed after X-irradiation, the human chromosomal fragments are usually present as translocations or insertions into hamster chromosomes. However, some cells contain a fragment consisting entirely of human chromosomal material with a human centromere. Because a dose of 80.0 Gy of X-rays results in cell death, the irradiated donor cells were fused with HPRT-deficient hamster recipient cells and hybrids selected as before on HAT medium. Non-selective retention of human chromosomal fragments seems to be a general phenomenon under these fusion conditions. In total, 103 independent somatic cell hybrid clones were isolated and assayed by Southern blotting for the retention of 14 DNA markers. Analysis of the results enabled the 14 markers to be mapped to a 20 Mb region of chromosome 21 and this map order was confirmed by PFGE analysis.

Although radiation hybrid mapping was originally developed for the human genome it has been extended to a number of other vertebrate genomes including the mouse, rat, dog, pig, cow, horse, baboon and zebra fish (for references see Geisler *et al.* 1999). Once a radiation hybrid panel is available, one can map any physical marker (STS, EST, simple sequence length polymorphism (SSLP), etc.) by looking for linkage to existing markers on the map (see Box 4.1).

HAPPY mapping

This method of mapping was developed by Dear & Cook (1993) and is analogous to radiation hybrid mapping. It uses haploid equivalents of DNA and the PCR, hence 'HAPPY mapping'. On this method, DNA from any source is broken randomly by γ-irradiation or shearing. Markers then are segregated by diluting the resulting fragments to give aliquots containing approximately one haploid genome equivalent. Markers are detected using PCR and linked markers tend to be found together in an aliquot. The map order of markers, and the distance between them, are deduced from the frequency with which they

Box 4.1 Theory of radiation hybrid mapping

The possible outcomes in a radiation hybrid mapping experiment are shown schematically in Fig. B4.1. In this figure, P is the *probability of retention*, i.e. the probability that a DNA segment is present in an RH clone or the proportion of clones containing a specific DNA segment. The value of P is a function of the radiation dose and the cell lines used and usually ranges from 10–50% with maximum mapping power occurring when $P = 50\%$. It is assumed that P is constant over small regions or the entire chromosome and that segments are lost or retained independently of each other.

The *breakage probability* (θ) is the probability that two markers are separated by one or more breaks. The closer the linkage between two markers, the closer to zero is the value of θ whereas for unlinked markers the value is 1. Note that breakage probability is a function of the radiation dose and the cell lines used.

In radiation hybrid mapping distances are expressed in Rays or centiRays where 1 cR equals a 1% frequency of breakage. Distances are calculated from the expression,

Map distance $(D) = -\log_e (1 - \theta)$ Rays

and assume that the number of breaks is determined by a Poisson distribution with mean D. This is analogous to Haldane's map function for genetic mapping. In the above equation, if θ is 0.1 then the distance is 11 cR and if it is 0.3 the distance is 36 cR. Distances determined by radiation hybrid mapping can be converted to physical distances if the physical distance between any two markers is known. For example, the q arm of human chromosome 11 is 86 Mb and 1618 cR. Therefore, 1 cR is equal to 53 kb. However, the mapping distance is dependent on the radiation dose applied as shown in the table below.

Radiation dose (rad)	Physical distance of 1 cR (kb)
3000	300
9000	50–70
10 000	25
50 000	4

A	B	Probability
+	+	$P(1 - \theta) + P^2\theta$
+	–	$PQ\theta$
–	+	$PQ\theta$
–	–	$Q(1 - \theta) + Q^2\theta$

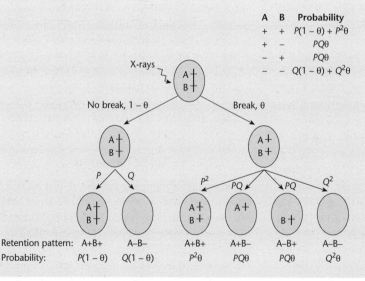

Retention pattern:	A+B+	A–B–	A+B+	A+B–	A–B+	A–B–
Probability:	$P(1 - \theta)$	$Q(1 - \theta)$	$P^2\theta$	$PQ\theta$	$PQ\theta$	$Q^2\theta$

Fig. B4.1 Probability of different outcomes in a radiation hybrid experiment.

co-segregate. The advantages of this method are that it is fast and accurate and is not subject to the distortions inherent in cloning, meiotic recombination or hybrid cell formation. The technique has been used with particular success with a number of protist genomes (Piper *et al.* 1998; Konfortov *et al.* 2000) whose high AT content causes problems in cloning and sequencing.

Table 4.2 Categorization of map objects. (Reproduced with permission from Cox *et al*. 1994, © American Association for the Advancement of Science.)

Mapping method	Experimental resource	Breakpoints	Markers
Meiotic	Pedigrees	Recombination sites	DNA polymorphisms
Radiation hybrid	Hybrid cell lines	Radiation-induced chromosome breaks	STSs
In situ hybridization	Chromosomes	Cytological landmarks	DNA probes
STS content	Library of clones	End points of clones	STSs
Clone-based fingerprinting	Library of clones	End points of clones	Genomic restriction sites

Integration of different mapping methods

Each of the mapping methods described above has its advantages and disadvantages and no one method is ideal. The size and complexity of the genome being analysed can greatly influence the methodologies employed in map construction. Nor is it uncommon for different research groups to use different mapping methods even when working on the same genome. Ultimately, the maps generated by the different methods need to be integrated. This is happening. For example, the BAC clones used to construct the human cytogenetic map (Cheung *et al.* 2001) have STSs that reference the radiation hybrid and linkage maps. Similarly, Chen *et al.* (2001) have devised a mapping method for rapid assembly and ordering of BAC clones on a radiation hybrid panel using STSs and PCR. Similarly, integrated maps are being constructed in other species, e.g. the rat (Bihoreau *et al.* 2001).

Many genomic mapping projects involve ordering two classes of objects relative to one another. These are *breakpoints* and *markers* (Table 4.2). Breakpoints, so called because they represent subdivisions of the genome, are defined by a specific experimental resource. Markers consist of unique sites in the genome and should be independent of any particular experimental resource. Although both types of object are essential for map construction, the map itself should be defined in terms of markers, especially those based on DNA sequence. One reason for this is that markers are permanent and easily shared. They can be readily stored as DNA sequence information and distributed in this fashion. By contrast, breakpoints are defined by experimental resources that tend to be transient and cumbersome to distribute. The most important reason to use sequence-based markers is that they can be easily screened against any DNA source. Thus they can be used to integrate maps constructed by diverse methods and investigators. Such integration is crucial to the assembly and assessment of maps.

The breakpoints divide the genome into 'bins' corresponding to the regions between breakpoints. In assessing mapping progress a first step is to determine the number of 'bins' that are occupied by markers and the distribution of these markers within each bin. Although some investigators report only the total number of makers used to construct the map, it is the number of occupied bins that provides the measure of progress. The distribution of the number of markers per bin is important because the goal is to have markers evenly, or at least randomly, spread rather than clustered.

The second step in assessing progress is to identify those occupied bins that are ordered relative to one another and to estimate the confidence in the ordering. Note that assignment of markers to bins can proceed throughout a mapping project but the ordering of bins is only possible as a project matures. Thus ordering is a good indication of the degree of completion. Finally, the distance in kilobases between ordered markers in a map needs to be measured.

As noted in Table 4.2, different experimental resources are used to construct the different kinds of maps. If consistency is to be achieved then it follows that different groups must use *identical* experimental material. Thus a key part of mapping is the construction of genomic libraries and cell lines by one research group for distribution to everyone who needs them. This has huge cost and logistical problems. Nevertheless, such libraries and cell lines are being made available. For example, Osoegawa *et al.* (2000, 2001) have constructed BAC libraries of the human and mouse genomes that are the universal reference material. Similarly, the BACs used for the cytogenetic mapping of the human genome (Cheung *et al.* 2001) are available from various stock centres.

In the context of resources, the work of the Centre d'Etude du Polymorphisme Humain (CEPH) deserves special mention. This organization was set up as a reference source for studies on human genetics. The need for such a reference source stems from the fact that directed matings in humans are not acceptable and because the human breeding cycle is too long to be experimentally useful. Consequently, CEPH maintains cell lines from three-generation human families, consisting in most cases of four grandparents, two parents and an average of eight children (Dausset *et al.* 1990). Originally cell lines from 40 families were kept but the number now is much larger. Such families are ideal for genetic mapping because it is possible to infer which allele was inherited from which parent (see Fig. 1.3 for example). The CEPH distributes DNAs from these families to collaborating investigators around the world.

Suggested reading

Bender W., Spierer P. & Hogness D.S. (1983) Chromosome walking and jumping to isolate DNA from the *Ace* and *rosy* loci and the bithorax complex in *Drosophila melanogaster*. *Journal of Molecular Biology* **168**, 17–33. *This is a classic paper and presents the principles of chromosome walking.*

Breen M. *et al.* (2001) Chromosome-specific single locus FISH probes allow anchorage of an 1800-marker integrated radiation-hybrid/linkage map of the domestic dog genome to al chromosomes. *Genome Research* **11**, 1784–1795. *This paper is an excellent example of how different mapping methods*

can be combined to develop a comprehensive physical and genetic map.*

Brookes A.J. (1999) The essence of SNPs. *Gene* **234**, 177–186. *This review covers the origin and genetical importance of SNPs in human DNA.*

Chee M. *et al.* (1996) Accessing genetic information with high-density arrays. *Science* **274**, 610–614. *This paper was the first detailed report of the use of DNA microchips to detect mutations in human DNA.*

Donis-Keller H. *et al.* (1987) A genetic linkage map of the human genome. *Cell* **51**, 319–337. *This is another classic paper that presented the first physical map of the human genome. It describes the identification and mapping of a large number of RFLPs.*

Fauth C. & Speicher M.R. (2001) Classifying by colors: FISH-based genome analysis. *Cytogenetics and Cell Genetics* **93**, 1–10. *This review details the different applications of FISH and the simultaneous use of fluorochromes.*

Weier H.U. (2001) DNA fiber mapping techniques for the assembly of high-resolution physical maps. *Journal of Histochemistry and Cytochemistry* **49**, 939–948. *This review summarizes different approaches to DNA fiber mapping and recent achievements in mapping ESTs and DNA replication sites.*

Useful websites

http://www.ncbi.nim.nih.gov/dbEST/
This website hosts the database on ESTs. Other specialized EST databases exist but this is the primary reference source.

http://snp.well.ox.ac.uk/
This site, and a mirror image in the USA, hosts the database on human SNPs.

http://schwartzlab.biotech.wisc.edu/
This website contains a wealth of information about optical mapping.

http://www.genethon.fr/english/
The details of the CEPH families and DNA databanks can be found on this site.

http://www.ebi.ac.uk/RHdb/RHdb.html
This site contains details of all radiation hybrid maps.

CHAPTER 5

Sequencing methods and strategies

Basic DNA sequencing

The first significant DNA sequence to be obtained was that of the cohesive ends of phage λ DNA (Wu & Taylor 1971) which are only 12 bases long. The methodology used was derived from RNA sequencing and was not applicable to large-scale DNA sequencing. An improved method, plus and minus sequencing, was developed and used to sequence the 5386 bp phage ΦX174 genome (Sanger *et al.* 1977a). This method was superseded in 1977 by two different methods, that of Maxam and Gilbert (1977) and the chain-termination or dideoxy method (Sanger *et al.* 1977b). For a while the Maxam and Gilbert method, which makes use of chemical reagents to bring about base-specific cleavage of DNA, was the favoured procedure. However, refinements to the chain-termination method meant that by the early 1980s it became the preferred procedure. To date, most large sequences have been determined using this technology, with the notable exception of bacteriophage T7 (Dunn & Studier 1983). For this reason, only the chain-termination method will be described here.

The chain-terminator or dideoxy procedure for DNA sequencing capitalizes on two properties of DNA polymerases: (i) their ability to synthesize faithfully a complementary copy of a single-stranded DNA template; and (ii) their ability to use 2′, 3′-dideoxynucleotides as substrates (Fig. 5.1). Once the analogue is incorporated at the growing point of the DNA chain, the 3′ end lacks a hydroxyl group and no longer is a substrate for chain elongation. Thus, the growing DNA chain is terminated, i.e. dideoxynucleotides act as chain terminators. In practice, the Klenow fragment of DNA polymerase is used because this lacks the 5′→3′ exonuclease activity associated with the intact enzyme. Initiation of DNA synthesis requires a primer and usually this is a chemically synthesized oligonucleotide which is annealed close to the sequence being analysed.

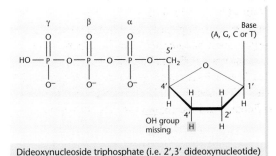

Normal deoxynucleoside triphosphate (i.e. 2′ deoxynucleotide)

Dideoxynucleoside triphosphate (i.e. 2′,3′ dideoxynucleotide)

Fig. 5.1 Dideoxynucleoside triphosphates act as chain terminators because they lack a 3′-OH group. Numbering of the carbon atoms of the pentose is shown (primes distinguish these from atoms in the bases). The α, β and γ phosphorus atoms are indicated.

DNA synthesis is carried out in the presence of the four deoxynucleoside triphosphates, one or more of which is labelled with ^{32}P, and in four separate incubation mixes containing a low concentration of one each of the four dideoxynucleoside triphosphate analogues. Therefore, in each reaction there is a population of partially synthesized radioactive DNA molecules, each having a common 5′-end, but each varying in length to a base-specific 3′-end (Fig. 5.2). After a suitable incubation period, the DNA in each mixture is denatured and electrophoresed in a sequencing gel.

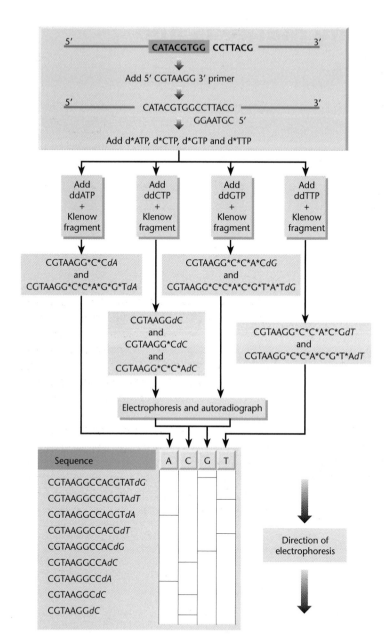

Fig. 5.2 DNA sequencing with dideoxynucleoside triphosphates as chain-terminators. In this figure asterisks indicate the presence of ^{32}P and the prefix 'd' indicates the presence of a dideoxynucleotide. At the top of the figure the DNA to be sequenced is enclosed within the box. Note that unless the primer is also labelled with a radioisotope the smallest band with the sequence CGTAAGG*dC* will not be detected by autoradiography as no labelled bases were incorporated.

A sequencing gel is a high-resolution gel designed to fractionate single-stranded (denatured) DNA fragments on the basis of their size and which is capable of resolving fragments differing in length by a single base pair. They routinely contain 6–20% polyacrylamide and 7 M urea. The function of the urea is to minimize DNA secondary structure which affects electrophoretic mobility. The gel is run at sufficient power to heat up to about 70°C. This also minimizes DNA secondary structure. The labelled DNA bands obtained after such electrophoresis are revealed by autoradiography on large sheets of X-ray film and from these the sequence can be read (Fig. 5.3).

To facilitate the isolation of single strands, the DNA to be sequenced may be cloned into one of the

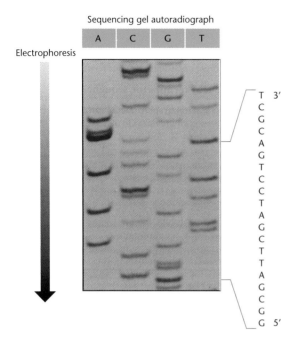

Sequencing gel autoradiograph

Electrophoresis

T 3'
C
G
C
A
G
T
C
C
T
A
G
C
T
T
A
G
C
G
G 5'

Fig. 5.3 Enlarged autoradiograph of a sequencing gel obtained with the chain terminator DNA sequencing method.

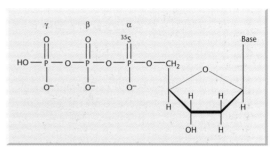

Fig. 5.5 Structure of an α-^{35}S-deoxynucleoside triphosphate.

Modifications of chain-terminator sequencing

Since its first description, the Sanger method of sequencing has been greatly improved in terms of read length, accuracy and convenience by a whole series of changes to the basic protocol. These changes include the choice of label, enzymes (Box 5.1) and template, the development of automated sequencing and a move from slab gel electrophoresis to capillary electrophoresis. These modifications are discussed in more detail below.

The sharpness of the autoradiographic images can be improved by replacing the ^{32}P-radiolabel with the much lower energy ^{33}P or ^{35}S. In the case of ^{35}S, this is achieved by including an α-^{35}S-deoxynucleoside triphosphate (Fig. 5.5) in the sequencing reaction. This modified nucleotide is accepted by

clustered cloning sites in the *lac* region of the M13 mp series of vectors (Fig. 5.4). A feature of these vectors is that cloning into the same region can be mediated by any one of a large selection of restriction enzymes but still permits the use of a single sequencing primer.

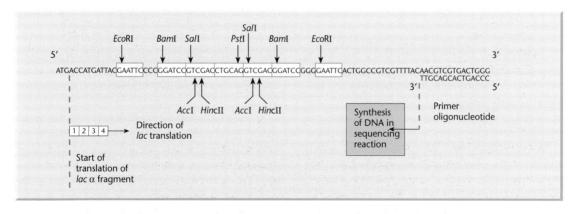

Fig. 5.4 Sequence of M13 mp7 DNA in the vicinity of the multipurpose cloning region. The upper sequence is that of M13 mp7 from the ATG start codon of the β-galactosidase α-fragment, through the multipurpose cloning region, and back into the β-galactosidase gene. The short sequence at the right-hand side is that of the primer used to initiate DNA synthesis across the cloned insert. The numbered boxes correspond to the amino acids of the β-galactosidase fragment.

Box 5.1 DNA polymerases used for Sanger sequencing

Many DNA polymerases have 5′→3′ and 3′→5′ exonuclease activities in addition to their polymerase activity. The 5′→3′ nuclease activity is detrimental to sequencing Regardless of whether the label for detection is on the 5′ end of the sequencing fragment, incorporated into the fragment as an internal label, or is on the terminator. In some cases, the domain of the polymerase that has the 5′→3′ exonuclease activity is absent, as in the case of T7 DNA polymerase. In others, the 5′→3′ nuclease domain can be removed by protease cleavage as first demonstrated by Klenow & Henningsen (1970) with *E. coli* DNA polymerase I. Alternatively, the domain can be removed by deletion, although some enzymes with deletions lose processivity. By contrast, *Taq* DNA polymerases with such deletions have greater thermostability and greater fidelity than full-length enzymes.

The 3′→5′ exonuclease activity is undesirable for sequencing applications because it hydrolyses the single-stranded sequencing primers. In most DNA polymerases this activity can be destroyed by point mutations or small deletions but some DNA polymerases, like those from *Thermus* species, naturally lack the activity.

When Sanger sequencing was first developed the enzyme used was the Klenow fragment of DNA polymerase I. A disadvantage of this enzyme is a sequence-dependent discrimination of dideoxy nucleotides (ddNTPs) for ordinary nucleotides

(dNTPs). This leads to a variation of the amount of fragments for each base in the sequencing reaction and hence uneven band intensities in autoradiographs or uneven peak heights if fluorescent labels are used (see p. 75). This leads to a requirement for high concentrations of ddNTPs, which are expensive, and increases the background noise with fluorescent labels. Problems also are encountered when using native *Taq* DNA polymerase (from *Thermus aquaticus*).

The discrimination of ddNTPs can be greatly reduced by replacing the Klenow polymerase with T7 DNA polymerase. The reason for this is that native *E. coli* DNA polymerase and *Taq* DNA polymerase have a phenylalanine in their active sites as compared with tyrosine in T7 DNA polymerase. Exchanging the phenylalanine residue with a tyrosine residue in the *E. coli* and *Taq* enzymes decreases the discrimination of ddNTPs by a factor of 250–8000 (Tabor & Richardson 1995).

Native *Taq* DNA polymerase exhibits another undesirable characteristic: uneven incorporation of ddNTPs with a strong bias in favour of ddGTP incorporation. This bias results from a strong interaction between ddGTP and an arginine residue at position 660. When this arginine residue is replaced with aspartate, serine, leucine, phenylalanine or tyrosine, the bias is eliminated. This results in more even band intensities, greater accuracy and longer read lengths (Li *et al.* 1999).

DNA polymerase and incorporated into the growing DNA chain. Non-isotopic detection methods also have been developed with chemiluminescent, chromogenic or fluorogenic reporter systems. Although the sensitivity of these methods is not as great as with radiolabels, it is adequate for many purposes.

The combination of chain-terminator sequencing and M13 vectors to produce single-stranded DNA is very powerful. Very good quality sequencing is obtainable with this technique, especially when the improvements given by ^{35}S-labelled precursors and T7 DNA polymerase are exploited. Further modifications allow sequencing of 'double-stranded' DNA, i.e. double-stranded input DNA is denatured by

alkali, neutralized, and one strand then is annealed with a specific primer for the actual chain-terminator sequencing reactions. This approach has gained in popularity as the convenience of having a universal primer has grown less important with the widespread availability of oligonucleotide synthesizers. With this development, Sanger sequencing has been liberated from its attachment to the M13 cloning system; e.g. PCR-amplified DNA segments can be sequenced directly. One variant of the double-stranded approach, often employed in automated sequencing, is 'cycle sequencing'. This involves a *linear* amplification of the sequencing reaction using 25 cycles of denaturation, annealing of a specific primer to one strand only, and extension in the presence of *Taq* DNA polymerase plus labelled dideoxynucleotides. Alternatively, labelled primers can be used with unlabelled dideoxynucleotides.

There are some important differences between M13 phages, plasmids and PCR products as templates in sequencing reactions. The single-stranded M13 phages can be sequenced only on one strand and share a limitation in size with PCR products to a practical maximum of 2–3 kb. Longer PCR products can be obtained under certain conditions but have not been used routinely for sequencing. Plasmids can harbour up to 10 kb fragments if low copy number vectors are used and this is a major advantage for linking sequences when very long stretches of DNA need to be sequenced (see p. 82). They also give better representation of DNA sequences from higher organisms with less tendency to eliminate repeated sequences (Chissoe *et al.* 1997; Elkin *et al.* 2001). A disadvantage of PCR sequencing is 'polymerase slippage', a term that refers to the inability of the *Taq* DNA polymerase to incorporate the correct number of bases when copying runs of 12 or more identical bases.

Automated DNA sequencing

In manual sequencing, the DNA fragments are radiolabelled in four chain-termination reactions, separated on the sequencing gel in four lanes, and detected by autoradiography. This approach is not well suited to automation. To automate the process it is desirable to acquire sequence data in real-time by detecting the DNA bands within the gel during the electrophoretic separation. However, this is not trivial as there are only about 10^{-15}–10^{-16} moles of DNA per band. The solution to the detection problem is to use fluorescence methods. In practice, the fluorescent tags are attached to the chain-terminating nucleotides. Each of the four dideoxynucleotides carries a spectrally different fluorophore. The tag is incorporated into the DNA molecule by the DNA polymerase and accomplishes two operations in one step: it terminates synthesis and it attaches the fluorophore to the end of the molecule. Alternatively, fluorescent primers can be used with non-labelled dideoxynucleotides. By using four different fluorescent dyes it is possible to electrophorese all four

chain-terminating reactions together in one lane of a sequencing gel. The DNA bands are detected by their fluorescence as they electrophorese past a detector (Fig. 5.6). If the detector is made to scan horizontally across the base of a slab gel, many separate sequences can be scanned, one sequence per lane. Because the different fluorophores affect the mobility of fragments to different extents, sophisticated software is incorporated into the scanning step to ensure that bands are read in the correct order. This is the principle on which the original Applied Biosystems (ABI) instruments operate.

An alternative to the four-dye system is to start with a single fluorescent-labelled primer which is used in all four sequencing reactions. The resulting fluorescent-labelled DNA strands are separated in four different lanes in the electrophoresis system. This is the basis of the Amersham Pharmacia Biotech ALF sequencer. It has a fixed argon laser which emits light that passes through the width of the gel and is sensed by detectors in each of the lanes.

Another variation is provided by the LI-COR two-dye near-infrared DNA analysis system. It can detect the products of two different sequencing reactions in parallel, enabling pooling reactions and simultaneous bidirectional sequencing. Sequencing both directions on a template by combining forward and reverse primers in the same direction produces twice the data from each reaction prepared.

Automated DNA sequencers offer a number of advantages that are not particularly obvious. First, manual sequencing can generate excellent data but even in the best sequencing laboratories poor autoradiographs frequently are produced that make sequence reading difficult or impossible. Usually the problem is related to the need to run different termination reactions in different tracks of the gel. Skilled DNA sequencers ignore bad sequencing tracks but many laboratories do not. This leads to poor quality sequence data. The use of a single-gel track for all four dideoxy reactions means that this problem is less acute in automated sequencing. Nevertheless, it is desirable to sequence a piece of DNA several times, and on both strands, to eliminate errors caused by technical problems. It should be noted that long runs of the same nucleotide or a high G + C content can cause compression of the bands on a gel, necessitating manual reading of the data, even with an automated system. Note also that

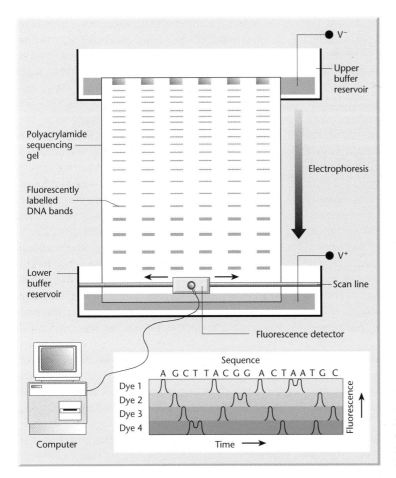

Fig. 5.6 Block diagram of an automated DNA sequencer and idealized representation of the correspondence between fluorescence in a single electrophoresis lane and nucleotide sequence.

multiple tandem short-repeats, which are common in the DNA of higher eukaryotes, can reduce the fidelity of DNA copying, particularly with *Taq* DNA polymerase. The second advantage of automated DNA sequencers is that the output from them is in machine-readable form. This eliminates the errors that arise when DNA sequences are read and transcribed manually.

DNA sequencing by capillary array electrophoresis

There are two problems with high volume sequencing in slab gels. First, the preparation of the gels is very labour-intensive and is a significant cost in large-scale sequencing centres. Secondly, DNA sequencing speed is related to the electric field applied to separate the fragments. To a first approximation,

the sequencing speed increases linearly with the applied electric field but application of a voltage across a material with high conductivity results in the generation of heat ('Joule heating'). Because of their thickness, slab gels cannot efficiently radiate heat and Joule heating limits the maximum electric field that can be applied.

Capillary electrophoresis is undertaken in high-purity fused silica capillaries with an internal diameter of 50 μm. Because of their small diameter, these capillaries are not prone to Joule heating even when high electric fields are applied in order to obtain rapid separation of DNA fragments. In addition, silica capillaries are very flexible, are easily incorporated into automated instruments and can be supplied pre-filled with a gel matrix. This reduces the hands-on operator time for 1000 samples per day from 8 h with a slab gel system to 15 min with a capillary gel system (Venter *et al.* 1998). For this

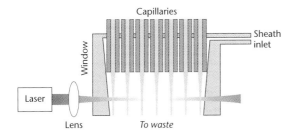

Fig. 5.7 Capillary linear-array sheath-flow cuvet instrument. Sheath fluid draws the analyte into thin streams in the centre of the flow chamber, with a single stream produced downstream from each capillary. A laser beam is focused beneath the capillary tips on the sample streams. A lens collects the fluorescence emission signal, which is then spectrally filtered and detected with either an array of photodiodes or with a changed couple device (CCD) camera. (Redrawn with permission from Dovichi & Zhang 2001.)

reason, capillary gels are very rapidly replacing slab gels in automated sequencing instruments. The current generation of capillary sequencers has arrays of up to 96 capillaries although there is one report of a 384 capillary instrument (Shibata *et al.* 2000).

In practice, DNA sequencing by capillary electrophoresis is very simple (Dovichi & Zhang 2001). Instead of placing the product of a sequencing reaction in the well of a slab gel it is applied to the top of a capillary gel. A number of different materials can be used for the gel matrix but it usually is linear (i.e. non-crosslinked) polyacrylamide. The labelled DNA fragments migrate through the capillary and emerge at the end in a vertical stream where they are detected (Fig. 5.7). Consideration of the detection method shown in Fig. 5.7 reveals another advantage of capillary gels over slab gels. With slab gels, the DNA fragments do not always run in a straight line and this makes automated reading more complicated but with capillaries this problem does not occur.

Basecalling and sequence accuracy

As DNA fragments pass the detector of a DNA sequencer they generate a signal. Information about the identity of the nucleotide bases is provided by the base-specific dye attached to the primer or dideoxy chain-terminating nucleotide. Additional steps include lane tracking and profiling, if slab gels

have been used, and trace processing. The latter involves the production of a set of four traces of signal intensities corresponding to each of the four bases over the length of the sequencing run. Using algorithms the four traces then are converted into the actual sequence of nucleotides. This process is known as *basecalling*.

The accuracy of the basecalling algorithms directly impacts the quality of the resulting sequence. In an ideal world the traces would be free of noise and peaks would be evenly spaced, of equal height and have a Gaussian shape. In reality, the peaks have variable spacing and height and there can be secondary peaks underneath the primary peaks. As a consequence, basecalling is error prone and for accurate sequence assembly it is essential to provide an estimate of quality for each assigned base (Buetow *et al.* 1999; Altshuler *et al.* 2000). The algorithm that gives the best estimate of error rates for basecalling with slab-gel based sequencing is *PHRED* (Ewing & Green 1998; Ewing *et al.* 1998). A PHRED quality score of X corresponds to an error probability of approximately $10^{-X/10}$. Thus a PHRED score of 30 corresponds to a 99.9% accuracy for the basecall. An improved basecalling algorithm, *Life Trace*, which is particularly suitable for capillary sequencing, has been described by Walther *et al.* (2001).

High throughput sequencing

The theoretical sequencing capacity of an automated DNA sequencer is easy to calculate. For a four-dye slab gel system, the capacity is the number of sequencing reactions that can be loaded on each gel, times the number of bases read from each sample, times the number of gels that can be run at once, times the number of days this can be carried out per year. For a 24-channel sequencer the capacity calculated in this way is 2.7 million bases per year. To use just one such instrument to sequence the human genome would require over 1000 years for single pass coverage and this clearly is not a practicable proposition.

To meet the demands of large-scale sequencing, 96-channel instruments have become commonplace and at least one 384-channel instrument has been developed (Shibata *et al.* 2000). By switching from slab gels to capillary systems, the electrophoresis run time is greatly reduced and nine runs can be achieved per 24 h period. Various other improvements to the biochemistry of the sequencing reactions and the chemistry of the gel matrix mean that the read length can be extended from the usual 500–600 bases to 800 bases. As a consequence it now is possible to generate 1–6 million bases of sequence per machine per month (Meldrum 2000b; Elkin *et al.* 2001) and Amersham Pharmacia Biotech claim 450 000 bases per day with their Megabace 1000 instrument. By comparison, 5 years previously the best throughput achieved was only (!) 40 thousand bases per month (Fleischmann *et al.* 1995). Those laboratories engaged in sequencing large genomes have large numbers of sequencing machines and can generate millions of base sequences per day, e.g. in excess of 18 million (Elkin *et al.* 2001).

To achieve the levels of sequence data quoted above it is not sufficient to have sequencing instruments with a high capacity. There are many manipulative steps required before samples are loaded on gels in preparation for electrophoresis. For example, DNA has to be isolated, fragmented and then cloned or amplified. The DNA then has to be re-isolated and subjected to the various sequencing reactions described earlier. Each of these procedures is labour intensive. Not surprisingly, there now are machines that can automate every one of them and details of these machines have been provided by Meldrum (2000a).

Sequencing strategies

Overview

Basically, there are three applications of sequencing:
1 sequencing of short regions of DNA to identify mutations of interest or single nucleotide polymorphisms (SNPs);
2 sequencing of complete genes and associated upstream and downstream control regions; and
3 sequencing of complete genomes.
For each application there is a choice of strategies. There also is a difference in the resource requirements. Almost any laboratory group can do sequencing of short regions of DNA and simple genes as found in prokaryotes and lower eukaryotes.

However, as the size of the DNA to be sequenced increases there is a need for more and more automation of the sequencing procedure and more and more computing power to sort the raw data into finished sequence. For this reason, genome sequencing is restricted to major sequencing centres.

Sequencing short stretches of DNA

When the objective is the identification of mutations or SNPs the requirement often is to analyse the same short stretch of DNA (10–20 nucleotides) from a large number of different samples. In essence, one *re-sequences* the same short stretch of DNA many times. This can be done by Sanger sequencing but the method generates many more data (500 bases of sequence) than are required. A number of alternative methods to Sanger sequencing have been developed that are much more appropriate for re-sequencing (pyrosequencing, hybridization microarrays) and these are described later (pp. 85–92).

Sequencing genes

The strategy for sequencing a gene depends on its size. In prokaryotes and lower eukaryotes genes are seldom more than a few kilobases in size because introns are rare and there is little or no repetitive DNA. In this instance, all that is required is to sequence a series of overlapping gene fragments. By contrast, in higher eukaryotes a gene may span several hundred kilobases or even several megabases. Sequencing DNA of this size requires a different approach particularly because 10 Mb of data are required to generate 1 Mb of confirmed sequence. This is the equivalent of sequencing an entire bacterial genome and the strategy for this is discussed in the next section. A simple more approach would be to analyse only the cDNA, which might be only 2 kb in length compared to 1 Mb for the intact gene. However, if the cDNA sequence is established first there is little incentive to sequence the genomic DNA at a later date! If one has only the genomic sequence, then the cDNA sequence is needed to identify unambiguously the introns and exons. A variant of the cDNA sequencing approach has been adopted by a number of groups associated with the human genome project. In this, clones are selected at random from a cDNA library and a small amount

Fig. 5.8 Schematic representation of hypothetical nested deletions obtained using a transposon-based cosmid vector. (a) Depiction of a hypothetical DNA clone. The green portion represents cloned DNA, the filled arrowheads the transposon sequences and the green half-arrows represent locations of primer binding sites. (b) Deletions extending into various sites within the cloned fragment resulting from selection for resistance to sucrose, caused by loss of sacB, and tetracycline. (c) Deletions resulting from selection for resistance to streptomycin, caused by loss of strA, and kanamycin. R/r indicates resistance and S/s indicates sensitivity. (Redrawn from Krishnan *et al.* 1995, by permission of Oxford University Press.)

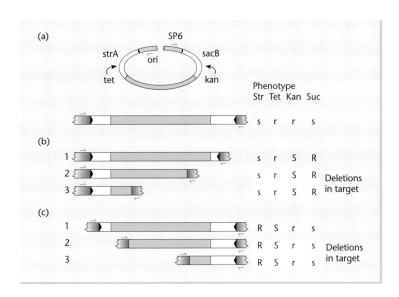

of sequence (ESTs, p. 51) determined for each, e.g. 200–300 nucleotides, a length that can be obtained in a single sequencing run. This information is incomplete as the full sequence of a cDNA usually comprises 1000–10 000 nucleotides. Furthermore, a low rate of errors in the sequence is not important, for the sequence acts merely as a 'signature'. These signatures are used to scan databases to determine whether they correspond to known genes. If so, no further work is necessary.

While cDNA sequencing reduces the workload, e.g. expressed genes constitute only 5% of the human genome, it is still technically demanding. The problems with this approach are fourfold. First, control elements such as enhancers and promoters as well as splice sites would not be sequenced. Secondly, many genes are expressed at very low levels or for very short periods and so may not be represented in cDNA libraries. Thirdly, there is some debate as to what actually constitutes a gene and therefore there is no agreement as to how many genes there are in each genome (for discussion see Fields *et al.* 1994; Bird 1995). Finally, much of the DNA not sequenced, so-called 'junk' DNA, could have functions that are as yet unknown.

A third approach to sequencing the genes of higher eukaryotes is to use feature mapping (Krishnan *et al.* 1995) to help guide the choice of regions to be subjected to detailed analyses. In feature mapping, large DNA fragments are cloned

into a transposon-based cosmid vector designed for generating nested deletions by *in vivo* transposition and simple bacteriological selection. These deletions place primer sites throughout the DNA of interest at locations that are easily determined by plasmid size (Fig. 5.8). In this way, Krishnan *et al.* (1995) generated 70 informative deletions of a 35 kb human DNA fragment. DNA adjacent to the deletion end points was sequenced and constituted the foundation of a feature map by: (i) identifying putative exons and positions of *Alu* elements; (ii) determining the span of a gene sequence; and (iii) localizing evolutionarily conserved sequences.

Genome sequencing

Two different strategies have been adopted for the sequencing of whole genomes: the 'clone by clone' approach and 'whole genome shotgun sequencing'. The choice of strategy has engendered much debate within the genome sequencing community. Some understanding of this debate is necessary to appreciate some of the key genome sequencing publications (Box 5.2).

With Sanger sequencing accurate read length is limited to 500–600 base pairs. Given this limitation, researchers employ a shotgun sequencing approach in which a random sampling of sequencing reads is collected from a larger target DNA sequence. With sufficient oversampling, the sequence of the target

Box 5.2 Some definitions used in genome sequencing projects

Contig   An overlapping series of clones or sequence reads that corresponds to a contiguous segment of the source genome

Coverage   The average number of times a genomic segment is represented in a collection of clones or sequence reads. Coverage is synonymous with redundancy

Minimal tiling path   A minimal set of overlapping clones that together provides complete coverage across a genomic region

Sequence-ready map   An overlapping bacterial clone map with sufficiently redundant clone coverage to allow for the rational selection of clones for sequencing

Finished sequence   The complete sequence of a clone or genome with a defined level of accuracy and contiguity

Full-shotgun sequence   A type of pre-finished sequence with sufficient coverage to make it ready for sequence finishing

Prefinished sequence   Sequence derived from a preliminary assembly during a shotgun-sequencing project

Working draft sequence   A type of prefinished sequence with sufficient coverage (8- to 10-fold) to make it ready for sequence finishing

---

can be inferred by piecing the sequence reads together into an assembly. For a given level of oversampling, the number of unsampled regions or gaps increases linearly with target size, as does the number of interspersed repetitive sequences that tend to confound assembly. After computer assembly of contigs, any gaps between contigs are closed experimentally and any misassembly is resolved. Because the order of the contigs and the size of the gaps is unknown, the assembly problem becomes harder as the size of the target DNA increases. For this reason it was generally assumed that cosmid targets represented the limit of the shotgun approach. Thus, whole genomes would be sequenced by first developing a set of overlapping cosmids that already had been ordered by physical mapping (Fig. 5.9). This strategy represents the conventional clone by clone or 'map-based' approach and it has been used successfully to generate the complete genome sequence of *Saccharomyces cerevisiae* (Goffeau *et al.* 1996) and the nematode *Caenorhabditis elegans* (*C. elegans* Sequencing Consortium 1998).

The view that cosmids represented the size limit for shotgun sequencing was destroyed when Fleischmann *et al.* (1995) determined the 1 830 137 bp sequence of the bacterium *Haemophilus influenzae* without prior mapping. The starting point was the preparation of genomic DNA, which then was

mechanically sheared and size fractionated. Fragments between 1.6 and 2.0 kb in size were selected, this narrow range being chosen to minimize variation in growth of clones. In addition, with a maximum size of only 2.0 kb the number of complete genes on a DNA fragment is minimized thereby reducing the chance of their loss through expression of deleterious gene products. The selected fragments were ligated to a sequencing vector and again size fractionated to minimize contamination from double-insert chimaeras or free vector. Finally, all cloning was undertaken in host cells deficient in all recombination and restriction functions to prevent deletion and rearrangement of inserts.

In all, 28 643 sequencing reactions were undertaken on inserts and this took eight individuals using 14 automated sequencers a total of 3 months. The collected sequence data were then assembled into contigs using sophisticated algorithms. Starting with an initial fragment a second candidate fragment was chosen with the best overlap based on oligonucleotide content. The contig was extended by the fragment only if strict match criteria were met. These match criteria included the minimum length of overlap, the maximum length of an unmatched end and the minimum percentage match. A total of 24 304 sequences were deemed useful and were assembled into 140 contigs after

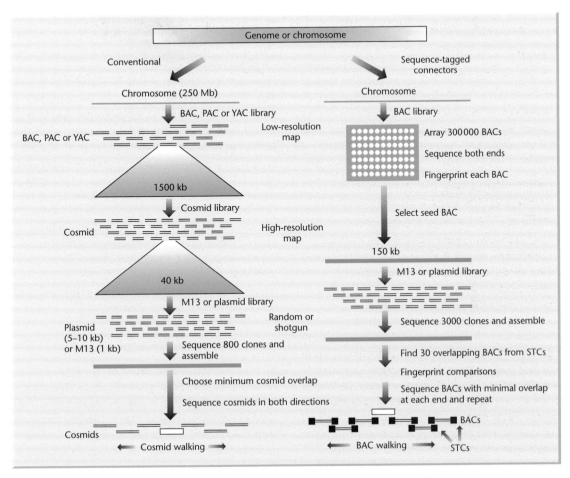

Fig. 5.9 Comparison of the conventional sequence approach with that proposed by Venter *et al.* (1996) (see text for full details). BAC, bacterial artificial chromosone; STC, sequence-tagged connector. (Redrawn from Venter *et al.* 1996 with permission from *Nature*, © Macmillan Magazines Ltd.)

30 h of central processing unit time. The 140 contigs were separated by 42 physical gaps (no template DNA for the region) and 98 sequence gaps (template available for gap closure). The sequence gaps and physical gaps were closed using the methods described in the next section.

The shotgun sequencing method for the whole *H. influenzae* genome has since been utilized many times with a wide range of other bacterial genomes. The success of the method led Venter *et al.* (1996) to propose a method for sequencing the entire human genome without first constructing a physical map. Rather than using yeast artificial chromosomes (YACs) and cosmids it would use bacterial artificial chromosomes (BACs) (Fig. 5.9) which can accept inserts up to 350 kb in length. A BAC library is pre-

pared which has an average insert size of 150 kb and a 15-fold coverage of the genome in question. The individual clones making up the library are arrayed in microtitre wells for ease of manipulation. Starting at the vector-insert points, both ends of each BAC clone are then sequenced to generate 500 bases from each end. These BAC end-sequences will be scattered approximately every 5 kb across the genome and make up 10% of the sequence. These 'sequence-tagged connectors' (STCs) will allow any one BAC clone to be connected to about 30 others.

Each BAC clone would be fingerprinted using one restriction enzyme to provide the insert size and detect artefactual clones by comparing the fingerprints with those of overlapping clones (Marra *et al.* 1997; see p. 48). A seed BAC of interest is sequenced and checked against the database of STCs to identify the 30 or so overlapping BAC clones. The two BAC clones showing internal consistency among the fingerprints and minimal overlap at either end then would be sequenced. In this way the entire human genome could be sequenced with just over 20 000 BAC clones. This proposal was not universally accepted but subsequently a modification of it was used to sequence chromosome 2 (19.6 Mb) of *Arabidopsis thaliana* (Lin *et al.* 1999b) and the entire human genome (International Human Genome Sequencing Consortium 2001).

Whereas Venter *et al.* (1996) proposed a hierarchical shotgun approach in which end sequences (STCs) are used to provide long-range continuity across the genome, Weber & Myers (1997) proposed a total shotgun approach to the sequencing of the human genome. In this approach, DNA would be sheared and size-selected before being cloned in *E. coli*. Cloned inserts would fall into two classes: long inserts of size 5–20 kb and short inserts of size 0.4–1.2 kb. Sequencing read lengths would be of sufficient magnitude so that the two sequence reads from the ends of the short inserts overlap. Both ends of the long inserts also would be sequenced and, because their spacing and orientation is known, they can be used to create a scaffold on which the short sequences can be assembled. This approach was not well received (Green 1997; Marshall & Pennisi 1998) but Myers *et al.* (2000) were able to show the validity of the approach by applying it to the 120 Mb euchromatic portion of the *Drosophila melanogaster* genome. Venter *et al.* (2001) then applied the method to the sequencing of the human genome. However, in both these cases, end sequences also were determined for 50 kb inserts in BACs to provide additional scaffolding information. Also, in the case of the human sequence, any sequence-tagged sites (STSs) that were detected helped to locate the fragment in the context of the overall genome STS map. An indication of the workload associated with sequencing the human genome in this way can be obtained by consideration of the data in Table 5.1.

There is no doubt that assembling a complete genome sequence is relatively easy if one has a collection of sequence-ready clones that already are ordered on a physical map. However, constructing the map is very laborious and takes much longer than sequencing. By contrast, whole-genome shotgun sequencing is very fast but sequence assembly could be more prone to errors. So, how comparable are sequence data produced by the two methods? The first comparison to be published is that for a 2.6 Mb sequence from *D. melanogaster* (Benos *et al.* 2001). Both methods identified a common set of 275 genes. An additional 15 genes were predicted from sequence data produced by one or other method but in each case the data are ambiguous. At the level of the protein sequence encoded by these 275 genes, significant differences were found in only 24. For 10 of these sequences it was not possible to tell which method had generated the (more) correct sequence. For the remaining 14 sequences the clone by clone approach yielded the best results. So, overall, the two different sequencing strategies generated data that were in remarkable agreement.

Closing sequence gaps

No matter what sequencing strategy is used there always are gaps at the assembly stage. These gaps fall into two categories: sequence gaps where a template exists and physical gaps where no template occurs. Sequence gaps can be closed using a primer-directed walking strategy as shown in Fig. 5.10. Physical gaps are much harder to close. In the case of the shotgun sequencing of the *H. influenzae* genome, a number of techniques were used. For example, oligonucleotide primers were designed and synthesized from the end of each contig. These primers were used in hybridization reactions based on the premise that labelled oligonucleotides homologous to the ends of adjacent contigs should hybridize to common DNA restriction fragments (Fig. 5.11a). Links were also made by searching each contig end against a peptide database. If the ends of two contigs matched the same database sequence, then the two contigs were tentatively assumed to be adjacent (Fig. 5.11b). Finally, two λ libraries were constructed from genomic *H. influenzae* DNA and were probed with the oligonucleotides designed from the ends of each contig. Positive plaques then were used to prepare templates and the sequence was determined from each end of the λ clone insert. These sequence fragments were searched against a

Table 5.1 Sequencing statistics for the shotgun sequencing of the human genome as undertaken by Venter *et al.* (2001). Note that the DNA sequenced was derived from five different individuals (A, B, C, D and F).

	Individual	Number of reads for different insert libraries				Total number of base pairs
		2 kbp	10 kbp	50 kbp	Total	
No. of sequencing reads	A	0	0	2 767 357	2 767 357	1 502 674 851
	B	11 736 757	7 467 755	66 930	19 271 442	10 464 393 006
	C	853 819	881 290	0	1 735 109	942 164 187
	D	952 523	1 046 815	0	1 999 338	1 085 640 534
	F	0	1 498 607	0	1 498 607	813 743 601
	Total	13 543 099	10 894 467	2 834 287	27 271 853	14 808 616 179
Fold sequence coverage	A	0	0	0.52	0.52	
(2.9 Gb genome)	B	2.20	1.40	0.01	3.61	
	C	0.16	1.17	0	0.32	
	D	0.18	0.20	0	0.37	
	F	0	0.28	0	0.28	
	Total	2.54	2.04	0.53	5.11	
Fold clone coverage	A	0	0	18.39	18.39	
	B	2.96	11.26	0.44	14.67	
	C	0.22	1.33	0	1.54	
	D	0.24	1.58	0	1.82	
	F	0	2.26	0	2.26	
	Total	3.42	16.43	18.84	38.68	
Insert size (mean)	Average	1951 bp	10 800 bp	50 715 bp		
Insert size (SD)	Average	6.10%	8.10%	14.90%		

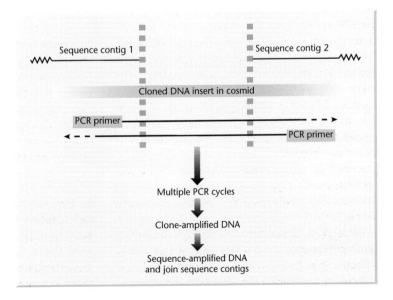

Fig. 5.10 Linking DNA sequence contigs by walking.

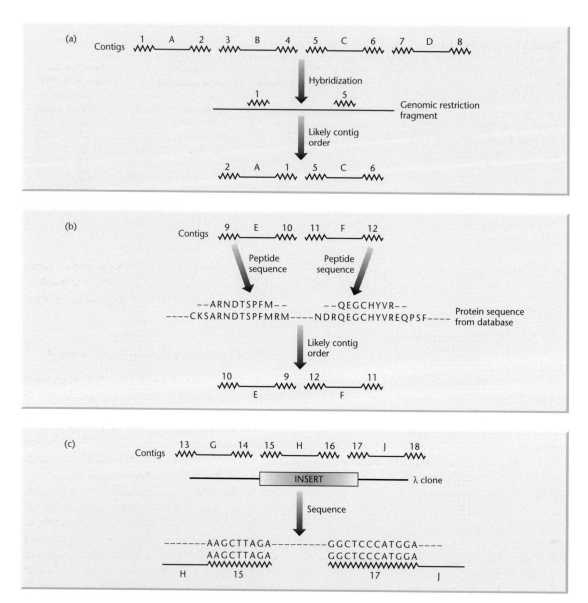

Fig. 5.11 The three methods for closing the physical gaps in sequencing by the method of Fleischmann *et al.* (1995). In each case the sequences at the ends of contigs are shown as wavy lines and individual sequences are given separate numbers. Contigs are denoted by large capital letters. In (b) the individual amino acids are represented by the standard single letter code. See text for a detailed description of each method.

database of all contigs and two contigs that matched the sequence from the opposite ends of the same λ clone were ordered (Fig. 5.11c). The λ clone then provided the template for closure of the sequence gap.

Another method for closing gaps is to use representational difference analysis (Lisitsyn *et al.* 1993).

In this technique one undertakes a subtractive hybridization of the library DNA from the total genomic DNA. In this way, Frohme *et al.* (2001) were able to close 11 out of 13 gaps in the sequence of the bacterium *Xylella fastidiosa*. Although this method is useful for isolating sequences that fall within gaps, any sequences isolated will not be

Table 5.2 Sequencing statistics for some eukaryotic genomes.

Organism	Year	Millions of bases sequenced	Total coverage (%)	Coverage of euchromatin (%)
Saccharomyces cerevisiae	1996	12	93	100
Caenorhabditis elegans	1998	97	99	100
Drosophila melanogaster	2000	116	64	97
Arabidopsis thaliana	2000	115	92	100
Human chromosome 21	2000	34	75	100
Human chromosome 22	1999	34	70	97
Human genome (consortium)	2001	2693	84	90
Human genome (Venter *et al.*)	2001	2654	83	88–93

useful if the sequences cannot be assembled into a contig that is anchored on at least one end of the gap.

Direct cloning of DNA that is missing in libraries is possible using transformation-associated recombination (TAR). As originally described (Larionov *et al.* 1996), one starts with a YAC containing *Alu* sequences. This vector is cleaved to generate two fragments, one consisting of *Alu*–telomere and the other of *Alu*–centromere–telomere. If these fragments are mixed with high-molecular-weight human DNA containing *Alu* sequences, recombination occurs during transformation to generate new YACs containing large human DNA inserts. Essentially, the *Alu* sequences act as 'hooks' and Noskov *et al.* (2001) have shown that the minimal length of sequence homology required is 60 bp. The hooks need not be *Alu* sequences but could be sequences derived from the ends of contigs that are used to trap DNA spanning sequence gaps.

Some of the gaps in complete sequences arise because of the difficulty in cloning DNA containing centromeres and telomeres. The absence of telomeres is easily explained: the absence of restriction sites in the $(TTAGGG)_n$ repeat of telomeres means that they are unlikely to be inserted into cloning vectors. The solution to this problem is to use half-YACs. These are circular yeast vectors containing a single telomere that have a unique restriction site at the end of the telomere (Riethman *et al.* (1989). On cleavage with the appropriate restriction enzyme, a linear molecule containing a single telomere at one end (half-YAC) is generated. This molecule is incapable of replicating in a linear form in yeast unless another telomere is added. Using such half-YACs, Riethman *et al.* (2001) were able to link 32 telomere regions to the draft sequence of the human genome.

How complete is a sequence assembly?

No genome of a eukaryote has been sequenced completely. The reason for this is the existence of regions, usually containing repetitive DNA, that are difficult to clone and/or sequence. The extent of these regions varies widely in different species and each sequencing consortium has made a pragmatic decision as to what constitutes a sufficient level of coverage for the genome on which they are working (Table 5.2). For example, about one-third of the overall sequence of *Drosophila* was not stable in the cloning systems used and was not sequenced but 97% of the euchromatin was sequenced. For the human genome, one definition of 'finished' is that fewer than one base in 10 000 is incorrectly assigned, more than 95% of the euchromatin is sequenced, and each gap is smaller than 150 kb. On this basis, only 25% of the human genome sequence is complete. However, this in no way decries the achievements to date.

Alternative DNA sequencing methodologies

Over the last 20 years many research groups have developed alternatives to the Sanger sequencing method. Initially, the impetus for developing

Fig. 5.12 The general principle of pyrosequencing. A polymerase catalyses incorporation of nucleotides into a nucleic acid chain. As each nucleotide is incorporated a pyrophosphate (PPi) molecule is released and incorporated into ATP by ATP sulfurylase. On addition of luciferin and the enzyme luciferase, this ATP is degraded to AMP with the production of light.

$$(\text{Oligonucleotide})_n + \text{nucleotide} \xrightarrow{\text{Polymerase}} (\text{Oligonucleotide})_{n+1} + \text{PPi}$$

$$\text{PPi} + \text{APS} \xrightarrow[\text{Sulfurylase}]{\text{ATP}} \text{ATP} + \text{sulphate}$$

$$\text{ATP} + \text{luciferin} + \text{O}_2 \xrightarrow{\text{Luciferase}} \text{AMP} + \text{PPi} + \text{oxyluciferin} + \text{LIGHT}$$

alternative methods was to increase the throughput of sequence in terms of bases read per day per individual. The advent of multichannel automated DNA sequencers, particularly the newer capillary sequencers, obviated the need for alternative methods for *de novo* sequencing of DNA molecules larger than a few megabases. Today, the rationale for using these alternative methodologies is not the sequencing of large stretches of DNA. Rather, they are used for applications where only short stretches of DNA need to be sequenced, e.g. analysis of SNPs, mutation detection, resequencing of disease genes, partial expressed sequence tag (EST) sequencing and microbial typing by analysis of 16S rRNA genes (Ronaghi 2001). Although Sanger sequencing is the 'gold standard', it generates much more information than is necessary and is time-consuming because it involves electrophoresis.

Although many different methods have been proposed as alternatives to Sanger sequencing, most have generated much initial excitement and then disappeared without trace. The exception is sequencing by hybridization which now is performed using microarrays ('gene chips'). Two newer methods of sequencing short stretches of DNA that are likely to be widely adopted are pyrosequencing and massively parallel signature sequencing (MPSS). Pyrosequencing most closely resembles Sanger sequencing in that it involves DNA synthesis, whereas MPSS has similarities to sequencing by hybridization. The continued refinement of two methods for producing gas-phase ions, electrospray ionization and matrix-assisted laser desorption ionization, has resulted in methods for the rapid characterization of oligonucleotides by mass spectrometry (Limbach *et al.* 1995). This methodology has not been widely adopted for sequencing and will not be discussed further here.

Pyrosequencing

Pyrosequencing is a DNA sequencing method that involves determining which of the four bases is incorporated at each step in the copying of a DNA template. As DNA polymerase moves along a single-stranded template, each of the four nucleoside triphosphates is fed sequentially and then removed. If one of the four bases is incorporated then pyrophosphate is released and this is detected in an enzyme cascade that emits light (Fig. 5.12). There are two variants of the pyrosequencing technique (Fig. 5.13). In solid-phase pyrosequencing (Ronaghi *et al.* 1996), the DNA to be sequenced is immobilized and a washing step is used to remove the excess substrate after each nucleotide addition. In liquid-phase sequencing (Ronaghi *et al.* 1998b) a nucleotide degrading enzyme (apyrase) is introduced to make a four enzyme system. Addition of this enzyme has eliminated the need for a solid support and intermediate washing thereby enabling the pyrosequencing reaction to be performed in a single tube. However, without the washing step, inhibitory substances can accumulate. The output from a typical pyrogram is shown in Fig. 5.14. It is worth noting that because the light emitted by the enzyme cascade is directly proportional to the amount of pyrophosphate released, it is easy to detect runs of 5–6 identical bases. For longer runs of a single base it may be necessary to use software algorithms to determine the exact number of incorporated nucleotides.

Template preparation for pyrosequencing is very easy. After generation of the template by the PCR, unincorporated nucleotides and PCR primers are removed. Two methods have been developed for this purification step. In the first, biotinylated PCR product is captured on magnetic beads, washed and denatured with alkali (Ronaghi *et al.* 1998a). In the

(a)

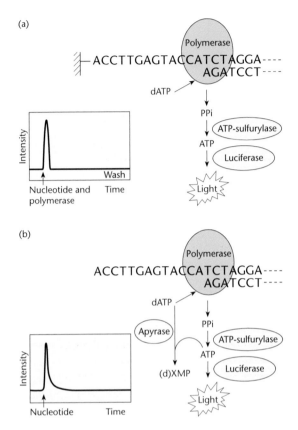

(b)

Fig. 5.13 The two types of pyrosequencing. (a) Schematic representation of the progress of the enzyme reaction in solid-phase pyrosequencing. The four different nucleotides are added stepwise to the immobilized primed DNA template and the incorporation event is followed using the enzyme ATP sulfurylase and luciferase. After each nucleotide addition, a washing step is performed to allow iterative addition. (b) Schematic representation of the progress of the enzyme reaction in liquid-phase pyrosequencing. Primed DNA template and four enzymes involved in liquid-phase pyrosequencing are placed in a well of a microtitre plate. The four different nucleotides are added stepwise and incorporation is followed using the enzyme ATP sulfurylase and luciferase. The nucleotides are continuously degraded by nucleotide-degrading enzyme allowing addition of subsequent nucleotide. dXTP indicates one of the four nucleotides. (Redrawn with permission from Ronaghi 2001.)

second method, akaline phosphatase or apyrase and exonuclease I are added to the PCR product to destroy the nucleotides and primers, respectively. The sequencing primer then is added and the mixture rapidly heated and cooled. This inactivates the enzymes, denatures the DNA and enables the

primers to anneal to the templates (Nordstrom *et al.* 2000a,b).

The acceptable read length of pyrosequencing currently is about 200 nucleotides, i.e. much less than is achieved with Sanger sequencing. However, many modifications are being made to the reaction conditions to extend the read length. For example, the addition of ssDNA-binding protein to the reaction mixture increases read length, facilitates sequencing of difficult templates and provides flexibility in primer design. The availability of automated systems for pyrosequencing (Ronaghi 2001) is facilitating the use of the technique for high-throughput analyses. A key benefit of the technique is the absence of a requirement to label the test sequence.

Sequencing by hybridization and the development of microarrays

The principle of sequencing by hybridization can best be explained by starting with a simple example. Consider the tetranucleotide CTCA, whose complementary strand is TGAG, and a matrix of the whole set of $4^3 = 64$ trinucleotides. This tetranucleotide will specifically hybridize only with complementary trinucleotides TGA and GAG, revealing the presence of these blocks in the complementary sequence. From this the sequence TGAG can be reconstructed. If instead of using trinucleotides, $4^8 = 65\ 536$ octanucleotides were used, it should be possible to sequence DNA fragments up to 200 bases long (Bains & Smith 1988; Lysov *et al.* 1988; Southern 1988; Drmanac *et al.* 1989). The length of the target that can be analysed is approximately equal to the square root of the number of oligonucleotides in the array (Southern *et al.* 1992). Two different experimental configurations have been developed for the hybridization reaction. Either the target sequence may be immobilized and oligonucleotides labelled or the oligonucleotides may be immobilized and the target sequence labelled. Each method has advantages over the other for particular applications. It is an advantage to label the oligonucleotides to analyse a large number of target sequences for fingerprinting. On the other hand, for applications that require large numbers of oligonucleotides of different sequence, it is advantageous to immobilize oligonucleotides and use the target sequence as the labelled probe.

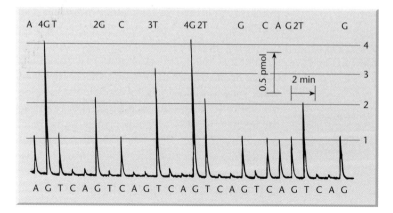

Fig. 5.14 Pyrogram of the raw data obtained from liquid-phase pyrosequencing. Proportional signals are obtained for one, two, three and four base incorporations. Nucleotide addition, according to the order of nucleotides, is indicated below the pyrogram and the obtained sequence is indicated above the pyrogram. (Redrawn with permission from Ronaghi 2001.)

Drmanac *et al.* have pioneered the first approach. For example, exons 5–8 of the *TP53* gene were PCR-amplified from 12 samples, spotted onto nylon filters and individually probed with 8192 non-complementary radiolabelled 7-base oligos (Drmanac *et al.* 1998). In this way, 13 distinct homozygous or heterozygous mutations in these 12 samples were detected by determining which oligonucleotide probes hybridized to the immobilized targets. The downside of this approach is the need to perform thousands of separate hybridization reactions.

The feasibility of the second approach (Fig. 5.15) was first demonstrated by Southern *et al.* (1992). They developed technology for making complete sets of oligonucleotides of defined length and co-valently attaching them to the surface of a glass plate by synthesizing them *in situ*. A device carrying all octapurine sequences was used to explore factors affecting molecular hybridization of the tethered oligonucleotides and computer-aided methods for analysing the data were developed. It was quickly realized that the light-directed synthetic (photo-lithography) methods routinely used in the semi-conductor industry could be combined with standard oligonucleotide synthesis to prepare microarrays carrying hundreds of thousands of oligonucleotides of predetermined sequence (Fodor *et al.* 1993; Pease *et al.* 1994; Southern 1996).

Manufacturing a microarray is deceptively simple. In the first step a mercury lamp is shone through a standard computer-industry photolithographic mask onto the synthesis surface. This activates specific areas for chemical coupling with a nucleoside which itself contains a 5′ protecting group. Further ex-posure to light removes this group, leaving a 5′-hydroxyl group capable of reacting with another nucleoside in the subsequent cycle. The choice of which nucleoside to activate is thus controlled by the composition of the mask. Successive rounds of deprotection and chemistry can result in an exponential increase in oligonucleotide complexity on a chip for a linear number of steps. For example, it requires only $4 \times 15 = 60$ cycles to synthesize a complete set of ~1 billion different 15 mers. The space occupied by each specific oligonucleotide is termed a feature and may house at least 1 million identical molecules (Fig. 5.16). A standard 1.6 cm square chip can house more than 1 million different oligos with a spacing of 1 μm, cf. the computer industry which routinely achieves resolution at 0.3 μm!

Although microchips have not been used for *de novo* sequencing of a genome, they have been used for re-sequencing genomes as exemplified by human mitochondrial DNA (Chee *et al.* 1996). For this purpose a 4L tiled array is set up in which L corresponds to the length of the sequence to be analysed. The sequence is probed with a series of oligomers of length P which exactly match the target sequence except for one position which is systematically substituted with each of the four bases A, T, G or C. In the example shown in Fig. 5.17, a tiled array of 15 mers varied at position 7 from the 3′ end is used. This is known as a $P^{15,7}$ array. To use such an array, the DNA to be analysed is amplified by long-range PCR. Fluorescently labelled RNA is then prepared by *in vitro* transcription and this is hybridized to the array. The hybridization patterns are imaged with a high-resolution confocal scanner and a typical

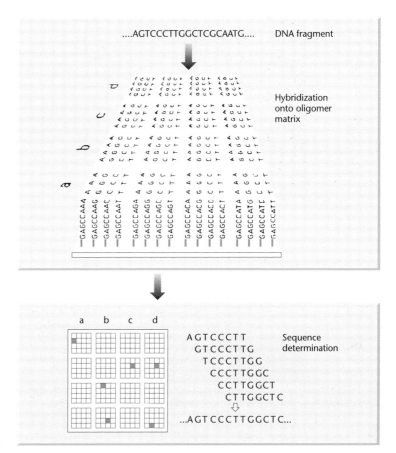

Fig. 5.15 Schematic representation of sequencing by hybridization where the sequence being examined is labelled and hybridized with immobilized oligomers. (Reprinted from Hoheisel 1994 by permission of Elsevier Science.)

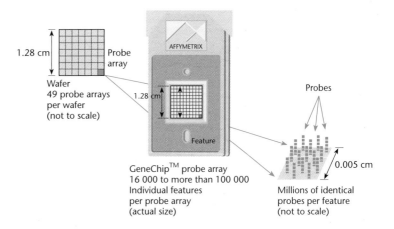

Fig. 5.16 Structure of a sequencing microchip. (Courtesy of Dr M. Chee, Affymetrix.)

result is shown in Fig. 5.17. For the purpose of re-sequencing the human mitochondrial genome (L = 16 569 bp) with a tiled array of $P^{15,7}$ probes then a total of 66 276 probes ($4 \times 16\,569$) of the possible $\sim 10^9$ 15 mers would be required.

Microchips are particularly useful for detecting mutations and polymorphisms, particularly SNPs, (Chee *et al.* 1996; Hacia *et al.* 1996, 1998; Kozal *et al.* 1996). As an example, consider again the mitochondrial genome. Chee *et al.* (1996) prepared a

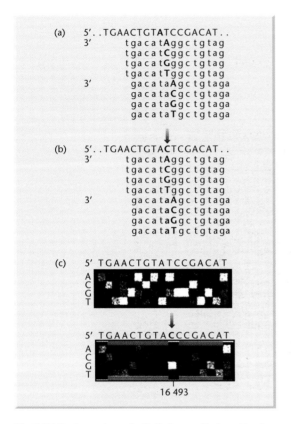

(c)   5′ TGAACTGTATCCGACAT

5′ TGAACTGTACCCGACAT

16 493

Fig. 5.17 Design and use of a 4L tiled array. Each position in the target sequence (uppercase letters) is queried by a set of four probes on the chip (lowercase letters), identical except at a single position, termed the substitution position, which is either A, C, G or T (bold black indicates complementarity, green a mismatch). Two sets of probes are shown, querying adjacent positions in target. (b) Effect of a change in the target sequence. The probes are the same as in (a), but the target now contains a single-base substitution (base C, shown in green and arrowed). The probe set querying the changed base still has a perfect match (the G probe). However, probes in adjacent sets that overlap the altered target position now have either one or two mismatches (green) instead of zero or one, because they were designed to match the target shown in (a). (c) Hybridization to a 4L tiled array and detection of a base change in the target. The array shown was designed to the mt1 sequence. (Top) Hybridization to mt1. The substitution used in each row of probes is indicated to the left of the image. The target sequence can be read 5′ to 3′ from left to right as the complement of the substitution base with the brightest signal. With hybridization to mt2 (bottom), which differs from mt1 in this region by a T → C transition, the G probe at position 16 493 is now a perfect match, with the other three probes having single-base mismatches (**A** 5, **C** 3, **G** 37, **T** 4 counts). However, at flanking positions, the probes have either single- or double-base mismatches, because the mt2 transition now occurs away from the query position. (Reprinted from Chee *et al.* 1996 by permission of the American Association for the Advancement of Science.)

P[25,13] tiling array consisting of 136 528 synthesis cells, each 35 μm square in size. In addition to a 4L tiling across the genome, the array contained a set of probes representing a single-base deletion at every position across the genome and sets of probes designed to match a range of specific mtDNA haplotypes. After hybridization of fluorescently labelled target RNA, 99% of the sequence could be read correctly simply by identifying the highest intensity in each column of four substitution probes. The array also was used to detect three disease-causing mutations in a patient with hereditary optic neuropathy (Fig. 5.18). A refinement to this method also has been developed. Ideally, the hybridization signals from the reference and test DNAs should be compared by hybridization to the same array. This can be carried out by using a two-colour labelling and detection scheme in which the reference is labelled with phycoerythrin (red) and the target with fluorescein (green). By processing the reference and target together, experimental variability during the fragmentation, hybridization, washing and detection steps is eliminated.

Variations of the microarray method

It should be noted that there are two variants of the hybridization-based method of DNA sequencing. The first approach is exemplified by the mitochondrial analysis described above. This is, oligos are used that are complementary to a significant subset of sequence changes of interest and one measures *gain of hybridization* signal to these oligos in test samples relative to reference samples. Relative 'gain' of signal by these oligos indicates a sequence change. In this respect, the microarray is being used in an analogous fashion to a conventional dot blot. This 'gain-of-signal' approach allows for a partial scan of a DNA segment for all possible sequence variations. An array designed to interrogate both strands of a target of length N for all possible single nucleotide substitutions would consist of 8N oligos of length 20–25 nucleotides, i.e. 80 000 oligos for a 10 kb sequence. To interrogate the same DNA target for all deletions of a particular length would require an additional 2N oligos, i.e. 100 000 oligos if screening for deletions of 1–5 bp. However, interrogating both target strands for insertions of length X requires $2(4^X)N$ oligos, i.e. 27 280 000 oligos if looking for all 1–5 bp insertions!

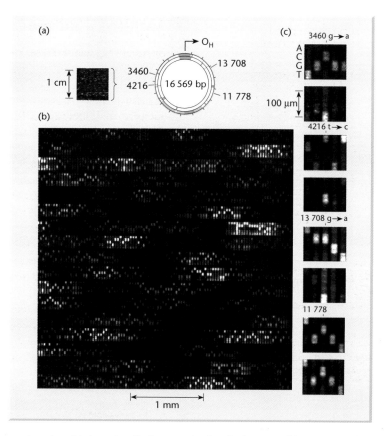

Fig. 5.18 Use of a sequencing microchip to analyse the human mitochondrial genome. (a) An image of the array hybridized to 16.6 kb of mitochondrial target RNA (L strand). The 16 569 bp map of the genome is shown, and the H strand origin of replication (O_H), located in the control region, is indicated. (b) A portion of the hybridization pattern magnified. In each column there are five probes, A, C, G, T and Δ, from top to bottom. The Δ probe has a single base deletion instead of a substitution and hence is 24 instead of 25 bases in length. The scale is indicated by the bar beneath the image. Although there is considerable sequence-dependent intensity variation, most of the array can be read directly. The image was collected at a resolution of ~100 pixels per probe cell. (c) The ability of the array to detect and read single base differences in a 16.6 kb sample is illustrated. Two different target sequences were hybridized in parallel to different chips. The hybridization patterns are compared for four positions in the sequence. Only the $P^{25,13}$ probes are shown. The top panel of each pair shows the hybridization of the mt 3 target, which matches the chip P^0 sequence at these positions. The lower panel shows the pattern generated by a sample from a patient with Leber's hereditary optic neuropathy (LHON). Three known pathogenic mutations, LHON 3460, LHON 4216, and LHON 13 708, are clearly detected. For comparison, the fourth panel in the set shows a region around position 11 778 that is identical in both samples. (Reprinted from Chee *et al.* 1996 by permission of the American Association for the Advancement of Science.)

In the loss-of-signal approach, sequence variations are scored by quantitating relative losses of hybridization signal to perfect match oligonucleotide probes in test samples relative to wild-type reference targets. Ideally, a homozygous sequence change results in a complete loss of hybridization signal to perfect match probes interrogating the region surrounding the sequence change (Fig. 5.19). For heterozygous sequence variations the signal loss theoretically is 50%. With this approach, an array designed to interrogate both strands of a 10 kb sequence for all possible sequences requires just 20 000 oligos. An added advantage is that multiple probes contribute to the detection of a sequence variation (see Fig. 5.19) thereby minimizing random sources of error caused by hybridization signal fluctuations – a problem with the gain-of-signal approach. The disadvantage of the loss-of-signal method is that the identity of the mutation cannot be determined without subsequent conventional sequencing.

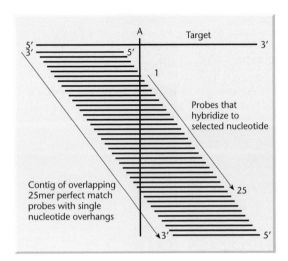

Fig. 5.19 Oligonucleotide probes used in loss of hybridization signal sequence analysis. Ideally each target nucleotide position contributes to hybridization to a set of N overlapping N-base perfect match probes in an oligonucleotide array. In this example, hybridization to 25 overlapping 25-base probes are affected by changes in a single target nucleotide. (Redrawn with permission from Hacia 1999.)

Availability of microarrays

Five different components are required for work with microarrays.
1 The chip itself with its special surface.
2 The device for producing the microarrays by spotting the oligos onto the chip or for their *in situ* synthesis.
3 A fluidic system for hybridization of the test DNA to the immobilized oligos.
4 A scanner to read the microarrays after hybridization.
5 Software programs to quantify and interpret the results.
All of the above are commercially available (Meldrum 2000b). In addition, microarrays carrying sets of oligos matched to particular sequences are also commercially available. However, it must be stressed that the value of the data obtained with a microarray depends critically on the quality of the arraying. Fortunately the laying down of arrays is becoming easier thanks to the use of bubble jet technology (Okamoto *et al.* 2000), maskless *in situ* synthesis of oligonucleotides (Singh-Gasson *et al.* 1999) and improvements in microarray surface chemistry (Beier & Hoheisel 2000).

Massively parallel signature sequencing

This sequencing method was developed by Brenner *et al.* (2000) to enable them to undertake a global analysis of gene regulation. This topic is covered in more detail later (p. 161) but essentially the relative abundance of different mRNAs is estimated by sequencing a large number of clones from representational cDNA libraries. Because the probability of finding a particular message is proportional to the number of clones sequenced, a large-scale sequencing project is required to detect very low abundance mRNAs if traditional Sanger sequencing is used. Instead, Brenner *et al.* (2000) have developed a method where a short 'signature' sequence can be determined *simultaneously* for every cDNA in a library. The methodology used is described on p. 161.

Suggested reading

Chee M. *et al.* (1996) Accessing genetic information with high-density arrays. *Science* **274**, 610–614. *This is a classic paper that describes in detail the use of DNA arrays to resequence the human mitochondrial genome and the identification of base changes in certain diseases.*

Fleischmann R.D. *et al.* (1995) Whole-genome random sequencing and assembly of *Haemophilus influenzae* Rd. *Science* **269**, 496–512. *This is a classic paper detailing the first complete sequence of a bacterial genome and details the principles involved in shotgun sequencing very large pieces of DNA.*

Green E.D. (2001) Strategies for the systematic sequencing of complex genomes. *Nature Reviews Genetics* **2**, 573–583. *An excellent review that covers sequencing strategy in more detail than possible in this chapter.*

Venter J.C. *et al.* (2001) The sequence of the human genome. *Science* **291**, 1304–1351.

International Human Genome Sequencing Consortium (2001) Initial sequencing and analysis of the human genome. *Nature* **409**, 860–933.
These two papers and the articles that accompany them give details of how two large sequencing groups undertook the massive task of sequencing the human genome. As well as describing the methodology that was used, they contain a wealth of interesting information.

Weber J.L. & Myers E.W. (1997) Human whole-genome shotgun sequencing. *Genome Research* **5**, 401–409.

Green P. (1997) Against whole-genome shotgun. *Genome Research* **5**, 410–417.
These two papers need to be read together as they give a good overview of the arguments for and against shotgun sequencing of large genomes. The first paper deals in detail with potential problems of repeated sequences, etc.

Useful websites

http://www.sciencemag.org/feature/plus/sfg/resources/res_sequence.shtml
This webpage is the entry point to the *Science* magazine resource guide to gene sequences.

http://www.tigr.org/tdb/mdb/mdb.html
This page provides entry to the database of sequenced microbial genomes maintained by the Institute for Genome Research.

http://www.ebi.ac.uk/genomes/mot/header.html
This site provides statistics on DNA sequencing projects and is updated daily.

CHAPTER 6

Genome annotation and bioinformatics

Introduction

Prior to the large-scale sequencing projects, genes and their functions were characterized on an individual basis. This often required laborious and painstaking work in terms of cloning, sequencing and the application of experimental methods to find genes embedded in large genomic fragments. In the current era of factory-style sequencing this has all changed and often individual genes are no longer the objective. The sequencing projects rapidly generate very large amounts of data. On their own, these data are not very informative and they have to be *mined* to extract relevant information.

The term *annotation* means obtaining biological information from unprocessed sequence data. *Structural annotation* means the identification of genes and other important sequence elements, and *functional annotation* means the determination of their functional roles in the organism. Rather than focusing on individual genes, it is now becoming more common to undertake the annotation of entire genomes. As more data become available, this can increasingly be carried out without experiments being performed at the bench. Raw genomic sequence can be annotated by comparison with databases of previously cloned genes and expressed sequence tags (ESTs), and sophisticated algorithms are available which allow gene prediction based on consensus features such as promoters, splice sites, polyadenylation sites and open-reading frames (ORFs). Annotation is achieved by the application of information technology to the management and analysis of biological data, a field known as *bioinformatics*. Originally the term applied to the computational manipulation and analysis of protein and DNA sequence data but has since been greatly expanded. For example, it now includes the prediction of three-dimensional structures, the prediction of the interactions of proteins with other molecules, and the analysis of the complete gene expression

pattern of tissues and organisms. The ultimate goal of bioinformatics is a detailed understanding of how organisms evolved and how, at a genetic and biochemical level, the different components of an organism develop and function. Some of these topics are explored in more detail in later chapters.

The science of bioinformatics has three components.

1 The development of new algorithms and statistics with which to assess the relationships among members of large data sets.

2 The use of these algorithms for the analysis and interpretation of various types of data including nucleotide and amino acid sequences, protein domain structures and protein interactions.

3 The development and implementation of tools (*databases*) that permit efficient access and management of different types of information.

In this chapter we focus on the last two topics. However, before covering these topics we review the more traditional methods of gene identification because they permitted some landmark advances in genetics.

Traditional routes to gene identification

It is hard to imagine in these days of sequencing projects and computer-based annotation the difficulty faced by researchers in the 1970s and 1980s in tracking down genes responsible for particular traits in higher eukaryotes. If some biochemical information was available about the gene or its product, it was sometimes possible to devise a cloning strategy that allowed direct isolation of the gene from a suitable cDNA or genomic library. This is the basis of *functional cloning* (Collins 1992) which is discussed at length in our sister text, *Principles of Gene Manipulation* (Primrose *et al.* 2001). For most traits, however, including thousands of inherited human diseases, no relevant biochemical information was available. In such cases, *positional cloning* strategies were developed in which the gene was first mapped to a particular candidate region, which was progressively narrowed down until the correct gene was identified. Positional cloning can be regarded as the earliest form of annotation because, once the candidate region was defined, the anonymous sequence had to be screened to identify candidate genes. The correct gene was then selected by searching for mutations present only in carriers and individuals affected by the trait.

In humans, one starts with a collection of pedigrees in which the disease gene is segregating. The disease is mapped against a panel of polymorphic markers until evidence of linkage is obtained. Clearly, the usefulness of linkage information increases as marker coverage becomes denser, but in most cases linkage analysis provides a candidate region in the order of 5–10 Mb of DNA. One way in which a candidate region can be further reduced in size is to screen genomic DNA from a large number of affected families and establish a collection of new polymorphic markers that provide dense coverage. In practice, however, there are often short cuts that can be exploited. Most of the early successes in human positional cloning were aided by the availability of individuals with chromosomal aberrations, e.g. deletions, inversions and translocations, which disrupted the gene and helped pinpoint its location. For example, the Duchenne muscular dystrophy gene was the first human gene to be successfully isolated by positional cloning, and the procedures that were used are discussed in Box 6.1. Positional cloning has been extended to other mammals, such as mice (e.g. Schumacher *et al.* 1996), and to plants, such as rice (Song *et al.* 1995) and sugar beet (Cai *et al.* 1997). However, in these species it is unnecessary to derive linkage information from pedigrees because directed breeding can be used. In mice, for example, high-resolution marker frameworks can be generated using interspecific backcrosses (e.g. *Mus musculus* vs. *Mus spretus*), allowing new phenotypes to be mapped rapidly and accurately (Copeland & Jenkins 1991). The high degree of synteny between the mouse and human genomes means that gene mapping in the mouse can often be used as a short cut to the isolation of the orthologous human gene (DeBry & Seldin 1996).

Once a candidate region was identified as tightly as possible, DNA clones were obtained spanning that region and these were searched for genes. Initially, a number of experimental *transcript mapping* strategies were used to identify genes. These include the following.

• *Hybridization of genomic clones to zoo blots.* This approach is based on the observation that coding sequences are strongly conserved during evolution, whereas non-coding DNA generally is not. DNA from a genomic clone that may contain a gene is hybridized to a Southern blot containing whole genomic DNA from a variety of species (a zoo blot). At reduced stringency, probes containing human genes, for example, will generate strong hybridization signals on genomic DNA from other animals (Monaco *et al.* 1986).

• *Hybridization of genomic clones to northern/reverse northern blots and cDNA libraries.* The major defining feature of a gene is that it is expressed, producing an RNA transcript. Therefore, if a genomic clone hybridizes to a northern blot (a blot containing only RNA) or a reverse northern blot (a blot containing cDNA, which is derived from RNA) or a cDNA library, it is likely that the genomic clone contains a gene. Unfortunately, this technique relies on the gene being expressed at a significant level in the tissue used to prepare the RNA or cDNA because the sensitivity is low. Part of the sensitivity problem reflects the small exons and large introns characteristic of higher eukaryote genes, which means that large genomic clones may only contain a few hundred base pairs of expressed DNA.

• *Identification of CpG islands.* CpG islands are short stretches of hypomethylated GC-rich DNA often found associated with vertebrate genes (see p. 22). Their function is unclear, but about 50% of human genes have associated CpG islands and these motifs can be exploited for gene identification. One approach is to search raw sequence data for CpG islands by computer (see below). However, the sequences found in CpG islands are scarce in bulk genomic DNA so certain rare-cutter restriction enzymes, such as *SacII* (which recognizes the site CCGCGG), generate small fragments which can indicate the presence of a gene (Cross & Bird 1995). An alternative PCR-based technique has also been used to identify CpG islands (Valdes *et al.* 1994).

• *cDNA selection and cDNA capture.* In the cDNA

Box 6.1 The Duchenne muscular dystrophy gene as an early example of positional cloning and transcript mapping

Isolation of the Duchenne muscular dystrophy (DMD) gene was the first successful positional cloning experiment, and the strategy benefited enormously from the availability of DMD individuals with characteristic chromosome aberrations allowing the candidate region to be precisely defined. It is important to emphasize that, at the time these experiments were carried out, there was absolutely no clue as to the biochemical basis of the disease and no suitable alternative cloning strategy. The disease was initially mapped using restriction fragment length polymorphisms and was thus localized to Xp21, a locus confirmed by the existence of a small number of DMD individuals with translocation breakpoints in this region. Two groups tackled the project using different approaches: in one case by enriching for DNA missing in a DMD individual carrying a deletion in the candidate region and in another by attempting to clone the translocation breakpoint.

Subtraction cloning

This approach (Kunkel *et al.* 1986) began with the identification of a young boy, known as 'BB' who suffered from four X-linked disorders, including DMD. Cytogenetic analysis showed that the boy had a chromosome deletion in the region Xp21, so a subtraction cloning procedure was devised to isolate the DNA sequences that were deleted in BB. Genomic DNA was isolated from BB and randomly sheared, generating fragments with blunt ends and non-specific overhangs. DNA was also isolated from an aneuploid cell line with four (normal) X chromosomes. This DNA was digested with the restriction enzyme *Mbo*I, generating sticky ends suitable for cloning. The *Mbo*I fragments were mixed with a large excess of the randomly sheared DNA from BB, and the mixture was denatured and then persuaded to reanneal extensively using phenol enhancement. The principle behind the strategy was that because the randomly fragmented DNA was present in a vast excess, most of the DNA from the cell line would be sequestered into hybrid DNA molecules that would be unclonable. However, those sequences present among the *Mbo*I fragments but absent from BB's DNA because of the deletion would only be able to reanneal to complementary strands

from the cell line. Such strands would have intact *Mbo*I sticky ends, and could therefore be ligated into an appropriate cloning vector. Using this strategy, Kunkel *et al.* generated a genomic library that was highly enriched for fragments corresponding to the deletion in BB. Clones from the library were tested by hybridization against normal DNA and DNA from BB to confirm they mapped to the deletion, resulting in the isolation of a single positive clone, pERT87. To confirm the genuine DMD gene had been isolated, three subclones of pERT87 were then tested against DNA from many other patients with DMD, revealing similar deletions in 6.5% of cases.

Cloning the translocation breakpoint

This approach (Ray *et al.* 1985) involved a teenage girl suffering from DMD, who presented a translocation involving the X-chromosome and the short arm of chromosome 21. This region of chromosome 21 was known to be rich in rRNA genes so a genomic library was prepared from a somatic cell hybrid line containing the reciprocal derivative chromosomes in a mouse background. This step was necessary to remove rRNA genes present on other human chromosomes. The library was screened to identify clones containing both rDNA and sequences specific to the X-chromosome. After a number of early setbacks, one such genomic clone was isolated corresponding to the translocation breakpoint. As above, this clone revealed deletions in a number of male patients with DMD.

Transcription mapping

The cDNA of the DMD gene was identified by Monaco *et al.* (1986) using a combination of zoo blotting, northern blotting and sequence comparison. Two subclones of the pERT87 clone hybridized to DNA from a variety of mammals and the chicken, and sequencing revealed an exon conserved between humans and mice. A northern blot of human fetal mRNA was screened with one of the subclones and a signal was obtained in muscle but not brain tissue. The same probe was used to isolate several overlapping cDNA fragments from a fetal muscle library.

selection approach (Lovett *et al.* 1991; Parimoo *et al.* 1991) an amplified cDNA library is hybridized to immobilized genomic clones covering the candidate genomic region. Of the cDNAs selected by this procedure, at least one should correspond to the desired gene. More recently, hybridization has been carried out in solution (cDNA capture) to enrich for cDNAs corresponding to a genomic clone. This technique relies on adequate expression of the target gene in the tissue used for cDNA preparation, but its sensitivity is much greater than other blot-based techniques. One further problem that has been encountered with this approach is the hybridization of cDNAs to (non-expressed) repetitive DNA elements and pseudogenes (see Lovett 1994).

• *Exon trapping (exon amplification).* This technique was independently devised by a number of research groups and involves an artificial splicing assay (Auch & Reth 1990; Duyk *et al.* 1990; Buckler *et al.* 1991; Hamaguchi *et al.* 1992). The advantage of this over the RNA/cDNA methods discussed above is that there is no need for the gene to be expressed. The general principle is that a genomic clone is inserted into an 'intron' flanked by two artificial exons within an expression vector. The vector is then introduced into mammalian cells by transfection and the recombinant expression cassette is transcribed and spliced to yield an artificial mRNA that can be amplified by RT-PCR. If the genomic clone does not contain an exon, the RT-PCR product will contain the two artificial exons in the vector and will be of a defined size. If the genomic clone does contain an exon, it will be spliced into the mature transcript and the RT-PCR product will be larger than expected (Church & Buckler 1999). Cosmid-based exon trap vectors can be used to trap multiple exons in one experiment (Datson *et al.* 1996; den Dunnen 1999).

Powerful as these methods are, they are limited by the fact that experiments have to be carried out at the bench on individual DNA clones. This is suitable where the goal is to identify individual genes, but for the high throughput annotation of entire genomes this is simply not fast enough to keep up with the rate at which sequence data accumulate. For predominantly this reason, bioinformatics has overtaken experimental approaches to gene identification. We now discuss how the use of computer databases and gene prediction tools have revolutionized genome annotation.

Databases

Databases are at the heart of bioinformatics. Essentially, they are electronic filing cabinets that offer a convenient and efficient method of storing vast amounts of information. There are many different database types, depending both on the nature of the information being stored (e.g. sequences, structures, gel images, etc.) and on the manner of data storage (flat-files, tables in relational databases, object-oriented databases, etc.). The number of different databases is growing very rapidly. During the year 2000, 55 new databases were created, bringing the total at the end of the year to 281! Each year, the first issue of the journal *Nucleic Acids Research* is devoted to an update on the status of the major databases (Baxevanis 2002) and an electronic version is available at http://www.nar.oupjournals.org. These databases are accessible to everyone and their web addresses can be found in the references just cited.

The primary nucleic acid and protein sequence databases are GenBank, held by the U.S. National Center for Biotechnology Information, the DNA Databank of Japan (DDBJ) and the Nucleotide Sequence Database maintained by the European Molecular Biology Laboratory (EMBL). New sequence data can be deposited with any one of these three groups because they automatically share the data on a daily basis. A typical DNA sequence record held by EMBL is shown in Fig. 6.1. For the purpose of searching for genes in DNA sequences derived from eukaryotes, dbEST is particularly useful. This is a database of ESTs (see p. 51) that have been generated by single pass sequencing of random clones from cDNA libraries. ESTs have been instrumental in generating gene maps both by hands-on experimentation and pure *in silico* analysis. For example, the first generation human gene map was generated by mapping ESTs onto radiation hybrid panels and yeast artificial chromosome (YAC) clones using PCR assays. Now genomic sequences are compared directly to the contents of dbEST in order to identify potential ORFs.

Overview of sequence analysis

When a long piece of DNA has been sequenced, the first task is to identify any genes that are present. In

```
EMBL (Release):RC22378
ID   RC22378   standard; RNA; PLN; 1440 BP.
XX
AC   U22378;
XX
SV   U22378.1
XX
DT   11-APR-1995 (Rel. 43, Created)
DT   04-MAR-2000 (Rel. 63, Last updated, Version 5)
XX
DE   Ricinus communis oleate 12-hydroxylase mRNA, complete cds.
XX
KW   .
XX
OS   Ricinus communis (castor bean)
OC   Eukaryota; Viridiplantae; Streptophyta; Embryophyta; Tracheophyta;
OC   Spermatophyta; Magnoliophyta; eudicotyledons; core eudicots; Rosidae;
OC   eurosids I; Malpighiales; Euphorbiaceae; Ricinus.
XX
RN   [1]
RP   1-1440
RX   MEDLINE; 95350145.
RA   van de Loo F.J., Broun P., Turner S., Somerville C.;
RT   "An oleate 12-hydroxylase from Ricinus communis L. is a fatty acyl
RT   desaturase homolog";
RL   Proc. Natl. Acad. Sci. U.S.A. 92(15):6743-6747(1995).
XX
RN   [2]
RP   1-1440
RA   Somerville C.R.;
RT   ;
RL   Submitted (08-MAR-1995) to the EMBL/GenBank/DDBJ databases.
RL   Chris R. Somerville, Plant Biology, Carnegie Institution of Washington, 290
RL   Panama Street, Stanford, CA 94305-4101, USA
XX
DR   AGDR; U22378; U22378.
DR   MENDEL; 10454; Ricco;1207;10454.
DR   SPTREMBL; Q41131; Q41131.
XX
FH   Key             Location/Qualifiers
FH
FT   source          1..1440
FT                   /db_xref="taxon:3988"
FT                   /organism="Ricinus communis"
FT                   /strain="Baker 296"
FT                   /tissue_type="developing endosperm"
FT   CDS             187..1350
FT                   /codon_start=1
FT                   /db_xref="SPTREMBL:Q41131"
FT                   /note="expressed only in developing endosperm of castor;
FT                   possible integral membrane protein of endoplasmic
FT                   reticulum; uses cytochrome b5 as intermediate electron
FT                   donor; fatty acid hydroxylase"
FT                   /product="oleate 12-hydroxylase"
FT                   /protein_id="AAC49010.1"
FT                   /translation="MGGGGRMSTVITSNNSEKKGGSSHLKRAPHTKPPFTLGDLKRAIP
FT                   PHCFERSFVRSFSYVAYDVCLSFLFYSIATNFFPYISSPLSYVAWLVYWLFQGCILTGL
 .
 .
 .
FT                   KPIMGEYYRYDGTPFYKALWREAKECLFVEPDEGAPTQGVFWYRNKY"
FT   polyA site      1440
FT                   /note="8 A nucleotides"
XX
SQ   Sequence 1440 BP; 367 A; 340 C; 321 G; 412 T; 0 other;
     gccaccttaa gcgagcgccg cacacgaagc ctcctttcac acttggtgac ctcaaatcaa        60
     acaccacacc ttataactta gtcttaagag agagagagag agagaggaga catttctctt       120
                                                                              .
                                                                              .
     ggcgttttct ggtaccggaa caagtattaa aaaagtgtca tgtagcctgt ttctttaaga      1380
     gaagtaatta gaacaagaag gaatgtgtgt gtagtgtaat gtgttctaat aaagaaggca      1440
//
```

Fig. 6.1 A typical DNA sequence database entry (dotted lines denote points at which, for convenience, material has been excised).

prokaryotes, gene density is generally high and most protein coding genes lack introns, so the task is relatively straightforward. However, if the DNA in question came from a higher eukaryote, a gene is much harder to recognize because it may be divided into many small exons and a similar number of larger introns. For example, in the human genome a typical exon is 150 bp and a typical intron is several kilobases and a complete gene can be hundreds of kilobases in length. Also, the mRNA from some genes can be edited in a number of different ways resulting in the generation of splice variants; i.e. different polypeptides are synthesized from a single gene. An added problem in gene finding is the signal to noise ratio. In bacterial genomes, genes make up 80–85% of the DNA. In the yeast *Saccharomyces cerevisiae* this figure drops to 70%. In the fruit fly and nematode this figure drops to 25% and in the human genome genes account for only 3–5% of the DNA. Thus, defining the precise start and stop position of a gene and the splicing pattern of its exons among all the non-coding sequence is exceedingly difficult.

Once a gene has been identified the nucleic acid sequence is converted into a protein sequence. The question then is, what is the function of this protein? By searching all the information contained in the various databases it may be possible to identify other proteins with a similar sequence and this may help to identify its function. Sequence comparisons also can be used to identify particular motifs in a protein such as ATP-binding or DNA-binding and these too can give information about function.

Detecting open-reading frames

If one starts with a length of DNA sequence and wants to identify possible genes then the first task is to identify the correct reading frame. Because there are three possible reading frames on each strand of a DNA molecule this is done by carrying out a *six-frame translation*. The result is six potential protein sequences (Fig. 6.2). The correct reading frame is assumed to be the longest frame uninterrupted by a stop codon (TGA, TAA or TAG). The longer this ORF, the more likely it is to be a gene because long ORFs are unlikely to occur by chance. Finding the end of such an ORF is easier than finding its beginning. The N-terminal amino acid of a protein usually

is methionine so the presence of an ATG codon might indicate the 5′ end of a gene. However, methionine is not always the first amino acid in a protein sequence and it can occur at other positions. Consequently, additional techniques are required to identify the start of an ORF.

In a mRNA molecule, the start codon may be flanked by a Kozak sequence (CCGCC<u>A</u>UGG) and finding this sequence helps to identify the 5′ end of the gene. Analysis of the codon usage can also be helpful because there are marked differences between coding and non-coding regions. Specifically, the use of codons for particular amino acids varies according to species (Table 6.1). These codon-use rules break down in regions of sequence that are not destined to be translated. Such untranslated regions often have an uncharacteristically high representation of rarely used codons. The identification of segments with a much higher than average GC content, and a higher than average frequency of the CpG dinucleotide, could be indicative of a CpG island. Such islands are found at the 5′ end of many vertebrate genes (Ioshikhes & Zhang 2000). As a final aid to finding ORFs, use can be made of the bias towards G or C as the third base in a codon. Although all of these tools can be applied manually to a DNA sequence, sophisticated computer programs are available (see below) that will do this for you.

There is an important caveat associated with the searching of nucleotide sequences for ORFs. If care is not taken at the sequencing stage then errors can creep into the finished sequence (see p. 27). Although incorrect basecalling is undesirable, its effects are fairly minimal unless it results in the erroneous creation or elimination of a termination codon. More importantly, an erroneous single base addition or deletion ('phantom indels') will disturb the reading frame and make correct identification of the ORF much more difficult.

Software programs for finding genes

The most important single development in genome annotation is the use of computers to predict the existence of genes in unprocessed genome sequence data (see reviews by Fickett 1996; Claverie 1997; Burge & Karlin 1998; Lewis *et al.* 2000; Gaasterland

```
Query Sequence:
            10          20          30          40          50
   0  TCCATTGAGC  CTTATACCAG  TAACATCTAC  ACTCGAAGAT  CTTGTCAGGG
  50  GAATTTCAGA  TTGTGAATCC  TCACTTACTG  AAAGATCTTA  CTGAGCGGGG
 100  CTTGTGGAAT  GAAGAGATGA  AAAATCAGAT  TATTGCATGC  AATGGCTCCA
 150  TTCAGTTTTC  CTTTTTCAGA  GCATACCAGA  AATTCCTGAT  GACCTGAAGC
 200  AACTCTATAA  GACCGTGTGG  GAAATCTCTC  AGAAGACTGT  TCTCAAGATG

Six-Frame Amino Acid Translation:

Forward 0
            10          20          30          40          50
   0  SIEPYTSNIY  TRRSCQGNFR  L!ILTY!KIL  LSGACGMKR!  KIRLLHAMAP
  50  FSFPFSEHTR  NS!!PEATL!  DRVGNLSEDC  SQD

Forward 1
            10          20          30          40          50
   0  PLSLIPVTST  LEDLVRGISD  CESSLTERSY  !AGLVE!RDE  KSDYCMQWLH
  50  SVFLFQSIPE  IPDDLKQLYK  TVWEISQKTV  LKM

Forward 2
            10          20          30          40          50
   0  H!ALYQ!HLH  SKILSGEFQI  VNPHLLKDLT  ERGLWNEEMK  NQIIACNGSI
  50  QFSFFRAYQK  FLMT!SNSIR  PCGKSLRRLF  SR

Reverse 0
            10          20          30          40          50
   0  HLENSLLRDF  PHGLIELLQV  IRNFWYALKK  EN!MEPLHAI  I!FFISSFHK
  50  PRSVRSFSK!  GFTI!NSPDK  IFECRCYWYK  AQW

Reverse 1
            10          20          30          40          50
   0  ILRTVF!EIS  HTVL!SCFRS  SGISGML!KR  KTEWSHCMQ!  SDFSSLHSTS
  50  PAQ!DLSVSE  DSQSEIPLTR  SSSVDVTGIR  LNG

Reverse 2
            10          20          30          40          50
   0  S!EQSSERFP  TRSYRVASGH  QEFLVCSEKG  KLNGAIACNN  LIFHLFIPQA
  50  PLSKIFQ!VR  IHNLKFP!QD  LRV!MLLV!G  SM
```

Fig. 6.2 A six-frame translation of an arbitrary DNA sequence. ! denotes a stop codon. (From Attwood & Parry-Smith 1999 © Pearson Education Limited 1999, reprinted by permission of Pearson Education Ltd.)

Codon	*Escherichia coli*	*Drosophila*	Human	Maize	Yeast
AGT	3	1	10	3	5
AGC	20	23	34	30	4
TCG	4	17	9	22	1
TCA	2	2	5	4	6
TCT	34	9	13	4	52
TCC	37	48	28	37	33

Table 6.1 Percentage use of the different serine codons in different organisms. (Reproduced with permission from Attwood & Parry-Smith 1999.)

& Oprea 2001). The advantage of computer-based prediction is its speed – annotation can be carried out concurrently with sequencing itself – but a disadvantage is its accuracy, particularly in the complex genome of higher eukaryotes. Essentially, two strategies are used for gene prediction: homology searching and *ab initio* prediction. Homology searching programs compare genomic sequence data to gene, cDNA, EST and protein sequences already present in databases. *Ab initio* prediction

algorithms search for gene-specific features such as promoters, splice sites and polyadenylation sites or for pertinent gene content, such as ORFs. Many of the currently available programs combine different search criteria, and their sensitivities vary widely (e.g. Burset & Guigó 1996; Rogic *et al.* 2001).

The identification of ORFs exceeding a certain length (usually about 300 nucleotides, equivalent to 100 amino acids) is sufficient to find most genes in prokaryote genomes. Genuine genes that are

Table 6.2 Types of algorithms used for searching for genes in DNA sequences.

Type of algorithm	Principle	Examples
Neural network (Uberacher & Mural 1991)	These are analytical techniques modelled on the processes of learning in cognitive systems. They use a data training set to build rules that can make predictions or classifications on data sets	GRAIL
Rule-based system	Uses an explicit set of rules to make decisions	GeneFinder
Hidden Markov model (Burge & Karlin 1997)	Represents a system as a set of discrete states and transitions between those states. Markov models are 'hidden' when one or more of the states cannot be directly observed. Each transition has an associated probability. Has the advantage of explicitly modelling how the individual probabilities of a sequence of features are combined into a probability estimate for the whole gene	GENSCAN GENIE HMMGene GeneMarkHMM FGENEH

smaller than this will be missed if such a threshold is rigorously applied, and there are also difficulties in identifying so-called *shadow genes*, i.e. overlapping ORFs on opposite strands. However, ambiguities arising from this type of genome organization can be resolved using algorithms that incorporate Markov models to highlight differences in base composition between genes and non-coding DNA, e.g. GEN-MARK (Borodovsky & McIninch 1993), a modified GeneScan algorithm (Ramakrishna & Srinivasan 1999) and Glimmer (Salzberg *et al.* 1999). As a result, it is now possible to identify all genes with near certainty in bacterial genomes. For example, Ramakrishna & Srinivasan (1999), using the improved GeneScan algorithm, reported a near 100% success rate in three microbial genomes: *Haemophilus influenze*, *Plasmodium falciparum* and *Mycoplasma genitalium*.

Several sophisticated software algorithms have been devised to handle gene prediction in eukaryotic genomes. Some of these gene predictors only predict a single feature, e.g. the exon predictors HEXON and MZEF. However, most attempt to use the output of several algorithms to generate a whole gene model in which a gene is defined as a series of exons that are coordinately transcribed. The principles used in these algorithms are summarized in Table 6.2. Most of these algorithms are available free of charge over the Internet, as listed in Box 6.2.

What features of eukaryotic genes are recognized by gene prediction programs? All protein-coding genes are transcribed and translated, so transcriptional and translational control signals, such as the TATA box, cap site, Kozak consensus and polyadenylation site would seem useful targets. Unfortunately, the diversity of eukaryotic promoters in combination with the small size of these target motifs detracts from their usefulness. For example, a TATA box is found in only about 70% of human genes (Fickett & Hatzigeorgiou 1997) while poly-adenylation signals can differ considerably from the consensus sequence AATAAA. Additionally, such signals identify only the first and last exons of a gene. Splice signals are much more useful because they define each exon and are almost invariant. Early gene-finding models assumed independence between positions in the 5′ and 3′ splice sites. More recently, however, dependencies between positions have been identified and have been built into gene-prediction algorithms (e.g. Burge & Karlin 1998). As well as these feature-dependent methods, differences in base composition between coding and non-coding DNA play an important part in gene prediction. Fickett and Tung (1992) compared a large number of base composition parameters in coding and non-coding DNA, and reached the conclusion that comparisons of hexamer base composition gave the best discrimination. Many of the currently used gene prediction programs incorporate Markov models to distinguish hexamer usage between coding and non-coding DNA.

It should be stressed that each of these gene-prediction algorithms needs to be 'trained' with data or else implanted with a set of rules. These activities are essential so that the algorithm can recognize key features that distinguish a gene or exon from

Box 6.2 Internet resources for genome annotation

Gene prediction software

http://genomic.sanger.ac.uk/gf/gfs.shtml FGENEH
http://www1.imim.es/geneid.html GENEID
http://www.fruitfly.org/seq_tools/genie.html GENIE
http://CCR-081.mit.edu/GENSCAN.html GENSCAN
http://www.cbs.dtu.dk/services/HMMgene HMMGene
http://genemark.biology.gatech.edu/GeneMark GeneMarkHMM
http://compbio.ornl.gov GRAIL
http://www.tigr.org/softlab/glimmer/glimmer.html GlimmerM

http://www.itba.mi.cnr.it/webgene GeneBuilder
http://www.sanger.ac.uk/Software/Wise2 Wise2/Genewise
http://blocks.fhcrc.org BLOCKS

Sites providing information on annotation

http://www.fruitfly.org/GASP1 Genome Annotation Assessment Project
http://www.geneontology.org Gene Ontology project
http://www.ebi.ac.uk/interpro InterPro

non-coding DNA. These features are not identical in all organisms. Thus an algorithm trained with nematode DNA will not perform satisfactorily with plant DNA without being re-trained. For example, GRAIL is one of the oldest gene prediction programs and can be used with human, mouse, *Arabidopsis*, *Drosophila* and *Escherichia coli* sequences but GENIE has been trained only on human and *Drosophila* sequences.

Even when an algorithm has been trained with data from a particular species it is not 100% accurate at identifying genes. This has been addressed in an ongoing international collaborative venture called the Genome Annotation Assessment Project (GASP). For example, Reese *et al.* (2000) selected two well-characterized regions of the *Drosophila* genome and presented the nucleotide sequence to the authors of the various algorithms for analysis. The best gene predictor had a sensitivity (detection of true positives) of 40% and a specificity (elimination of false positives) of 30% when required to predict entire gene structures. The errors generated included incorrect calling of exon boundaries, missed or phantom exons, or failure to detect entire genes. In a similar study using human DNA, Fortna & Gardiner (2001) found that the best results were obtained by running five different programs and counting consensus exons obtained from any two or more programs (Table 6.3).

Using homology to find genes

The algorithms described in Table 6.2 are known as *ab initio* programs because they attempt to predict genes from sequence data without the use of prior knowledge about similarities to other genes. Finding genes in long sequences can be facilitated by looking for matches with sequences that are known to be transcribed, e.g. a cDNA, an EST or even a gene in another species.

The pace of genome annotation changed radically with the growth of EST data, because genomic sequences can be rapidly screened for EST hits to identify potential genes. EST clones are derived from the 3′ ends of poly(A)$^+$ transcripts and contain 3′ untranslated sequences (p. 51). However, they often extend far enough towards the 5′ end to reach the coding sequence and thus overlap with predicted exons but they cannot be expected to identify all coding exons. It should be noted that not all ESTs can be assumed to be reliable indicators of a gene or a mature mRNA. In some cases they can be derived from unprocessed intronic sequences, primed from the genomic poly(A) tract, or from processed pseudogenes.

The most widely used programs for carrying out similarity searches of a query sequence with database sequences are the BLAST (Basic Local Alignment Search Tool) family (Altschul *et al.* 1990).

Table 6.3 Exon predictions using different exon search programs with and without expressed sequence tags (ESTs). For each gene the actual number of exons has been experimentally verified. (Reprinted from Fortna & Gardiner 2001 by permission of Elsevier Science.)

Accession no.	Gene	No. of exons	GRAIL			Genscan			MZEF			FGENES			FGENE H			≥2 Exons*			ESTs†			Exon + EST‡		
			TP	FP	ME	TP	FP	ME	TP	FP	ME	TP	FP	ME	TP	FP	ME	TP	FP	ME	TP	FP	ME	TP	FP	ME
AP001715	B3/GCFC	17	14	5	3	17	2	0	14	1	3	16	4	1	17	2	0	17	4	0	9	19	8	9	2	8
	ORF4	3	3	1	2	0	1	3	0	4	2	0	0	3	0	0	3	0	1	2	2	9	1	1	2	2
	B37	3	2	1	1	0	0	3	3	1	0	2	1	0	1	0	2	3	0	0	1	4	2	1	0	2
AP001717	PRKCBP2	1	1	0	0	1	0	0	0	0	1	1	0	0	1	0	0	1	0	0	1	0	0	1	0	0
	IFNGR2	7	6	4	1	4	1	3	2	3	5	4	3	3	3	3	4	5	1	2	7	8	0	6	2	1
	C21orf4	7	2	0	5	3	0	4	4	1	3	4	1	3	0	0	7	4	0	3	7	5	0	4	1	3
	C21orf55	2	0	0	2	0	0	2	0	1	2	0	0	2	0	0	2	0	0	2	1	0	1	0	0	2
	GART	22	19	2	3	15	0	7	19	4	3	19	0	3	14	1	8	20	0	2	22	9	0	21	2	1
	SON	11	9	2	2	10	1	1	9	4	2	7	1	4	10	1	1	11	1	0	11	23	0	10	11	1
	CRYZL1	13	7	6	6	7	1	6	7	8	6	8	1	5	7	0	6	8	1	5	12	14	1	9	2	4
	B17	9	4	0	5	4	0	5	7	2	2	5	1	4	8	1	1	7	1	2	9	6	0	8	1	1
AP001753	AGPAT3	8	7	8	1	7	6	1	6	1	2	7	5	1	7	7	1	7	5	1	8	5	0	7	1	1
	TMEM1	23	17	12	6	20	6	3	13	1	10	18	7	5	14	7	9	19	6	4	16	25	7	15	12	8
	PWP2H	21	17	2	4	20	4	1	14	1	7	19	1	2	19	6	2	21	3	0	18	4	3	18	3	3
	C21orf33	7	4	0	3	7	4	0	3	0	4	5	1	2	6	0	1	7	0	0	7	4	0	7	1	0
	KIAA0653	6	4	2	2	5	2	1	2	0	4	6	1	0	5	1	1	5	0	1	4	0	2	4	0	2
	DNMT3L	12	10	1	2	11	0	1	6	0	6	6	4	6	11	3	1	11	1	1	10	3	2	10	1	2
Total		172	124	47	48	131	28	41	110	32	62	127	31	45	123	32	49	147	24	25	145	138	27	131	41	41

* Consistent exon predictions from any two or more programmes.

† Exon predictions from ESTs alone.

‡ Consistent exon prediction from EST match and any one or more exon programmes.

TP, True positives; FP, false positives; ME, missed exons.

Box 6.3　The principles of similarity searching

A DNA or amino acid sequence on its own is not very informative; rather, it must be analysed by comparative methods against existing databases to develop hypotheses concerning relatives and function. The general approach is to use algorithms to compare a new or query sequence to all the other sequences in a database. These comparisons are made in a pairwise fashion. Each comparison is given a score reflecting the degree of similarity between the query and the sequence being compared. The higher the score, the greater the degree of similarity. Note that similarity is not the same as homology. Similarity is the extent to which two sequences are related, whereas homology is similarity resulting from having a common ancestor.

Sequence alignments can be global or local. A global alignment is an optimal alignment that includes all characters from each sequence, whereas a local alignment is an optimal alignment that includes only the most similar local region or regions. Discrimination between real and artefactual matches is carried out using an estimate of probability that the match might occur by chance. Most of the sequence comparison programs (e.g. BLAST) use algorithms that search sequence databases for optimal local alignments. They start by breaking the query and database sequences into fragments ('words') and initially seeking word matches. The initial search is for a word of length W that scores at least T when compared to the query using a given scoring matrix (see below). Word hits then are extended in either direction in an attempt to generate an alignment with a score exceeding the threshold of S. The T parameter dictates the speed and sensitivity of the search.

The quality of each pairwise alignment is represented as a score and the different scores are ranked. Scoring matrices are used to calculate the score of the alignment base by base or amino acid by amino acid. A unitary matrix is used for DNA pairs because each position can be given a score of $+1$ if it matches and a score of zero if it does not. Substitution matrices are used for amino acid alignments (see Box 8.1, p. 135). These are matrices in which each possible residue substitution is given a score reflecting the probability that it is related to the corresponding residue in the query. The alignment score is the sum of the scores for each position.

Positions at which a letter is paired with a null are called gaps. Gap scores are negative. Because a single mutational event may cause the insertion or deletion of more than one residue, the presence of a gap is frequently ascribed more significance than the length of the gap.

The significance of each alignment is computed as a P (probability) value or an E (expectation) value and these are just different ways of representing the significance of the alignment because

$$P = 1 - e^{-E}.$$

The most highly significant P values are those close to 1 whereas the lower the E value the greater its significance.

BLASTN queries a nucleotide sequence against a nucleotide database, BLASTX translates a nucleotide query into all six frames and searches a protein database, and BLASTP uses a protein sequence to search a protein database. Before undertaking similarity searches it is important to remove repetitive sequences and any vector sequences that might be present, and various programs (e.g. RepeatMasker and VecScreen) are available that do this. As well as being used for gene finding, the BLASTN and BLASTP algorithms are widely used in comparative genomics (see Chapter 7). A key aspect of sequence comparisons is an assessment of whether a given alignment constitutes evidence for homology; i.e. what is the probability that alignment can be expected purely from chance alone? The statistics of sequence similarity is a specialist topic in its own right (see Box 6.3) and a useful tutorial can be found at http://www.ncbi.nlm.nih.gov/BLAST/tutorial.

The current trend in gene prediction is to make as much use of sequence-similarity data as possible. The latest generation of gene prediction algorithms, such as Grail/Exp, GenieEST and GenomeScan (Yeh *et al.* 2001), combine *ab initio* predictions with

Box 6.4 How many genes in the human genome?

Two groups (International Human Genome Sequencing Consortium 2001; Venter *et al.* 2001) recently published a draft sequence of the human genome. The two groups used different gene-prediction methodologies and, not surprisingly, obtained different answers to the question of the number of genes (Bork & Copley 2001). Venter *et al.* used a proprietary package called Otto which gives sequence similarity the highest priority. Evidence for transcription was derived from similarities to sequences present in RefSeq (a library of well-characterized human genes), Unigene (a database of human ESTs) and SWISS-PROT (a primary database of protein sequences). Otto then used GENESCAN to find and refine the splicing pattern of the predicted gene. Using this approach the number of genes was estimated to be 39 000, but the evidence for 12 000 of these was considered to be weak.

The other group started with *ab initio* gene predictions using GENESCAN and then strengthened these predictions using nucleotide and protein-sequence similarities. The predicted genes then were further scrutinized with GenieEST and against the RefSeq library. This methodology suggested that the number of genes was about 32 000, of which 15 000 were known genes and 17 000 were predictions. However, these 32 000 genes are estimated to come from 24 500 actual genes, the remainder being pseudogenes or fragments of real genes. As an added complication, the sensitivity of prediction is only 60% so it is likely that there are another 6800 genes (40% of 17 000). This gives an estimate of the gene number as 31 300.

similarity data into a single probability model. Both Reese *et al.* (2000) and Fortna & Gardiner (2001) found that algorithms that take similarity data into account are better at predicting gene structure (see also Table 6.3). The problems encountered in searching complete genomes for the total inventory of genes is best illustrated by the human genome sequencing initiative (Box 6.4).

Analysis of non-coding RNA and extragenic DNA

In all cells, but particularly multicellular eukaryotes, there is much more to the genome than coding regions. For example, analysis of non-coding RNAs and regulatory regions can provide much useful information. Of the non-coding RNAs, rRNAs are the easiest to find and this usually is done by similarity searching. tRNAs can be found by using tRNAScan-SE, a program that includes searching for characteristic structural features such as the ability to form hairpins. The methods for identifying non-coding RNA have been reviewed by Eddy (1999).

Regulatory regions are particularly important sequence features of genomes. To date, a relatively

small number of transcriptional-factor-binding sites have been identified by classical experimental methods. The existence of such sites in a query sequence is suggestive of a regulatory region. However, these sequences typically are fairly short and could occur by chance. Better evidence for such regulatory regions is their conservation in the sequences upstream from the same gene in two related species, e.g. mouse and humans.

Identifying the function of a new gene

The simplest way to identify the function of new genes, or putative genes, is to search for homologues in the protein databases. If a putative protein encoded by an uncharacterized ORF shows statistically significant similarity to another protein of known function, this proves beyond doubt that the ORF in question is a bona fide new gene and often identifies its function. A search of this kind is usually undertaken with one of the BLAST tools (e.g. BLASTP or PSI-BLAST), FASTA or HMMER. BLAST programs are the fastest sequence alignment tools but compromise some degree of sensitivity in favour of speed. FASTA and HMMER are slower but more

```
SWALL:PAGT_HUMAN
ID  PAGT   HUMAN   STANDARD;   PRT;    559 AA.
AC  Q10472;
DT  01-OCT-1996 (Rel. 34, Created)
DT  01-OCT-1996 (Rel. 34, Last sequence update)
DT  01-MAR-2002 (Rel. 41, Last annotation update)
DE  Polypeptide N-acetylgalactosaminyltransferase (EC 2.4.1.41) (Protein-
DE  UDP acetylgalactosaminyltransferase) (UDP-GalNAc:polypeptide, N-
DE  acetylgalactosaminyltransferase) (GalNAc-T1).
GN  GALNT1.
OS  Homo sapiens (Human).
OC  Eukaryota; Metazoa; Chordata; Craniata; Vertebrata; Euteleostomi;
OC  Mammalia; Eutheria; Primates; Catarrhini; Hominidae; Homo.
OX  NCBI_TaxID=9606;
RN  [1]
RP  SEQUENCE FROM N.A.
RC  TISSUE=Salivary gland;
RX  MEDLINE=96115928; PubMed=8690719;
RA  Meurer J.A., Naylor J.M., Baker C.A., Thomsen D.R., Homa F.L.,
RA  Elhammer A.P.;
RT  "cDNA cloning, expression, and chromosomal localization of a human
RT  UDP-GalNAc:polypeptide, N-acetylgalactosaminyltransferase.";
RL  J. Biochem. 118:568-574(1995).
RN  [2]
RP  SEQUENCE FROM N.A.
RX  MEDLINE=96025800; PubMed=7592619;
RA  White T., Bennett E.P., Takio K., Soerensen T., Bonding N.,
RA  Clausen H.;
RT  "Purification and cDNA cloning of a human UDP-N-acetyl-alpha-D-
RT  galactosamine:polypeptide N-acetylgalactosaminyltransferase.";
RL  J. Biol. Chem. 270:24156-24165(1995).
CC  -!- FUNCTION: THIS PROTEIN CATALYZES THE INITIAL REACTION IN O-LINKED
CC      OLIGOSACCHARIDE BIOSYNTHESIS, THE TRANSFER OF AN N-ACETYL-D-
CC      GALACTOSAMINE RESIDUE TO A SERINE OR THREONINE RESIDUE ON THE
CC      PROTEIN RECEPTOR.
CC  -!- CATALYTIC ACTIVITY: UDP-N-acetyl-D-galactosamine + polypeptide =
CC      UDP + N-acetyl-D-galactosaminyl-polypeptide.
CC  -!- COFACTOR: MANGANESE AND CALCIUM.
CC  -!- PATHWAY: GLYCOSYLATION.
CC  -!- SUBCELLULAR LOCATION: Type II membrane protein. Golgi.
CC  -!- SIMILARITY: BELONGS TO THE GLYCOSYLTRANSFERASE FAMILY 2.
CC  -!- SIMILARITY: CONTAINS 1 RICIN B-TYPE LECTIN DOMAIN.
CC  ----------------------------------------------------------------------
CC  This SWISS-PROT entry is copyright. It is produced through a collaboration
CC  between the Swiss Institute of Bioinformatics and the EMBL outstation -
CC  the European Bioinformatics Institute. There are no restrictions on its
CC  use by non-profit institutions as long as its content is in no way
CC  modified and this statement is not removed. Usage by and for commercial
CC  entities requires a license agreement (See http://www.isb-sib.ch/announce/
CC  or send an email to license@isb-sib.ch).
CC  ----------------------------------------------------------------------
DR  EMBL; U41514; AAC50327.1; -.
DR  EMBL; X85018; CAA59380.1; -.
DR  MIM; 602273; -.
DR  InterPro; IPR001173; Glycos_transf_2.
DR  InterPro; IPR000772; Ricin_B_lectin.
DR  Pfam; PF00535; Glycos_transf_2; 1.
DR  Pfam; PF00652; Ricin_B_lectin; 3.
DR  SMART; SM00458; RICIN; 1.
DR  PROSITE; PS50231; RICIN_B_LECTIN; 1.
KW  Transferase; Glycosyltransferase; Transmembrane; Signal-anchor;
KW  Golgi stack; Glycoprotein; Manganese; Calcium; Lectin.
FT  PROPEP        1     40       REMOVED IN SOLUBLE POLYPEPTIDE
FT                               N-ACETYLGALACTOSAMINYLTRANSFERASE
FT                               (BY SIMILARITY).
FT  CHAIN        41    559       POLYPEPTIDE N-
FT                               ACETYLGALACTOSAMINYLTRANSFERASE, SOLUBLE
FT                               FORM.
FT  DOMAIN        1      8       CYTOPLASMIC (POTENTIAL).
FT  TRANSMEM      9     28       SIGNAL-ANCHOR (TYPE-II MEMBRANE PROTEIN)
FT                               (POTENTIAL).
```

```
FT   DOMAIN       29    559    LUMENAL, CATALYTIC (POTENTIAL).
FT   DOMAIN      439    559    RICIN B-TYPE LECTIN.
FT   CARBOHYD     95     95    N-LINKED (GLCNAC . . .) (POTENTIAL).
FT   CARBOHYD    117    117    O-LINKED (POTENTIAL).
FT   CARBOHYD    118    118    O-LINKED (POTENTIAL).
FT   CARBOHYD    119    119    O-LINKED (POTENTIAL).
FT   CARBOHYD    141    141    N-LINKED (GLCNAC . . .) (POTENTIAL).
FT   CARBOHYD    288    288    O-LINKED (POTENTIAL).
FT   CARBOHYD    541    541    N-LINKED (GLCNAC . . .) (POTENTIAL).
FT   CARBOHYD    552    552    N-LINKED (GLCNAC . . .) (POTENTIAL).
SQ   SEQUENCE    559 AA;    64219 MW;    CD68118CB201EE5B    CRC64;
     MRKFAYCKVV  LATSLIWVLL  DMFLLLYFSE  CNKCDEKKER  GLPAGDVLEP  VQKPHEGPGE
  .
  .
  .
     LCLDVSKLNG  PVTMLKCHHL  KGNQLWEYDP  VKLTLQHVNS  NQCLDKATEE  DSQVPSIRDC
     NGSRSQQWLL  RNVTLPEIF
 //
```

Fig. 6.3 An abbreviated version of a typical entry in the SWISS-PROT database (dotted lines denote points at which, for convenience, material has been excised).

sensitive. Although there are a number of different protein databases that can be searched, the best are SWISS-PROT and TrEMBL.

The SWISS-PROT database is not just a repository of protein sequences; rather, it is a collection of confirmed protein sequences that are extensively annotated with information such as function and biological role of the protein, protein family assignments, and bibliographical references (Fig. 6.3). The quality of the data in SWISS-PROT is very high because the content is actively managed (*curated*). Because of the speed limitations of full annotation, the less robust TrEMBL database has been developed. TrEMBL consists of entries in the same format as those in SWISS-PROT which are derived from translation of all coding sequences in the EMBL nucleotide sequence database and which are not in SWISS-PROT already. Note that, because of the redundancy in the genetic code, it is easier to compare amino acid sequences than nucleotide sequences. Unlike SWISS-PROT entries, those in TrEMBL are awaiting manual annotation. However, they are given a potential functional annotation. This is done by selecting proteins in SWISS-PROT that belong to the same group of proteins as the unannotated protein, extracting the annotation shared by all functionally characterized proteins of this group and assigning this common annotation to the unannotated protein.

Significant matches of a novel gene to another sequence may be in any of four classes (Oliver 1996a). First, a match may predict both the bio-chemical and physiological function of the novel gene. An example is ORF YCR24c, identified during the whole genome sequencing of *Saccharomyces cerevisiae*, which has a close sequence similarity to an Asn-tRNA synthetase from *Escherichia coli*. Secondly, a match may define the biochemical function of a gene product without revealing its cellular function. An example of this was five protein kinase genes found on yeast chromosome III whose biochemical function is clear (they phosphorylate proteins) but whose particular physiological function in yeast was unknown. Thirdly, a match may be to a gene from another organism whose function in that organism is unknown. For example, ORF YCR63w from yeast matched protein G10 from *Xenopus* and novel genes from *Caenorhabditis elegans* and humans but, at the time, the function of all of them was unknown. Finally, a match may occur to a gene of known function that merely reveals that our understanding of that function is superficial, e.g. yeast ORF YCL17c and the NifS protein of nitrogen-fixing bacteria. After similar sequences were found in a number of bacteria that do not fix nitrogen it was shown that the NifS protein is a pyridoxal phosphate-dependent aminotransferase.

Secondary databases of functional domains

During evolution, recombination between unrelated genes has led to the generation of novel proteins.

Table 6.4 Secondary protein databases that can be used to identify protein motifs and protein domains.

Database	Details
PFAM	A collection of profiles and alignments for common protein families
PRINTS	A compendium of short protein motifs that captures common protein folds and domains
PROSITE	A database of longer protein signatures known as profiles
ProDom	A collection of protein domains
SMART	A curated collection of protein domains
BLOCKS	A database of conserved protein regions and their multiple alignments

Often these proteins are novel combinations of functional domains. This needs to be borne in mind when doing protein sequence similarity searches. Indeed, any analysis of gene function should include a search against databases of functional domains. There are a number of such databases (Table 6.4) and details of most of them can be found in Attwood and Parry-Smith (1999).

These secondary databases have been constructed using different analytical methods such as motif recognition, fingerprints of collections of motifs, domain profiles and hidden Markov models. Consequently, each has different strengths and weaknesses and hence different areas of optimum application. This makes it difficult to interpret the results when a predicted protein hits entries in several of the databases. To resolve this issue, a cross-referencing system called InterPro has been developed (Apweiler *et al.* 2001a,b). InterPro permits a protein sequence to be screened against each of the secondary databases and then extracts all the relevant information (Fig. 6.4). An example of the use of InterPro can be found in the analysis of the proteome of *Drosophila* (Rubin *et al.* 2000). A related database is CluSTr which offers an automatic classification of SWISS-PROT and TrEMBL proteins into groups of related proteins. The use of protein structural information for functional annotation is discussed in more detail in Chapter 8.

Gene ontology

Gene ontology (Ashburner *et al.* 2000) is a classification system that enables protein function to be related to gross cellular or whole organism functions

such as central metabolism, nucleic acid replication, cell division, pathogenesis, etc. (Fig. 6.5). It combines the breadth required to describe biological functions among very diverse species with the specificity and depth needed to distinguish a particular protein from another member of the same family. All the organism-specific databases use the standard gene ontology vocabulary to annotate genes and proteins (Stein 2001).

Analyses not based on homology

Although homology searching plays a major part in converting gene structure to function, other analytical tools are available that give information about secondary structure, e.g. location of α-helices and β-strands, based on examination of the primary sequence. For example, a number of programs are available which produce hydropathy (hydrophobicity) profiles for predicting possible transmembrane domains. These are runs of hydrophobic amino acids about 20–25 residues in length. Because there are many different hydropathic rankings for the amino acids, there are different scales of hydrophobicity and this leads to differences in detail regarding the number and extent of the potential hydrophobic regions. Where there is consensus between the methods, there is a greater chance that the predictions are reliable. In addition to hydropathy profiles it is possible to identify putative amphipathic helices. These are helices that display a characteristic charge separation in terms of the distribution of polar and non-polar amino acid residues on opposite faces. This 'sidedness' allows such helices to sit comfortably at polar–apolar interfaces and its detection

```
InterPro Entry IPR000772
Ricin B lectin domain
Database      InterPro
Accession     IPR000772; Ricin_B_lectin (matches 253 proteins)
Name          Ricin B lectin domain
Type          Domain
Dates         08-OCT-1999 (created)
              12-MAR-2001 (last modified)
Signatures    PS50231; RICIN_B_LECTIN (241 proteins)
              PF00652; Ricin_B_lectin (195 proteins)
              SM00458; RICIN (183 proteins)
Children      IPR003558; Cytolethal distending toxin A (6 proteins)
[tree]        IPR003559; Cytolethal distending toxin C (4 proteins)
Abstract      Ricin is a legume lectin from the seeds of the castor bean plant, Ricinus communis. The
              seeds are poisonous to people, animals and insects and just one milligram of ricin can
              kill an adult.
              Primary structure analysis has shown the presence of a similar domain in many
              carbohydrate-recognition proteins like plant and bacterial AB-toxins, glycosidases or
              proteases [1, 2, 3]. This domain, known as the ricin B lectin domain, can be present in
              one or more copies and has been shown in some instance to bind simple sugars, such as
              galactose or lactose.
              The ricin B lectin domain is composed of three homologous subdomains of 40 amino
              acids (alpha, beta and gamma) and a linker peptide of around 15 residues (lambda).
              It has been proposed that the ricin B lectin domain arose by gene triplication from
              a primitive 40 residue galactoside-binding peptide [4, 5]. The most characteristic,
              though not completely conserved, sequence feature is the presence of a Q-W pattern.
              Consequently, the ricin B lectin domain as also been refered as the $(QxW)_3$ domain and
              the three homologous regions as the QxW repeats [2, 3]. A disulfide bond is also
              conserved in some of the QxW repeats [2].
              The 3D structure of the ricin B chain has shown that the three QxW repeats pack around
              a pseudo threefold axis that is stabilised by the lambda linker [4]. The ricin B lectin
              domain has no major segments of a helix or beta sheet but each of the QxW repeats
              contains an omega loop [5]. An idealized omega-loop is a compact, contiguous segment
              of polypeptide that traces a 'loop-shaped' path in three-dimensional space; the main
              chain resembles a Greek omega.
Examples      • P26514 XYNA_STRLI: Xylanase A (EC 3.2.1.8) (gene xInA), from Streptomyces.
              • P46084 HA33_CLOBO: 33-kD hemagglutinin component of the botulinum neurotoxin
                complex.
              • Q05308 SP1_RARFA: Serine protease I (RPI) (EC 3.4.21.-), from Rarobacter
                faecitabidus [6].
              • P49260 PA2R_RABIT: Phospholipase A2 receptor, from mammals.
              • P02879 RICI_RICCO: Ricin, from Ricinus communis (Castor bean) [4].
              • P22222 E13B_OERXA: Glucan endo-1, 3-beta-glucanase (EC 3.2.1.39), from Oerskovia
                xanthineolytica.
              • P09545 HLYA_VIBCH: The bacterial AHH1/ASH4/HlyA/VVHA family of hemolysins.
                View examples
References     1. Hirabayashi J., Dutta S.K., Kasai K.
                 Novel galactose-binding proteins in Annelida. Characterization of 29-kDa tandem
                 repeat-type lectins from the earthworm Lumbricus terrestris.
                 J. Biol. Chem. 273(23): 14450-14460(1998).
              2. Hazes B., Read R.J.
                 A mosquitocidal toxin with a ricin-like cell-binding domain.
                 Nat. Struct. Biol. 2(5): 358-359(1995).
              3. Hazes B.
                 The $(QxW)_3$ domain: a flexible lectin scaffold.
                 Protein Sci. 5(8): 1490-1501(1996).
              4. Rutenber E., Ready M., Robertus J.D.
                 Structure and evolution of ricin B chain.
                 Nature 326(6113): 624-626(1987).
              5. Rutenber E., Robertus J.D.
                 Structure of ricin B-chain at 2.5 A resolution.
                 Proteins 10(3): 260-269(1991).
              6. Shimoi H., Iimura Y., Obata T., Tadenuma M.
                 Molecular structure of Rarobacter faecitabidus protease I. A yeast-lytic serine
                 protease having mannose-binding activity.
                 J. Biol. Chem. 267(35): 25189-25195(1992).
              7. Harris N., Peters L.L., Eicher E.M., Rits M., Raspberry D., Eichbaum Q.G., Super
                 M., Ezekowitz R.A.
                 The exon-intron structure and chromosomal localization of the mouse macrophage
                 mannose receptor gene Mrc1: identification of a Ricin-like domain at the N-terminus
                 of the receptor.
                 Biochem. Biophys. Res. Commun. 198(2): 682-692(1994).
Database      PROSITE doc; PDOC50231
links         PROSITE predoc; QDOC50231
Matches       Table all Graphical all Condensed graphical view
```

Fig. 6.4 An example entry from the InterPro database.

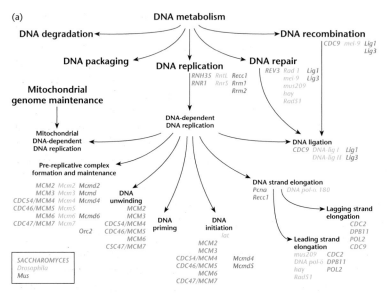

(a)

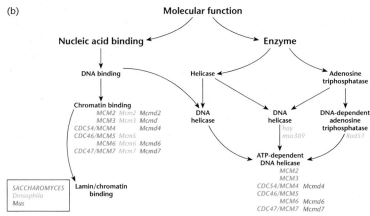

(b)

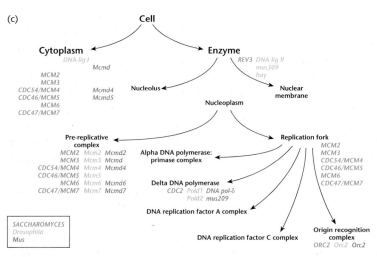

(c)

Fig. 6.5 Examples of gene ontology. Three examples illustrate the structure and style used by gene ontology to represent the gene ontologies and to associate genes with nodes within an ontology. The ontologies are built from a structured, controlled vocabulary. The illustrations are the products of work in progress and are subject to change when new evidence becomes available. For simplicity, not all known gene annotations have been included in the figures. (a) Biological process ontology. This section illustrates a portion of the biological process ontology describing DNA metabolism. Note that a node may have more than one parent. For example, 'DNA ligation' has three parents: 'DNA-dependent DNA replication'; 'DNA repair'; and 'DNA recombination'. (b) Molecular function ontology. The ontology is not intended to represent a reaction pathway, but instead reflects conceptual categories of gene-product function. A gene product can be associated with more than one node within an ontology, as illustrated by the MCM proteins. These proteins have been shown to bind chromatin and to possess ATP-dependent DNA helicase activity, and are annotated to both nodes. (c) Cellular component ontology. The ontologies are designed for a generic eukaryotic cell, and are flexible enough to represent the known differences between diverse organisms. (Reprinted from Ashburner *et al.* 2000 by permission of Nature Publishing Group, New York.)

gives clues as to the possible cellular role of a new protein.

Genome annotation

Many different genomes have been and are being completely sequenced. As the sequence data accumulates attempts are made, using the tools described above, to identify all the genes and ascribe functions to them. However, a predicted gene or predicted protein function is exactly that – a prediction. The reliability of that prediction depends on the experimental data to support it. Therefore, in annotating genes and genomes, biologists must use their own knowledge and intuition plus information from the literature to design experiments to support their interpretations. An example is provided by Pollack (2001) who found that examination of genomic or enzymatic data alone provided an incomplete picture of metabolic function in the bacterium *Ureaplasma urealyticum.*

The genome of *U. urealyticum* is 752 kb in length and has 613 protein-coding genes and 39 RNA-coding genes. Biological roles have been ascribed to 53% of the protein-coding genes, 19% have no known function although they are similar to other hypothetical genes, and 28% are hypothetical genes with no significant relationship to any other gene in any database (Glass *et al.* 2000). The metabolism of *Ureaplasma* has been studied for a long time and hence one would expect a good fit between annotated genes and known enzyme activities. Although in the majority of cases there is such agreement, a considerable number of examples were found where there were annotated genes without detectable activity (e.g. deoxyguanosine kinase, hypoxanthine-guanine phosphoribosyltransferase) or reported enzyme activity without annotation (e.g. pyruvate carboxylase, aspartate aminotransferase).

There are a number of explanations for the apparent disagreement between biological function and gene annotation. These include inaccurate sequencing, inaccurate annotation, mistakes in carrying out enzymatic assays, mutations in crucial residues that eliminate enzymatic activity and differences between strains. It could be that the annotation is correct but that post-translational modification renders the enzyme undetectable or inactive or the enzyme is only synthesized in certain media or environmental conditions. The best way of matching annotated genes to biological function is to inactivate the gene product by gene disruption, preferably by deletion of the entire ORF. A number of ways of doing this are available and are described in Chapter 10. While a strategy for annotation that involves individual knockouts of every gene is feasible with microbial genomes, it is not practical and/or ethical for the much larger genomes of higher eukaryotes. For the annotation of these larger genomes it is essential to bring biologists and bioinformaticians together and get them working in teams where they can both stimulate and challenge each other. Stein (2001) has reviewed the different organizational models that have been used for these annotation teams.

Given the importance of genome annotation, it is essential that different teams use similar methods. Fortunately, standard genome annotation languages are gaining general acceptance. GAME is particularly valuable for describing experimental evidence that supports an annotation, and DAS (distributed annotation system) is particularly useful for indexing and visualization. These are accompanied by a number of software tools (BioPerl 2001, BioPython 2001, BioJava 2001 and BioCORBA 2001) for storing, manipulating and visualizing these genome annotations.

Molecular phylogenetics

The main use of DNA and protein-sequence data is the analysis of cellular function. However, the data can also be used to investigate the evolution of genes and their protein products and this is known as molecular phylogeny. There are two approaches to building models of sequence evolution (Whelan *et al.* 2001): empirical models are built through comparisons of large numbers of observed sequences whereas parametric models are built on the basis of the chemical or biological properties of DNA and amino acids.

The parametric approach is favoured for studies on DNA evolution and the parameters used are base frequency, base exchangeability and rate heterogeneity. The base frequency parameter describes the frequency of the bases A, G, C and T averaged over all sequence sites and is influenced by the overall GC

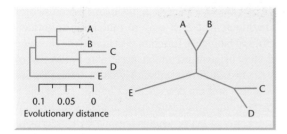

Fig. 6.6 Two ways of representing phylogenetic relationships. (From Attwood & Parry-Smith 1999 © Pearson Education Ltd. 1999, reprinted by permission of Pearson Education Ltd.)

content. Base exchangeability describes the relative tendencies for one base to be substituted for another and base transitions (purine–purine and pyrimidine–pyrimidine) are expected to be more frequent than base transversions (purine–pyrimidine). Rate heterogeneity is the variation in mutation rates along a stretch of DNA because of biochemical constraints, structural features, etc.

In contrast to phylogenetic studies with DNA, those on proteins use an empirical approach in which the number of amino acid substitutions is computed. On its own, the number of changes is not very revealing and so phylogenists use sophisticated statistical methods, such as maximum likelihood, to make inferences about the patterns and processes of evolution. Because proteins are encoded by DNA it might be assumed that protein and DNA phylogenies would be identical, but this is not necessarily so. The reason for this is that silent mutations occur in DNA because of the redundancy of the genetic code. Mutations at the DNA level that do not result in amino acid substitutions only get incorporated into DNA phylogenies.

Regardless of whether one is studying DNA or protein phylogeny, the relationships are usually represented graphically. There are two methods of

doing this (Fig. 6.6). In phylogenetic trees, evolutionary distance is measured in terms of horizontal branch length. In dendograms, evolutionary distance is measured along the length of the segments. Although the results of the two methods look different they depict exactly the same relationships.

Suggested reading

Attwood T.K. & Parry-Smith D.J. (1999) *Introduction to Bioinformatics*. Addison Wesley Longman, Harlow. *This short textbook gives a good overview of the methodology used in sequence comparisons and complements the material presented in this chapter.*

Goodman N. (2002) Biological data becomes computer literate: new advances in bioinformatics. *Current Opinion in Biotechnology* **13**, 68–71.

Stern L. (2001) Genome annotation: from sequence to biology. *Nature Reviews Genetics* **2**, 493–503.
Two excellent reviews that cover most of the material presented in this chapter.

Whelan S., Lio P. & Goldman N. (2001) Molecular phylogenetics: state-of-the-art methods for looking into the past. *Trends in Genetics* **17**, 262–272. *This paper presents a good overview of this specialist topic that will increase in importance as more and more genomes are sequenced.*

Useful websites

http://ww.ebi.ac.uk
European Bioinformatics Institute.

http://www.ncbi.nlm.nih.gov
National Centre for Biotechnology Information.

http://www.ddbj.nig.ac.jp
DNA Database of Japan.

By accessing any one of these three websites it is possible to access any publicly available programs and databases simply by following the links. Details of websites also can be found in the database summary that forms the first issue of each year of *Nucleic Acids Research* (see e.g. Baxevanis 2002).

CHAPTER 7

Comparative genomics

Introduction

Comparative genomics is the study of the differences and similarities in genome structure and organization in different organisms. For example, how are the differences between humans and other organisms reflected in our genomes? How similar are the number and types of proteins in humans, fruit flies, worms, plants, yeasts and bacteria? There are two drivers for comparative genomics. One is a desire to have a much more detailed understanding of the process of evolution at the gross level (the origin of the major classes of organism) and at a local level (what makes related species unique). The second driver is the need to translate DNA sequence data into proteins of known function. The rationale here is that DNA sequences encoding important cellular functions are more likely to be conserved between species than sequences encoding dispensable functions or non-coding sequences. The ideal species for comparison are those whose form, physiology and behaviour are as similar as possible, but whose genomes have evolved sufficiently that non-functional sequences have had time to diverge. Furthermore, peptides with a similar primary or secondary structure are likely to have similar functions.

Orthologues, paralogues and gene displacement

In order to compare genome organization in different organisms it is necessary to distinguish between *orthologues* and *paralogues*. Orthologues are homologous genes in different organisms that encode proteins with the same function and which have evolved by direct vertical descent. Paralogues are homologous genes within an organism encoding proteins with related but non-identical functions. Implicit in these definitions is that orthologues evolve simply by the gradual accumulation of muta-

tions, whereas paralogues arise by gene duplication followed by mutation accumulation. Good examples of paralogues are the protein superfamilies described in Chapter 2 (see Fig. 2.8 and Table 2.3).

There are many biochemical activities that are common to most or all living organisms, e.g. the citric acid cycle, the generation of ATP, the synthesis of nucleotides, DNA replication, etc. It might be thought that in each case the key proteins would be orthologues. Indeed, 'universal protein families' shared by all archae, eubacteria and eukaryotes have been described (Kyrpides *et al.* 1999). However, there is increasing evidence that functional equivalence of proteins requires neither sequence similarity nor even common three-dimensional folds (Galperin *et al.* 1998; Huynen *et al.* 1999). Such non-orthologous *gene displacements* may be rather common. For example, analysis of the citric acid cycle enzymes in bacteria and archaea indicates that for at least 25% of the *Escherichia coli* enzymes a displacement can be found in at least one other species.

At the opposite end of the spectrum from gene displacements, it is possible for proteins with very divergent functions and primary sequences to be related because they have similar three-dimensional structures (Brenner *et al.* 1998). Thus conventional sequence analysis may not detect homology. For example, the structures of D-alanine: D-alanine ligase, glutathione synthetase and the ATP-binding domains of carbamoyl phosphate synthetase and succinyl-CoA synthetase are so similar that they almost certainly evolved from the same ancestral protein. Nevertheless, pairwise comparisons of the primary sequences fail to confirm the homology.

Protein evolution by exon shuffling

Analysis of protein sequences and three-dimensional structures has revealed that many proteins are composed of discrete domains. These so-called mosaic proteins are particularly abundant in the metazoa. The majority of mosaic proteins are extracellular or constitute the extracellular parts of membrane-bound proteins and thus they may have played an important part in the evolution of multicellularity. The individual domains of a mosaic protein are often involved in specific functions which contribute to its overall activity. These domains are evolutionarily mobile which means that they have spread during evolution and now occur in otherwise unrelated proteins (Doolittle 1995). Mobile domains are characterized by their ability to fold independently. This is an essential characteristic because it prevents misfolding when they are inserted into a new protein environment. To date, over 60 mobile domains have been identified.

A survey of the genes that encode mosaic proteins reveals a strong correlation between domain organization and intron–exon structure (Kolkman & Stemmer 2001); i.e. each domain tends to be encoded by one or a combination of exons and new combinations of exons are created by recombination within the intervening sequences. This process yields rearranged genes with altered function and is known as *exon shuffling*. Because the average intron is much longer than the average exon and the recombination frequency is proportional to DNA length, the vast majority of crossovers occur in non-coding sequences. The large number of transposable elements and repetitive sequences in introns will facilitate exon shuffling by promoting mismatching and recombination of non-homologous genes. An example of exon shuffling is described in Box 7.1.

Although mosaic proteins are most common in the metazoa, they are found in unicellular organisms. Because a large number of microbial genomes have been sequenced, including representatives from the three primary kingdoms (Archaea, Eubacteria and Eukarya), it is possible to determine the evolutionary mobility of domains. With this in mind, Wolf *et al.* (2000a) searched the genomes of 15 bacteria, four archaea and one eukaryote for genes encoding proteins consisting of domains from the different kingdoms. They found 37 examples of proteins consisting of a 'native' domain and a horizontally acquired 'alien' domain. In several instances the genome contained the gene for the mosaic protein as well as a sequence encoding a stand-alone version of

the alien domain but more usually the stand-alone counterpart was missing.

Comparative genomics of prokaryotes

At the time of going to press (early 2002) complete genome sequences were available for 78 different prokaryotes. Simple analysis of the sequence data reveals two features of note. First, the genome sizes vary from 0.58 Mb (*Mycoplasma genitalium*) to nearly 8 Mb (*Saccharopolyspora erythraea*), i.e. a 14-fold difference. Secondly, the gene density is remarkably similar across all species and is about 1 gene per kilobase of DNA. This means that large prokaryotic genomes contain many more genes than smaller ones. By contrast, the human genome contains only twice as many genes as *Drosophila*. So how can we account for size diversity of prokaryotes?

When the different genomes are arranged in size order (Fig. 7.1) some interesting features emerge. First, the archaebacteria exhibit a very much smaller range of genome sizes. This could be an artefact of the small number of genomes examined but more probably reflects the fact that they each occupy a specialized environment and have little need for metabolic diversity. Secondly, the smallest eubacterial genomes are found in those organisms that normally are found associated with animals or humans, e.g. mycoplasmas, rickettsias, chlamydiae, etc. Those organisms that can occupy a greater number of niches have a larger genome size. Not surprisingly, there is a good correlation between genome size and metabolic and functional diversity as demonstrated by the size of the genomes of *Bacillus* and *S. erythraea* (formation of spores, synthesis of antibiotics) and *Pseudomonas aeruginosa* (degradation of a wide range of aromatic compounds).

The minimal genome

The genome of *M. genitalium* is 0.58 Mb and it is the smallest sequenced to date. This begs the question, what is the minimal genome that is consistent with a free-living cellular organism? If one ignores functionally important RNA molecules and non-coding sequences, the problem is one of defining the minimal protein set. The first attempt at identifying

Box 7.1 Haemostatic proteins as an example of exon shuffling

The process of blood coagulation and fibrinolysis involves a complex cascade of enzymatic reactions in which inactive zymogens are converted into active enzymes. These zymogens belong to the family of serine proteases and their activation is accompanied by proteolysis of a limited number of peptide bonds. Comparison of the amino acid sequences of the haemostatic proteases with those of archetypal

serine proteases such as trypsin shows that the former have large N-terminal extensions (Fig. B7.1). These extensions consist of a number of discrete domains with functions such as substrate recognition, binding of co-factors, etc. and the different domains show a strong correlation with the exon structure of the encoding genes.

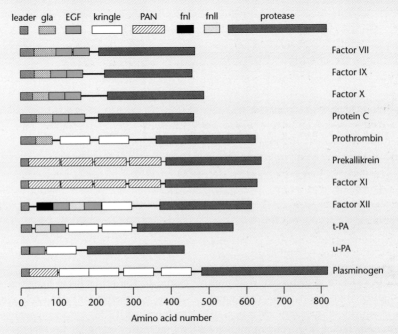

Fig. B7.1 Domain structures of the regulatory proteases of blood coagulation and fibrinolysis. The different domains: grey, serine protease domain; dark green, EGF-like domain; dotted, Gla domain; cross-hatch, PAN domain; light green, fibronectin type II domain (fn2); black, fibronectin type I domain (fn1). (Adapted from Kolkman & Stemmer 2001.)

this minimal set was made by compiling a list of orthologous proteins in *Haemophilus influenzae* and *M. genitalium* (Mushegian & Koonin 1996). The expectation was that this list would predominantly contain proteins integral for cell survival as both bacteria are essentially parasites and thus should have shed auxiliary genes. Altogether 244 orthologues were identified but this list is unlikely to be complete because of the occurrence of non-orthologous gene displacements. Some of these gene displacements can be inferred because both organisms appear to have key metabolic pathways that

are incomplete. In this way, Mushegian (1999) extended the minimal protein set to 256 genes.

The problem with the above approach is that if one is too strict in defining the degree of similarity between two proteins required to constitute orthologues then the minimal protein set is greatly underestimated. A variation of the above method is to identify orthologous groups; i.e. clusters of genes that include orthologues and, additionally, those paralogues where there has been selective gene loss following gene duplication. When this approach was taken with four eubacteria, one archaebacterium

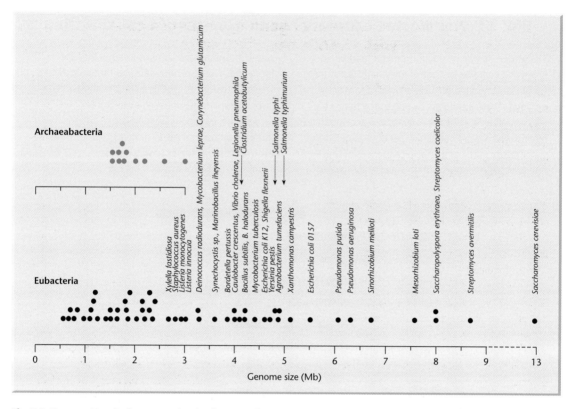

Fig. 7.1 Genome sizes of eubacteria and archaebacteria whose genomes have been completely sequenced.

and one yeast, 816 clusters of orthologous groups (COGs) were identified. Of these, 327 contained representatives of all three kingdoms (Mushegian 1999). Based on this set of 327 proteins it was possible to reconstruct all the key biosynthetic pathways. When the analysis was repeated with sequence data from an additional three archaebacteria and 12 eubacteria, the minimal protein set was slightly reduced to 322 COGs.

Analysis of larger genomes

Comparison of the *P. aeruginosa* (6.3 Mb) and *E. coli* (4.5 Mb) genomes indicates that the large genome of *P. aeruginosa* is the result of greater genetic complexity rather than differences in genome organization. Distributions of open-reading frame (ORF) sizes and inter-ORF spacings are nearly identical in the two genomes. If the larger genome of *P. aeruginosa* arose by recent gene duplication one would expect it to have a similar number of paralogous groups com-

pared to the other large bacterial genomes and a larger number of ORFs in each group. In fact, the number of ORFs in the paralogous groups in *Pseudomonas* is similar to the other genomes. Thus selection for environmental versatility (Box 7.2) has favoured genetic capability through the development of numerous small paralogous gene families whose members encode distinct functions. As a general rule, one would expect that as the size of the prokaryotic genome increases then the number of paralogues also would increase and this is what has been observed (Table 7.1). Furthermore, the biochemical bias in these paralogues reflects the biology of the host organism (Box 7.2).

Analysis of all the prokaryotic genomes sequenced to date has revealed two intriguing observations. First, almost half the ORFs identified are of unknown biological function. This suggests that a number of novel biochemical pathways remain to be identified. Secondly, approximately 25% of all ORFs identified are unique and have no significant sequence

Box 7.2 Correlation of genome sequence data with the biology of bacteria

Pseudomonas aeruginosa

Pseudomonas aeruginosa is a bacterium that is extremely versatile both ecologically and metabolically. It grows in a wide variety of habitats including soil, water, plant surfaces, biofilms and both in and on animals including humans. A major problem with *P. aeruginosa* is its resistance to many disinfectants and antibiotics. Pseudomonads are characterized by a limited ability to grow on carbohydrates but a remarkable ability to metabolize many other compounds including an astonishing variety of aromatics.

Analysis of the genome of *P. aeruginosa* (Stover *et al.* 2000) reveals a general lack of sugar transporters and an incomplete glycolytic pathway, both of which explain the poor ability to grow on sugars. By contrast, it has large numbers of transporters for a wide range of metabolites and a substantial number of genes for metabolic pathways not found in many other bacteria such as *E. coli*. As might be expected for an organism with great metabolic versatility, a high proportion of the genes (> 8%) are involved in gene regulation. The organism also has the most complex chemosensory system of all the complete bacterial genomes with four loci that encode probable chemotaxis signal-transduction pathways. Finally, sequencing revealed the presence of a large number of undescribed drug efflux systems (see Fig. 12.10) which probably account for the inherent resistance of the organism to many antibacterial substances.

Caulobacter crescentus

Caulobacter crescentus is a bacterium that is found in oligotrophic (very low nutrient) environments and is not capable of growing in rich media. Not surprisingly, genome sequencing (Nierman *et al.* 2001) has shown that the bacterium possesses a large number of genes for responding to environmental substrates. For example, 2.5% of the genome is devoted to motility, there are two

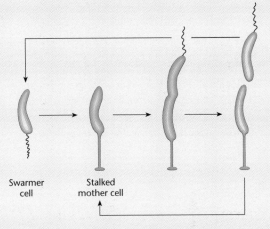

Fig. B7.2 The life cycle of *Caulobacter crecentus*.

chemotaxis systems and over 16 chemoreceptors. It also has 65 members of the family of outer membrane proteins that catalyse energy-dependent transport across the membrane. By contrast, the metabolically-versatile *P. aeruginosa* has 32 and other bacteria less than 10.

The bacterium also has an obligatory life cycle involving asymmetric cell division and differentiation (Fig .B7.2). Thus it comes as no surprise that genome sequencing reveals a very high number of two-component signal-transduction proteins, e.g. 34 histidine protein kinase (HPK) genes, 44 response regulator (RR) genes and 27 hybrid (HPK/RR) genes. In addition, the frequency of the GAnTC target site for DNA methylation was much less than would be expected if it occurred at random.

Deinococcus radiodurans

This bacterium is remarkable for its ability to survive extremely high doses of ionizing radiation. For example, it can grow in the presence of chronic radiation (6 kilorads/hour) and withstand acute exposures to 1500 kilorads. The organism also is resistant to dessication, oxidizing agents and ultraviolet radiation. These properties could be the

continued on p. 118

Box 7.2 *continued*

result of one or more of prevention, tolerance and repair. Genome sequencing (White *et al.* 1999; Makarova *et al.* 2001) has shown that systems for the prevention and tolerance of DNA damage are present but that the key mechanism of resistance is an extremely efficient DNA repair system. Although all of the DNA repair genes identified in *D. radiodurans* have functional homologues in other prokaryotes, no other species has the same high degree of gene redundancy. The bacterium also has

multiple genes for proteins involved in exporting oxidation products of nucleotides. Another important component may be the presence of DNA repeat elements scattered throughout the genome. These repeats satisfy several expected requirements for involvement in recombinational repair, including that they are intergenic, they are ubiquitous, and they occur at a frequency that is comparable to the number of double-stranded DNA breaks that can be tolerated.

Table 7.1 Relationship between paralogues and genome size.

Organism	Genome size relative to *E. coli*	Percentage of proteins belonging to paralogues
Pseudomonas aeruginosa	1.4	75
Escherichia coli	1	50
Caulobacter crescentus	0.88	48
Haemophilus influenzae	0.38	35
Mycoplasma genitalium	0.12	26

similarity to any other available protein sequence. Although this might be an artefact of the small number of bacterial species studied by whole-genome analysis, it does support the observation of incredible biological diversity between bacteria. More importantly, it indicates that there are large numbers of new protein families yet to be discovered, e.g. over 1000 proteins in each of *Bacillus subtilis*, *E. coli* and *Deinococcus radiodurans*!

Because the DNA and protein sequence databases are updated daily it pays to revisit them from time to time to determine if homologues to previously unidentified proteins have been found. It also pays to re-examine sequence data as new and more sophisticated bioinformatics tools are being developed. The benefits of this can be seen from the work of Robinson *et al.* (1994). They re-examined 18 Mb of prokaryotic DNA sequence and uncovered more than 450 genes that had escaped detection. A more specific example is that of Dandekar *et al.* (2000) who re-examined the sequence data for *Mycoplasma pneumoniae*. They identified an additional 12 ORFs and eliminated one identified previously and found

an additional three RNA genes. They also shortened eight protein reading frames and extended 16 others.

Horizontal gene transfer

Horizontal, or lateral, gene transfer is the occurrence of genetic exchange between different evolutionary lineages. It is generally believed that such gene transfer has occurred many times in the course of evolution but detecting it has been very difficult. Now that complete genome sequences are available the task should be much easier. Two methods that have been used are the detection of sequences with unusual nucleotide composition or the failure to find a similar gene in closely related species. For example, analysis of the genomes of two bacterial thermophiles indicated that 20–25% of their genes were more similar to genes in archaeabacteria than those of eubacteria (Aravind *et al.* 1998; Nelson *et al.* 1999). These archaeal-like genes occurred in clusters in the genome and had a markedly different nucleotide composition and could have arisen by horizontal gene transfer.

Garcia-Vallve *et al.* (2000) have developed a statistical procedure for predicting whether genes of a complete genome have been acquired by horizontal gene transfer. This procedure is based on analysis of G + C content, codon usage, amino acid usage and gene position. When it was applied to 24 sequenced genomes it suggested that 1.5–14.5% of genes had been horizontally transferred and that most of these genes were present in only one or two lineages. However, Koski *et al.* (2001) have urged caution in the use of codon bias and base composition to predict horizontal gene transfer. They compared the ORFs of *E. coli* and *Salmonella typhi*, two closely related bacteria that are estimated to have diverged 100 million years ago. They found that many *E. coli* genes of normal composition have no counterpart in *S. typhi*. Conversely, many genes in *E. coli* have an atypical composition and not only are also found in *S. typhi*, but are found at the same position in the genome, i.e. they are *positional* orthologues.

Karlin (2001) has defined genes as 'putative aliens' if their codon useage difference from the average gene exceeds a high threshold and codon useage differences from ribosomal protein genes and chaperone genes also are high. Using this method, in preference to variations in G + C content, he noted that stretches of DNA with anomalous codon useage were frequently associated with pathogenicity islands. These are large stretches of DNA (35–200 kb) that encode several virulence factors and are present in all pathogenic isolates of a species and usually absent from non-pathogenic isolates. Such pathogenicity islands may have been spread laterally, particularly by phages and/or plasmids (see also Chapter 12, p. 214).

Comparative genomics of related bacteria

Because bacteria exhibit such a marked metabolic and ecological diversity, relatively little information can be gained by comparing the gene and protein content of distantly related species. Exceptions to this are the identification of essential housekeeping genes and the elaboration of the minimal genome. By comparing gene content and gene order in closely related bacteria it is much easier to understand the evolutionary processes that have shaped the bacterial populations of today.

In evolutionary terms, the most closely related organisms are different isolates of the same species. Kato-Maeda *et al.* (2001) have undertaken an analysis of deletions in 19 clinical isolates of *Mycobacterium tuberculosis* and made three interesting observations.

1 All but one of the strains had deletions equivalent to ~0.3% of the genome.

2 Most of the DNA deleted contained ancestral genes whose functions are no longer required, e.g. phage-related genes, insertion sequences, etc.

3 As the amount of DNA deleted increased, the likelihood that the bacteria will cause pulmonary cavitation decreased.

This suggests that the accumulation of mutations tends to diminish pathogenicity.

Alm *et al.* (1999) undertook a comparison of the complete sequences of two strains of *Helicobacter pylori* and found that they were very similar with only about 6% of the genes being specific for each strain. One interesting aspect of *H. pylori* is that it encodes more than 20 putative restriction-modification (R-M) systems, a feature that requires more than 4% of the genome. Lin *et al.* (2001) found that less than one-third of these R-M systems were functional in either strain. Furthermore, although most of the R-M genes were common to both strains, the genes that were functional were different in the two strains. All the strain-specific genes were active, whereas most of the shared genes were inactive. These observations suggest that *H. pylori* is constantly acquiring new R-M systems and inactivating the old ones. Why this should be is not clear at present.

Comparison of three isolates of *E. coli*, one laboratory strain and two O157 pathogenic isolates, also has uncovered some interesting structural features (Hayashi *et al.* 2001; Perna *et al.* 2001). The genomic backbone comprises 4.1 Mb that clearly are homologous between the three isolates although an inversion was found around the replication terminus. However, the homology is punctuated by hundreds of lineage-specific islands of apparently introgressed DNA scattered throughout the genome. Both the O157 isolates have a genome that is more than 800 kb larger than the laboratory strain and this size difference is entirely caused by variations in the amount of island DNA. Many of the islands in the pathogenic isolates encode putative virulence factors and prophages or prophage-like elements.

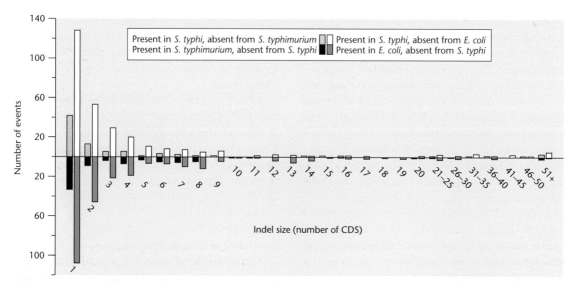

Fig. 7.2 Distribution of insertions and deletions in *Salmonella typhi* relative to *Escherichia coli* and *Salmonella typhimurium*. The graph shows number of insertion–deletion events plotted against the size of the inserted or deleted element (shown as number of genes), clearly indicating that most of the events involve a small number of genes. Values above the lines represent genes present in *S. typhi*; values below the line represent genes absent in *S. typhi*. Dark bars show the comparison with *S. typhimurium*; light bars with *E. coli*. (Redrawn with permission from Parkhill *et al.* 2001b.)

The next level of comparison is to assess the similarities and differences between different species of the same genus and this has been carried out for three different bacterial genera. Read *et al.* (2000) undertook a comparative analysis of the genome sequences of two strains of *Chlamydia trachomatis* and two strains of *Chlamydia pneumoniae*. Apart from a number of large chromosomal inversions, the genomes of the two species exhibited a high degree of synteny and little indication of horizontal gene transfer. However, these organisms are obligate intracellular parasites and thus may have little selective pressure exerted on them.

Two serovars of *Salmonella* (*S. typi* and *S. typhimurium*) have been completely sequenced (McClelland *et al.* 2001; Parkhill *et al.* 2001a) and comparisons have been made with another close relative, *E. coli* (Fig. 7.2), with which there is extensive synteny. As would be expected, the relationship between *S. typhi* and *S. typhimurium* is very much closer than between *S. typhi* and *E. coli*, although there still are significant differences. There are 601 genes (13.1%) that are unique to *S. typhi* compared with *S. typhimurium* and 479 genes (10.9%) unique to *S. typhimurium* relative to *S. typhi*. By contrast, there are 1505 genes (32.7%) unique to *S. typhi*

relative to *E. coli* and 1220 genes (28.4%) unique to *E. coli* relative to *S. typhi*. Another difference between *S. typhi* and *S. typhimurium* is the presence of 204 pseudogenes in the former and only 39 in the latter. In most cases these pseudogenes are relatively recent because they are caused by a single frameshift or stop codon. It is worth noting that complete sequencing of closely related genomes facilitates the detection of pseudogenes. This is because a frameshift or premature stop codon is only recognizable if the gene is colinear with a functional homologous gene in another genome. One biological difference between the two *Salmonella* serovars is that *S. typhi* only infects humans, whereas *S. typhimurium* can infect a wide range of mammals. This may be related to differences in pseudogene content because many of the pseudogenes in *S. typhi* are in housekeeping functions and virulence components.

A more interesting inter-species comparison (Cole *et al.* 2001) is that of *Mycobacterium tuberculosis* and *Mycobacterium leprae*, the causative organisms of tuberculosis and leprosy (Table 7.2). Of the 1604 ORFs in *M. leprae*, 1439 had homologues in *M. tuberculosis*. Most of the 1116 pseudogenes were translationally inert but also had functional counterparts in *M. tuberculosis*. Even so, there has still

Table 7.2 Comparison of the genomes of two *Mycobacterium* spp. (Reproduced from Cole *et al.* 2001.)

Feature	*Mycobacterium leprae*	*Mycobacterium tuberculosis*
Genome size	3 268 203	4 411 532
G + C (%)	57.79	65.61
Protein coding (%)	49.5	90.8
Protein coding genes (No.)	1604	3959
Pseudogenes (No.)	1116	6
Gene density (bp per gene)	2037	1114
Average gene length (bp)	1011	1012
Average unknown gene length (bp)	338	653

been a massive gene decay in the leprosy bacillus. Genes that have been lost include those for part of the oxidative respiratory chain and most of the microaerophilic and anaerobic ones plus numerous catabolic systems. These losses probably account for the inability of microbiologists to culture *M. leprae* outside of animals. At the genome organization level, 65 segments showed synteny but differ in their relative order and distribution. These breaks in synteny generally correspond to dispersed repeats, tRNA genes or gene-poor regions and repeat sequences occur at the junctions of discontinuity. These data suggest that genome rearrangements are the result of multiple recombination events between related repetitive sequences.

Common themes in different bacteria

As more and more bacterial genomes are sequenced certain structural themes start to emerge. Eisen *et al.* (2000) and Suyama & Bork (2001) have noted that in closely related bacteria chromosomal inversions are most likely to occur around the origin or terminus of replication. Another observation is the large number of prophages and prophage remnants that litter many genomes. The exact significance of these features is not known at present.

The systematic comparison of gene order in bacterial and archaeal genomes has shown that there is very little conservation of gene order between phylogenetically distant genomes. A corollary of this is that whenever statistically significant conservation of gene order is observed then it could be indicative of organization of the genes into operons. Wolf *et al.* (2001) undertook a comparison of gene order in all

the sequenced prokaryotic genomes and found a number of potential operons. Most of these operons encode proteins that physically interact, e.g. ribosomal proteins and ABC-type transporter cassettes. More important, this analysis enabled functions to be assigned to genes based on predictions of operon function.

Comparative genomics of organelles

Comparisons of mitochondrial genomes

As far as we know, mtDNA has the same fundamental role in all eukaryotes that contain it: it encodes a limited number of RNAs and proteins essential for formation of a functional mitochondrion. Although the genetic role of mtDNA appears to be universally conserved, this genome exhibits remarkable variation in conformation and size (Fig. 7.3) as well as in actual gene content, arrangement and expression (Tables 7.3 and 7.4). Like a typical bacterial genome, many mtDNAs map as circular molecules but linear molecules also exist. Whereas the mtDNA of animals and fungi are relatively small (15–20 kb), those of plants are very large (200–2000 kb). Plant mitochondria rival the eukaryotic nucleus, and especially the plant nucleus, in terms of the C-value paradox they present; i.e. larger plant mitochondrial genomes do not appear to contain more genes than smaller ones, but simply have more spacer DNA. This paradox extends to plant–animal comparisons where the *Arabidopsis* mtDNA is 20 times larger than human mtDNA but has less than twice the number of genes.

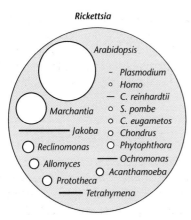

Fig. 7.3 Size and gene content of mitochondrial genomes compared with an α-Proteobacterial (*Rickettsia*) genome. Circles and lines represent circular and linear genome shapes, repectively. (Reprinted from Gray *et al.* 1999 by permission of the American Association for the Advancement of Science.)

Table 7.3 Comparative features of mtDNA.

Property	Range
Conformation	Circular or linear
Size	< 6 to > 2000 kb
Non-coding DNA (%)	< 10 to > 80
Number of genes	5–97

As a result of the steady accumulation of sequence data it now is evident that mtDNAs come in two basic types. These have been designated as 'ancestral' and 'derived' (Gray *et al.* 1999) and their characteristics are summarized in Table 7.4. It is generally believed that mitochondria are the direct descendants of a bacterial endosymbiont that became established in a nucleus-containing cell and an ancestral mitochondrial genome is one that has retained clear vestiges of this eubacterial ancestry. The prototypal ancestral mtDNA is that of *Reclinomonas americana*, a heterotrophic flagellated protozoon. The mtDNA of this organism contains 97 genes including all the protein-coding genes found in all other sequenced mtDNAs. Derived mitochondrial genomes are ones that depart radically from the ancestral pattern. In animals and many protists this is accompanied by a substantial reduction in overall size and gene content. In plants, and particularly angiosperms, there has been extensive gene loss but size has increased as a result of frequent duplication of DNA and the capture of sequences from the chloroplast and nucleus (Marienfeld *et al.* 1999).

If mitochondria are derived from a bacterium, what is the closest relative of that bacterium that exists today? The current view is that it is *Rickettsia prowazekii*, the causative agent of epidemic typhus. This organism favours an intracellular lifestyle that could have initiated the endosymbiotic evolution of the mitochondrion. The genome of *R. prowazekii* has been sequenced and the functional profile of its genes shows similarities to mitochondria (Andersson *et al.* 1998). The structure, organization and gene content of the bacterium most resembles that of the mtDNA of *Reclinomonas americana*.

Intracellular gene transfer

The principal function of the mitochondrion is the

Table 7.4 Properties of ancestral and derived mtDNAs. (Reprinted from Marienfeld *et al.* 1999 by permission of Elsevier Science.)

Ancestral mtDNA	Derived mtDNA
1 Many extra genes compared with animal mtDNA	1 Extensive gene loss
2 rRNA genes that encode eubacteria-like 23S, 16S and 5S rRNAs	2 Marked divergence in rDNA and rRNA structure
3 Complete, or almost complete, set of tRNA genes	3 Accelerated rate of sequence divergence in both protein-coding and rRNA genes
4 Tight packing of genetic information with few or no introns	4 Highly biased use of codons including, in some cases, elimination of certain codons
5 Eubacterial-like gene clusters	5 Introduction of non-standard codon assignments
6 Use standard genetic code	

generation of ATP via oxidative phosphorylation. At least 21 genes encode proteins critical for oxidative phosphorylation and one would expect all of these genes to be located in the mtDNA. Similarly, a mtDNA location would be expected for the genes encoding the 14 ribosomal proteins that are required to translate mtRNA. However, sequence data indicate that many mitochondrial genomes lack a number of key genes (Table 7.5) and the missing genes can be found in the nucleus. Functional transfer of mitochondrial genes to the nucleus has stopped in animals, hence their consistency in size. Part of the reason for this is that further transfer is blocked by changes in the mitochondrial genetic code. However, this gene transfer continues to occur in plants and protists because there is no genetic code barrier to transfer.

In the case of the mitochondrial *cox2* gene, transfer to the nucleus is still on-going in the case of the legumes (Palmer *et al.* 2000). Analysis of 25 different legumes identified some genera in which the *cox2* gene was located in the mitochondrion, some in which it was nuclear, and some where it was present in both genomes. In most cases where two copies of the gene are present, only one gene is transcriptionally active, although at least one genus was found in which both genes are transcribed.

Adams *et al.* (2000a) studied the distribution of the *rps10* gene in 277 angiosperms and identified 26 cases where the gene has been lost from the mtDNA. In 16 of these loss lineages, the nuclear gene was characterized in detail. To be active in the nucleus, a gene acquired from mtDNA must be inserted into the nuclear genome in such a way that a mature translatable mRNA can be produced. Moreover, the resulting protein is made in the cytoplasm and must be targeted to and imported into mitochondria. What emerged was that in some cases pre-existing copies of other nuclear genes have been parasitized with the *rps10* coding sequence. In several instances a mitochondrial targeting sequence has been coopted to provide entry for the RPS10 protein back into the mitochondrion but different nuclear genes provide this sequence in different plants. In other cases, the RPS10 protein is imported despite the absence of an obvious targeting sequence. These results, and similar findings for other mitochondrial genes (Adams *et al.* 2001), provide confirmation that nuclear transfer is on-going and has happened on many separate occasions in the past. Nor is nuclear transfer confined to mitochondrial genes for Millen *et al.* (2001) have made similar observations with chloroplast genes. Henze & Martin (2000) have reviewed the mechanisms whereby this transfer can occur.

Horizontal gene transfer

In the previous section we discussed intracellular horizontal evolution whereby genes move between the mitochondrion, the chloroplast and the nucleus. However, cross-species acquisition of DNA by plant mitochondrial genomes also has been detected (Palmer *et al.* 2000). The DNA in question is a homing group I intron. These introns encode site-specific endonucleases with relatively long target sites that catalyse their efficient spread from intron-containing alleles to intron-lacking alleles of the same gene in genetic crosses. This intron has been detected in the mitochondrial *cox1* gene of 48 angiosperms out of 281 tested. Based on sequence data for the intron and the host genome, it appears that this intron has been independently acquired by cross-species horizontal transfer to the host plants on many separate occasions. What is not clear are the identities of the donor and recipient in each individual case. By contrast with this group I intron, the 23 other introns in angiosperm mtDNA belong to group II and all are transmitted in a strictly vertical manner.

Comparative genomics of eukaryotes

Minimal eukaryotic genomes

In determining the minimal genome we are seeking to answer a number of different questions. What is the minimal size of the genome of a free-living unicellular eukaryote and how does it compare with the minimal bacterial genome? That is, what are the fundamental genetic differences between a eukaryotic and a prokaryotic cell? Next, what additional genetic information does it require for multicellular coordination? In animals, what are the minimum sizes for a vertebrate genome and a mammalian genome? Finally, what is the minimum size of genome for a flowering plant? Given that many

Table 7.5 Comparison of the gene content in completely sequenced land plant, green algal and red algal mitochondrial genomes. (Reproduced with permission from Gray *et al*. 1999.)

	Ath	Mpo	Pwi	Nol	Pmi	Ceu	Ppu	Cme	Ram
Complex I									
nad1	■	■	■	■	■	■	■	■	■*
nad2	■	■	■	■	■	■	■	■	■*
nad3	■	■	■	■	■	○	■	■	■*
nad4	■	■	■	■	■	■	■	■	■*
nad4L	■	■	■	■	■	○	■	■	■*
nad5	■	■	■	■	■	■	■	■	■*
nad6	■	■	■	■	■	■	■	■	■*
nad7	■	□	■	■	○	○	○	○	■*
nad8	○	○	○	○	○	○	○	○	■
nad9	■	■	■	■	○	○	○	○	■*
nad10	○	○	○	■	○	○	○	○	■*
nad11	○	○	○	○	○	○	○	○	■
Complex II									
sdh2	○	○	○	○	○	○	■	■	■*
sdh3	○	■	○	○	○	○	■	■	■*
sdh4	□	■	○	○	○	○	■	■	■*
Complex III									
cob	■	■	■	■	■	■	■	■	■*
Complex IV									
cox1	■	■	■	■	■	■	■	■	■*
cox2	■	■	■	■	○	○	■	■	■*
cox3	■	■	■	■	○	○	■	■	■*
Complex V									
atp1	■	■	■	■	○	○	○	○	■*
atp3	○	○	○	○	○	○	○	○	■*
atp6	■	■	■	■	■	○	■	■	■*
atp8	■	■	■	■	■	○	■	■	■*
atp9	■	■	■	■	○	○	■	■	■*
Cytochrome *c* biogenesis									
yejR	■	■	○	○	○	○	○	■	■*
yejU	■	■	○	○	○	○	○	■	■*
yejV	■	■	○	○	○	○	○	■	■*
yejW	○	○	○	○	○	○	○	■	■*
SSU ribosomal proteins									
rps1	○	■	○	○	○	○	○	○	■*
rps2	○	■	■	■	○	○	○	○	■*
rps3	■	■	■	■	○	○	■	■	■*
rps4	■	■	■	■	○	○	○	■	■*
rps7	■	■	■	■	○	○	○	○	■*
rps8	○	■	○	■	○	○	○	■	■*
rps10	○	■	■	■	○	○	○	○	■*
rps11	○	■	■	■	○	○	■	■	■*
rps12	■	■	■	■	○	○	■	■	■*
rps13	○	■	■	■	○	○	○	○	■*
rps14	□	■	■	■	○	○	○	■	■*
rps19	□	■	■	■	○	○	○	○	■*

(continued)

Table 7.5 (*continued*)

	Ath	Mpo	Pwi	Nol	Pmi	Ceu	Ppu	Cme	Ram
LSU ribosomal proteins									
rpl1	O	O	O	O	O	O	O	O	■
rpl2	■	■	O	O	O	O	O	O	■*
rpl5	■	■	■	■	O	O	O	■	■*
rpl6	O	■	■	■	O	O	O	■	■*
rpl10	O	O	O	O	O	O	O	O	■
rpl11	O	O	O	O	O	O	O	O	■
rpl14	O	O	O	■	O	O	O	■	■*
rpl16	■	■	■	■	O	O	■	■	■*
rpl18	O	O	O	O	O	O	O	O	■
rpl19	O	O	O	O	O	O	O	O	■
rpl20	O	O	O	O	O	O	O	■	■*
rpl27	O	O	O	O	O	O	O	O	■
rpl31	O	O	O	O	O	O	O	O	■
rpl32	O	O	O	O	O	O	O	O	■
rpl34	O	O	O	O	O	O	O	O	■

Abbreviations: Ath, *Arabidopsis thaliana*; Mpo, *Marchantia polymorpha*; Pwi, *Prototheca wickerhamii*; Nol, *Nephroselmis olivacea*; Pmi, *Pedinomonas minor*; Ceu, *Chlamydomonas eugametos*; Ppu, *Porphyra purpurea*; Cme, *Cyanidioschyzon merolae*; Ram, *Reclinomonas americana*.
Complex I, NADH : ubiquinol oxidoreductase; Complex II, succinate : ubiquinone oxidoreductase; Complex III, ubiquinol : cytochrome *c* oxidoreductase; Complex IV, cytochrome *c* oxidase; Complex V, ATP synthase.
■, gene present; O, gene absent; □, pseudogene.
*, Gene assumed to have been present in the mtDNA of the common ancestor of red and green algae by virtue of its presence in one or more of the green plant/red algal mitochondrial genomes listed in the table.

eukaryotic genomes contain large amounts of non-coding DNA these questions have to be answered by considering both genome size and the number of proteins that are encoded.

In contrast to the situation with bacterial genomes, only a very limited number of genomes from microbial eukaryotes have been sequenced. Thus it is difficult to specify what constitutes the minimal genome size for a free-living eukaryote. However, a eukaryotic parasite (*Encephalitozoon cuniculi*) has a genome of only 2.9 Mb and that of a close relative, *Encephalitozoon intestinalis*, is estimated to be 2.3 Mb. These genomes contain very little repetitive DNA other than for rDNA and probably contain less than 2000 genes (Vivares & Metenier 2000). This is still 7–8 times more than for the minimal prokaryotic genome and it is hard to predict how many of these genes will be involved with mitosis and related events. Nevertheless, the change from a prokaryote to a eukaryote does not require a major increase in genome size.

Of the multicellular organisms whose genomes have been sequenced, *Arabidopsis*, *Caenorhabditis* and *Drosophila* encode approximately similar numbers of proteins (11 000–18 000). This may be the minimal complexity required by extremely diverse multicellular eukaryotes to execute development and respond to their environment (*Arabidopsis* Genome Initiative 2000). Among the vertebrates, the Japanese puffer fish (*Fugu rubripes*) has the smallest genome identified to date but has a similar gene repertoire to other vertebrates such as humans. Whereas about 35 000 genes are spread over 3000 Mb of DNA in the human genome, in *Fugu* these same genes are restricted to just 400 Mb. Analysis of the *Fugu* genome reveals that its genes are densely packed. It has only short intergenic and intronic sequences that are devoid of repetitive elements (Venkatesh *et al.* 2000). Nevertheless, the splice points (intron–exon boundaries) are remarkably similar in the two species. At the top of the vertebrate evolutionary tree are the mammals and the smallest mammalian genomes identified to date are those of *Miniopterus*, a species of bat. Their genome size is 50% of that of humans but it is not known if this smaller size is a result of genome compaction or loss of genes.

Comparisons of the major sequenced genomes

The eukaryotic genomes that have been completely sequenced are from organisms that are so distant in evolutionary terms that comparisons can be made only at the gross level. A good starting point is to compare the numbers and types of repetitive elements in the different genomes because it is these, rather than the numbers of genes, that account for the major differences in genome sizes. Table 7.6 shows that the euchromatic portion of the human genome has a much higher density of transposable element copies than the euchromatic DNA of the other three multicellular organisms examined. Furthermore, long and short interspersed nuclear elements (LINEs and SINEs) account for 75% of the repetitive DNA in the human genome, whereas the other genomes have no dominant families. The age of repetitive DNA can be defined in terms of percentage of nucleotide substitution from the consensus sequence: the more substitution then the older the repeat. Using this definition, the human genome is filled with copies of ancient transposons, whereas the transposons in the other genomes tend to be of more recent origin. The accumulation of old repeats is likely to be determined by the rate at which organisms engage in 'housecleaning' through genomic deletion.

At the level of the gene, differences in intron and exon structure can be seen in the different eukaryotic genomes with *Arabidopsis* being significantly different (Table 7.7). The conservation of preferred exon size across the three animal genomes suggests a conserved exon-based component of the splicing machinery. The data on intron size are somewhat deceptive because the figures for average length are skewed by a small number of very long introns.

The number of coding genes in the human sequence is estimated to be about 35 000. This compares with 6000 for a yeast cell, 13 000 for a fly, 18 000 for a worm and 26 000 for a plant. None of the numbers for the multicellular organisms is particularly accurate. Nevertheless, humans do not gain their undoubted complexity over worms and plants by using many more genes. Although there is a higher degree of 'alternative splicing' in human than other species, allowing more proteins to be encoded per gene, the numbers of different proteins do not account for the physical and behavioural differences between species.

Many eukaryotic proteins are composed of a number of different domains. Analysis of the genome sequences reveals that over 90% of the domains that can be identified in human proteins are found in *Drosophila* and *Caenorhabditis* proteins as well. Thus, vertebrate evolution has required the invention of few new domains. Novel combinations of domains, in some cases involving many different domains (Table 7.8), generate novel proteins. Despite the fact that many proteins are mosaics of domains, a high percentage of the proteins encoded by any one of the

Table 7.6 Number and nature of interspersed repeated DNA sequences in different eukaryotic genomes. For details of the different kinds of repeated sequences see p. 27. (Table reproduced with permission from International Human Genome Sequencing Consortium 2001.)

	Human		**Drosophila**		**Caenorhabditis**		**Arabidopsis**	
	Percentage of bases	Approximate number of families	Percentage of bases	Approximate number of families	Percentage of bases	Approximate number of families	Percentage of bases	Approximate number of families
LINE/SINE	33.40	6	0.70	20	0.40	10	0.50	10
LTR	8.10	100	1.50	50	0.00	4	4.80	70
DNA	2.80	60	0.70	20	5.30	80	5.10	80
Total	44.40	170	3.10	90	6.50	90	10.50	160

Table 7.7 Comparison of exons and introns in different eukaryotic genomes.

	Human	Drosophila	Caenorhabditis	Arabidopsis
Average length of coding sequence (bp)	1340	1497	1311	2013
Average intron length (bp)	3300	487	267	170
Most common intron length (bp)	87	59	47	
Average exon length (bp)	120–140	120–140	120–140	250

Table 7.8 Numbers of proteins with multiple, but different, domains in three eukaryotes.

Unique domains per protein	No. of proteins in		
	Drosophila	Caenorhabditis	Saccharomyces
2	1474	1248	402
3	413	335	95
4	156	114	23
5	52	38	4
6	8	9	1
7 or more	4	3	0

sequenced genomes have orthologues in the other sequenced genomes. For example, 60% of the predicted human proteins have sequence similarities to yeast, fly, nematode or plant proteins and 61% of fly proteins, 43% of nematode proteins and 46% of yeast proteins have human counterparts. As has been found in prokaryotic genome sequencing projects, a significant proportion of the genes identified in eukaryotic genomes have, as yet, no counterparts in the databases.

Venter *et al.* (2001) identified 1523 human proteins that had strict orthologues in both *Drosophila* and *Caenorhabditis*. The distribution of the functions of these conserved protein sets is shown in Fig. 7.4. Not surprisingly, this set is not distributed in the same way as the whole human set. Compared with the whole human set, there are several categories that are over-represented in the conserved set by a factor of 2 or more and these are the ones expected to be conserved over evolution. The functions that differ significantly between humans and the other sequenced genomes are those involved in acquired immunity, neural development, intercellular and intracellular signalling, haemostasis and apoptosis.

Figure 7.5 shows a comparison of the functional categories of plant (*Arabidopsis*) proteins versus those found in the other sequenced genomes. About 48–60% of the *Arabidopsis* proteins involved in protein synthesis have counterparts in the other species, reflecting highly conserved gene functions. By contrast, only 8–23% of the plant proteins involved in transcription have counterparts in the other species, suggesting independent evolution of many plant transcription factors (*Arabidopsis* Genome Initiative 2000). Two features stand out when comparing the *Arabidopsis* complement of transcriptional regulators with that of other organisms. First, less than 20% of the plant transcription factors are zinc-coordinating proteins compared with *Drosophila* (51%), yeast (56%) and *Caenorhabditis* (64%). Secondly, no single family of transcription factors dominates in plants, whereas the opposite is true in the other eukaryotes (Riechmann *et al.* 2000).

Striking differences are seen in the extent of parology found in the different genomes sequenced to date (Table 7.9). The absolute number of singletons (non-paralogous genes) and gene families is in the same range in the different multicellular organisms. However, the number of gene families with more

(a)

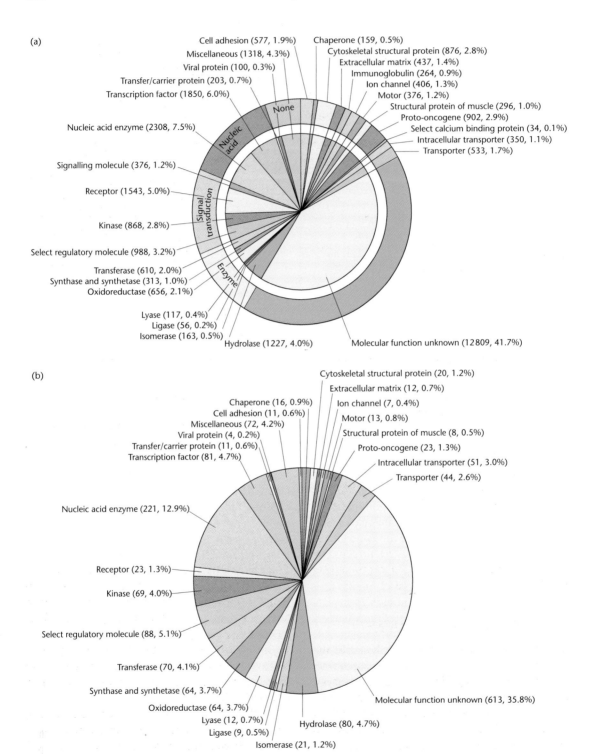

Cell adhesion (577, 1.9%)
Miscellaneous (1318, 4.3%)
Viral protein (100, 0.3%)
Transfer/carrier protein (203, 0.7%)
Transcription factor (1850, 6.0%)

Chaperone (159, 0.5%)
Cytoskeletal structural protein (876, 2.8%)
Extracellular matrix (437, 1.4%)
Immunoglobulin (264, 0.9%)
Ion channel (406, 1.3%)
Motor (376, 1.2%)
Structural protein of muscle (296, 1.0%)
Proto-oncogene (902, 2.9%)
Select calcium binding protein (34, 0.1%)
Intracellular transporter (350, 1.1%)
Transporter (533, 1.7%)

Nucleic acid enzyme (2308, 7.5%)

Signalling molecule (376, 1.2%)

Receptor (1543, 5.0%)

Kinase (868, 2.8%)

Select regulatory molecule (988, 3.2%)

Transferase (610, 2.0%)
Synthase and synthetase (313, 1.0%)
Oxidoreductase (656, 2.1%)

Lyase (117, 0.4%)
Ligase (56, 0.2%)
Isomerase (163, 0.5%)
Hydrolase (1227, 4.0%)

Molecular function unknown (12 809, 41.7%)

None
Nucleic acid
Signal transduction
Enzyme

(b)

Cytoskeletal structural protein (20, 1.2%)
Extracellular matrix (12, 0.7%)
Ion channel (7, 0.4%)
Motor (13, 0.8%)
Structural protein of muscle (8, 0.5%)
Proto-oncogene (23, 1.3%)
Intracellular transporter (51, 3.0%)
Transporter (44, 2.6%)

Chaperone (16, 0.9%)
Cell adhesion (11, 0.6%)
Miscellaneous (72, 4.2%)
Viral protein (4, 0.2%)
Transfer/carrier protein (11, 0.6%)
Transcription factor (81, 4.7%)

Nucleic acid enzyme (221, 12.9%)

Receptor (23, 1.3%)

Kinase (69, 4.0%)

Select regulatory molecule (88, 5.1%)

Transferase (70, 4.1%)

Synthase and synthetase (64, 3.7%)

Oxidoreductase (64, 3.7%)
Lyase (12, 0.7%)
Ligase (9, 0.5%)
Isomerase (21, 1.2%)

Hydrolase (80, 4.7%)

Molecular function unknown (613, 35.8%)

Fig. 7.4 Functional analysis of orthologues in the human, *Drosophila* and *Caenorhabditis* genomes. (a) Distribution of the molecular functions of 26 383 putative human genes. Each slice lists the numbers and percentages of human gene functions assigned to a given category of molecular function. (b) Functions of 1523 putative orthologues common to human, fruit fly and nematode genomes. (Reprinted from Venter *et al.* 2001 by permission of the American Association for the Advancement of Science.)

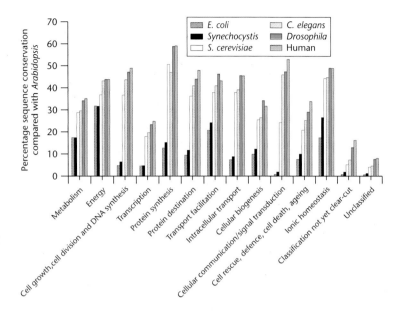

Fig. 7.5 Comparison of functions of genes in *Arabidopsis* with those in other organisms whose genomes have been sequenced. (Redrawn with permission from the *Arabidopsis* Genome Initiative 2000.)

Table 7.9 Proportion of genes in different organisms present as either singletons or in paralogous families. (Reproduced with permission from the *Arabidopsis* Genome Initiative 2000.)

	No. of singletons and distinct gene families	Percentage unique	Percentage of gene families containing				
			2 members	3 members	4 members	5 members	> 5 members
Haemophilus influenzae	1587	88.8	6.8	2.3	0.7	0.0	1.4
Saccharomyces cerevisiae	5105	71.4	13.8	3.5	2.2	0.7	8.4
Drosophila melanogaster	10 736	72.5	8.5	3.4	1.9	1.6	12.1
Caenorhabditis elegans	14 177	55.2	12.0	4.5	2.7	1.6	24.0
Arabidopsis thaliana	11 601	35.0	12.5	7.0	4.4	3.6	37.4

than two members is considerably more pronounced in *Arabidopsis* than other eukaryotes.

Comparisons of gene order

Comparing gene order in different organisms is just one of the many tools that can be used to develop a molecular phylogeny. However, the current set of completely sequenced eukaryotic genomes (human, *Drosophila*, *Caenorhabditis*, *Arabidopsis* and *Saccharomyces*) is too divergent to reveal the dynamics of gene-order evolution. For this reason, Seoighe *et al.* (2000) chose to compare gene order in two yeast species, *S. cerevisiae* and *Candida albicans*. Many examples of inversions of gene order were found but about half of them were single gene inversions. For

example, 103 of the 298 pairs of genes that occur as neighbours in both yeasts had different orientations. By contrast, in prokaryotes inversions tend to involve much bigger sections of the genome and are centred on the origin or terminus of replication (see p. 17). The rate of inversions is much higher in eukaryotes than prokaryotes and the direction of transcription has little effect on gene orientation (Table 7.10).

At the opposite end of the eukaryotic spectrum, Dehal *et al.* (2001) undertook a comparison of gene order between human chromosome 19 (HSA19) and related regions in the mouse genome. HSA19 exists as a single conserved linkage group in most primates and linkage within each chromosome arm is highly conserved (syntenic) in dogs, cattle

Table 7.10 Comparison of gene-order conservation between two yeasts and two bacteria. (Reproduced with permission from Seoighe *et al.* 2000).

	Candida albicans– *Saccharomyces cerevisiae*	*Haemophilus influenzae–* *Escherichia coli*
No. genes (genome A/genome B)	9168/5800	1709/4289
Elongation factor 1 α identity (percentage)*	91	93
No. shared orthologues	3960 (68%)	1330 (78%)
Conserved pairs (percentage)†	9	36.2
Conserved pairs including gene orientation (percentage)	6	35.7
Gene orientation of conserved pairs (→→/←→/→←)‡	1/0.76/0.99	1/0.11/0.0

* The protein sequence conservation within the pairs of species, as measured by the sequence identity of elongation factor 1α, is similar.

† A conserved gene pair is defined as an adjacent pair of genes in genome A that is both present and adjacent in genome B. In determining whether two genes are adjacent, genes that are not shared between the two species are ignored.

‡ The largest fraction of conserved pairs with conserved direction of transcription was set to one, the other fractions are relative to this. The data clearly show that prokaryotes show a higher degree of gene-order conservation in general than *C. albicans* and *S. cerevisiae*, specifically regarding the conservation of the orientation of genes in conserved pairs.

and sheep. By contrast, the HSA19-homologous sequences in mouse are dispersed over four chromosomes but there still is extensive conservation of gene order. However, despite identical gene content, several HSA19 regions are substantially larger than the related intervals in mouse. The principal reason is differences in the number of repetitive elements, particularly SINEs.

The syntenic relationship between human and other mammalian genomes is not too surprising. More surprising is that a similar relationship exists for the human and zebrafish genomes (Barbazuk *et al.* 2000). A major benefit of synteny is that information on gene location in a highly mapped organism can be used to place the corresponding gene in a poorly mapped relative.

Other aspects of comparative genomics

Lateral gene transfer from bacteria

Detailed examination of the complete human DNA sequences suggested that between 113 and 223 genes have been transferred from bacteria to humans or one of our vertebrate ancestors (Inter-national Human Genome Sequencing Consortium 2001). The evidence for such lateral gene transfer is the presence in the human genome of sequences very similar to bacterial genes but their absence from non-vertebrate genomes. Lateral gene transfer is not the only explanation for these results. Random independent loss of genes in different lineages could yield similar distribution patterns. When the number of non-vertebrate genomes screened against the human genome is increased there is a decrease in the number of possible examples of lateral gene transfer (Salzberg *et al.* 2001).

An alternative way to re-assess the possibility of lateral gene transfer from bacteria is to construct phylogenetic trees for each candidate gene (Andersson *et al.* 2001). If a vertebrate gene sequence is nested within a robust cluster of bacterial sequences then the most probable explanation is that the gene in question was laterally transferred from bacteria. One possible example of such lateral gene transfer is a putative *N*-acetylneuraminate lyase gene. One version of this gene has been shown to have been transferred from bacteria into the protozoan parasite *Trichomonas vaginalis*. The vertebrate version of this gene clusters unequivocally with genes from the bacteria *Yersinia pestis* and *Vibrio cholerae* (Fig. 7.6). This is indicative of lateral

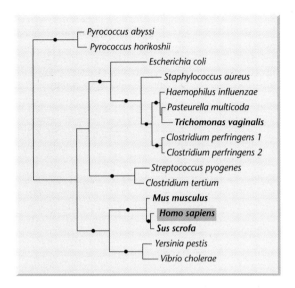

Fig. 7.6 Phylogeny of the gene encoding *N*-acetylneuraminate lyase among prokaryotes, a protozoan (*Trichomonas vaginalis*) and three vertebrates (mouse, human and pig). (Reprinted from J.O. Andersson *et al.* 2001 by permission of the American Association for the Advancement of Science.)

gene transfer involving bacteria and eukaryotes that is independent of the bacteria to *Trichomonas* transfer.

Phylogenetic footprinting

Most of the preceding discussion has been concerned with the comparison of protein sequences. The reason for this is that the redundancy in the genetic code means that protein sequences evolve at a much slower rate than DNA sequences and hence it is much easier to detect orthologues and paralogues at the protein level. Nevertheless, comparisons of DNA sequences also can be informative because gene sequences tend to evolve at a much slower rate than non-coding DNA sequences. The reason for this is that gene products are subject to selective pressure. Comparative DNA sequence analysis exploits this local difference in mutation rates to identify functional elements such as genes, regulatory sequences, splice sites and binding sites. This is accomplished by detecting orthologous DNA sequences, a process known as *phylogenetic footprinting*.

When selecting species for comparative sequence analysis, one challenge is that the species selected should be sufficiently diverged that functional elements stand out from less conserved non-functional sequences but be sufficiently close that orthologous functional elements have not been lost by evolution. An additional issue is that regulatory elements are quite short and can be difficult to detect by alignment amongst the 'noise' generated by the excess of diverged non-functional sequences. Cliften *et al.* (2001) undertook comparative DNA analysis with sequence data derived from *S. cerevisiae* and seven other *Saccharomyces* spp. Not only were they able to detect regulatory elements and RNA genes that do not encode proteins but they also identified small protein-coding genes that had gone undetected in previous sequence analyses. However, to achieve this they needed to compare four species simultaneously and use a motif discovery algorithm as well as an alignment algorithm.

Blanchette (cited in Tompa 2001) has taken phylogenetic footprinting one step further by abandoning alignment altogether, and using specialized algorithms that find the most conserved motifs among the input sequences. This permitted the detection of 9 bp regulatory elements upstream of the *rbcS* gene in 10 plant species (three monocots, seven dicots). The algorithms were also able to detect regulatory sequences in diverse vertebrate species.

Shabalina *et al.* (2001) have analysed the sequences of 100 pairs of orthologous intergenic regions from the human and mouse genomes and found that approximately 17% of the nucleotides are invariant. This is about threefold higher than for *Caenorhabditis* intergenic regions and suggests that intergenic DNA might be functionally important.

Where do new genes come from?

Analysis of the sequenced eukaryotic genomes shows that they encode thousands of genes that have no counterparts in prokaryotes. What are the sources of these genes? Many 'new' genes are likely to be old ones that have been modified beyond recognition. Examples of this type include domains involved in protein–protein interactions such as von Willebrand A, fibronectin type III, immunoglobulin and SH3 modules (Ponting *et al.* 2000). These domains show extensive proliferation in higher eukaryotes but have only a distant relationship to homologues in prokaryotes and lower eukaryotes.

A clue to another source of new genes could be the observation that many of the new eukaryotic domains are α-helical. Such domains could have evolved from the condensed coiled structures that are present in prokaryotes but especially abundant in eukaryotes (Koonin *et al.* 2000). The origin of some other proteins is totally unexpected as exemplified by hedgehog, a key regulator of eukaryotic development. Hedgehog appears to have evolved by fusion of an intein with an extensively modified metalloprotease domain. However, the source of many other genes remains a mystery.

One source of new genes could be transposable elements. When Nekrutenko & Li (2001) examined 13 799 human gene sequences they found that 533 (4%) protein-coding regions contained transposable elements or fragments of them. These transposable elements could have been inserted directly into an exon or could have been inserted into an intron and subsequently recruited as a new exon. A more detailed examination of those genes showed that only 10% of them contained the transposable element. In the remainder, the transposable element was recruited as a novel exon, an event made possible by the presence on transposable elements of potential splice-sites. Thus, insertion of transposable elements might be one cause of the high frequency of alternative splicing in human protein-coding genes. More interesting, Nekrutenko & Li (2001) found two mouse genes that have one or many transposable elements in the protein-coding regions and that have no human or rat orthologues. Thus, insertion of transposable elements could create new genes.

Suggested reading

Gray M.W. (1999) Evolution of organellar genomes. *Current Opinion in Genetics and Development* **9**, 678–687.

Gogarten J.P. & Olendzenski L. (1999) Orthologs, paralogs and genome comparisons. *Current Opinion in Genetics and Development* **9**, 630–636.

These two reviews give a good overview of the use of genomics information to understand the evolutionary process.

International Human Genome Sequencing Consortium (2001) Initial sequencing and analysis of the human genome. *Nature* **409**, 860–921.

Venter J.C. *et al.* (2001) The sequence of the human genome. *Science* **291**, 1304–1351.

Arabidopsis Genome Initiative (2000) Analysis of the genome sequence of the flowering plant *Arabidopsis thaliana*. *Nature* **408**, 796–813.

These three papers contain a wealth of information on comparative genomics, including relevant data from earlier eukaryotic genome sequencing projects.

Bennetzen J. (2002) Opening the door to comparative plant biology. *Science* **296**, 60–78. *The draft sequence of the rice genome was published after this book went to print. This article provides a commentary on comparative cereal genomics and makes comparisons with the* Arabidopsis *genome.*

Wood V. *et al.* (2002) The genome sequence of *Schizosaccharomyces pombe*. *Nature* **415**, 871–80. *This paper was published after this book went to print. As well as presenting an analysis of the genome of* S. pombe *(a fission yeast) it provides a comparison with the genome of* S. saccharomyces *(a budding yeast).*

Read T.D. *et al.* (2002) Comparative genome sequencing for discovery of novel polymorphisms in *Bacillus anthracis*. *Science* **296**, 2028–33. *This fascinating paper describes how comparative genomics has been used to identify the origin of the strains of anthrax used in recent acts of bioterrorism.*

Copeland N.G., Jenkins N.A. and O'Brien S.J. (2002) Mmu 16 – comparative genomic highlights. *Science* **296**, 1617–18. *This short paper provides a useful comparative summary of the human and mouse genomes based on a number of recent publications.*

Shankar N., Baghdayan A.S. and Gilmore M.S. (2002) Modulation of virulence within a pathogenicity island in vancomycin-resistant *Enterococcus faecalis*. *Nature* **417**, 746–50. *A fascinating paper describing how a commensal organism can become a pathogen through the acquisition of new traits.*

Useful websites

http://news.bmn.com/alerts
At this site it is possible to subscribe to an on-line Comparative Genomics Newsletter that provides short reviews on many of the topics in this chapter.

http://www.ornl.gov/hgmfs/faq/compgen.html
http://www.llnl.gov/str/Stubbs.html
These two websites give a general overview of the comparative genomics of mice and men.

http://pga.lbl.gov/overview.html
This site provides a useful entry into a different aspect of comparative genomics: its use to discover the role of *cis*-regulating elements in human disease.

http://www.infobiogen.fr/services/dbcat
This site presents a catalogue of the many different databases that are available.

CHAPTER 8

Protein structural genomics

Introduction

Following the structural annotation of a genome, the next step is the *functional annotation* of each new gene. As discussed in Chapter 6, one way in which tentative functions can be assigned to new genes is to compare their sequences to genes, cDNAs and expressed sequence tags (ESTs) already stored in databases. However, a large number of genes fail to match any previously identified sequence, and in these cases alternative methods are required for functional annotation. One way in which functions can be assigned to anonymous genes (also known as *orphans* or *ORFans* because they do not belong to any known gene family) is to examine the structure of the encoded protein, known as a *hypothetical protein*, and compare it to protein structures that have already been determined. This is because three-dimensional protein structures are much more strongly conserved in evolutionary terms than primary amino acid sequences, and it is the three-dimensional protein structure (rather than the primary sequence) which actually carries out the biochemical function of the molecule.

The determination of protein structure was once regarded as the ultimate stage in any biological study. Such analysis would only be undertaken when there was a clear and thorough understanding of the function of the protein, and the desire was to understand the functional implications of its structure and interactions at the atomic level. *Structural genomics* (also known as *structural proteomics*)[1] is a relatively new field which has turned this idea on its head. The aim of structural genomics is to determine the three-dimensional structures of large numbers of proteins, often before any other biological information is available, providing data that will reveal distant evolutionary relationships between orphan genes and those whose functions are known. This brings the determination of protein structure to the beginning of the investigative process (Burley 2000; Skolnick & Kolinski 2000; Brenner 2001; Norin & Sundstrom 2002).

Determining gene function by sequence comparison

As the sequencing phase of each genome project nears completion, potentially all the protein coding sequences in the organism become available (Chapter 6). The quickest way to assign a probable function to a new gene or cDNA is to carry out *pairwise comparisons* to all other sequences existing in databases using alignment tools such as BLAST (Altschul *et al.* 1990) or FASTA (Pearson & Lipman 1988). Conserved sequence often indicates conserved function, so if cDNA X is very similar in sequence to cDNA Y, and cDNA Y encodes a protein phosphatase, then it is likely that cDNA X also encodes a protein phosphatase. As discussed in Chapter 6, such analysis cannot usually reveal the precise function of a gene in a cellular or whole organism context, only its biochemical category. Further experiments would be necessary for a more in-depth functional understanding, e.g. by determining the expression pattern of the gene and mutant phenotype (see later chapters). Even so, the knowledge that the gene product is a phosphatase would help in developing experimental strategies to investigate its function more thoroughly.

Powerful though tools such as BLAST and FASTA are, they are only really useful when there is > 30% sequence identity between the query sequence and entries in the sequence databases. As the

---

[1] The term 'structural genomics' was initially used to describe the structural phase of the genome projects, i.e. physical mapping, clone contig assembly and sequencing (McKusick 1997). More recently, the term has been adopted by the structural biology community to describe genome annotation initiatives based on protein structure determination (Teichmann *et al.* 1999). We use the term in the latter sense in this chapter.

evolutionary relationship between sequences becomes more distant, pairwise comparisons tend to
become less reliable. Thus, where there is 20–30%
identity, over half of all evolutionary relationships
fail to be detected by these standard alignment tools
(Brenner *et al.* 1998). One way to address this problem is to carry out multiple sequence comparisons,
which use the characteristics of sets of related
sequences as a search query. PSI-BLAST (Position-
Specific Iterated BLAST) is one commonly used
algorithm (Altschul *et al.* 1997). Initially, a standard
BLAST search is carried out with an orphan gene as
the query. This reliably finds matches to any homologues with > 30% identity and may also find some
more distantly related sequences. However, the program then generates a profile of the set of collected
sequences and carries out a second search, allowing
more homologues to be collected. The process can be
repeated indefinitely, and has been shown to detect
three times as many homologous relationships as
pairwise comparisons (Martin *et al.* 1998; Park *et al.*
1998). As the power of these tools increases, however, it becomes more important to screen matches
for false positives, which can be generated by the
most common structural domains in proteins. These
are known as low complexity regions (LCRs). For
example, amino acid sequences corresponding to
transmembrane helices and coiled-coil domains are
so common that they are found in many unrelated
proteins (Huynen *et al.* 1998b). Another potential
downfall is pollution in sequence databases. It is
known that all databases contain a small but significant proportion of errors, so even a confident hit
generated by an anonymous gene or cDNA can be
misleading. It is worthwhile considering that annotating a new gene solely on the basis of someone's
annotation of an existing gene can propagate and
reinforce any errors that have been made (Bork &
Bairoch 1996; Zhang & Smith 1997; Brenner 1999).

Determining gene function through conserved protein structure

Sequence comparison is a very powerful method,
but even with sophisticated multiple alignment tools,
< 50% of newly discovered genes can be matched to
previously identified genes with known functions.

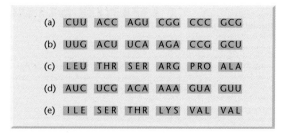

Fig. 8.1 Degeneracy in nucleotide and amino acid sequences.
Up to six codons may specify the same amino acid, with the
result that nucleotide sequences a and b, which are conserved
at only eight out of a possible 18 positions, specify exactly the
same peptide (c). Nucleotide sequence d, which matches
sequence a at only four positions, encodes a different peptide
(e), but one in which the amino acids at each position have
very similar chemical properties to those in c. These two
peptides would be likely to fold in similar ways.

Although the success rate is likely to improve as
more information accumulates in the databases, it
is still a significant bottleneck to functional annotation. For example, it was shown that even 2 years
after the complete genome sequence of the yeast
Saccharomyces cerevisiae became available, the number of orphan open-reading frames had not been
significantly reduced (Fischer & Eisenberg 1999a).
This is despite the massive increase in available
sequence data from other organisms over the same
time period, and suggests that a substantial number
of gene families exist with no associated functional
information in any of the sequenced genomes (Jones
2000). It is here that the determination of three-
dimensional protein structure can help. Structure
is much more strongly conserved than sequence
because of *degeneracy*, a concept that is most often
applied to nucleotide sequences but is also relevant
to protein structures. In the genetic code, most
amino acids are specified by more than one codon,
and in some cases up to six codons are used. Therefore, a certain amount of divergence in nucleotide
sequence can occur without altering the primary
amino acid sequence of a protein (Fig. 8.1). Similarly, chemically similar amino acids such as
leucine, isoleucine and valine can be regarded as
conserved residues rather than mismatches when
comparing protein sequences, because they would
be expected to influence the folding of the polypep-

Box 8.1 Protein substitution matrices

Nucleotide sequences are compared on the basis of identity, with gaps being introduced where necessary to achieve alignment (see Box 6.3, p. 104). With amino acid sequences, sequence alignment is more complex because replacing one amino acid with another can have a range of effects. In conservative changes, one amino acid is replaced with another that has similar chemical properties and this would be expected to have a minor effect on the folding of the polypeptide backbone and thus conserve the structure of the protein. On the other hand, replacing a small amino acid such as valine with another carrying a large polar side chain, such as aspartate, would have a more significant effect. This would be known as a non-conservative substitution. To take this into account, amino acid sequences are compared not only on the basis of percentage identity, but also on the basis of percentage similarity.

Similarity is calculated using protein substitution matrices, which are based on calculations of the relative mutability of amino acids. In general, conservative changes are more common in

evolutionary terms, so the pattern of changes observed in collections of related amino acid sequences reveals how likely it is for one amino acid to be substituted for another. Originally, a set of calculations were made by Dayhoff (1978), known as the percentage of accepted point mutations (PAM). These were based on mutational preferences in aligned proteins with a certain level of sequence identity. This matrix is rarely used nowadays because it was based on a small collection of proteins within which globular and other soluble proteins were over-represented. More recently, a substitution matrix has been calculated using a larger and more representative protein set (Gonnet *et al*. 1992). The most widely used set of matrices is BLOSUM (Henikoff & Henikoff 1993) which uses blocks of aligned amino acid sequences. For example, BLOSUM80 is a matrix in which blocks of amino acids showing 80% sequence identity are aligned. Most of the standard homology search programs, including the BLAST and FASTA derivatives for protein sequence comparisons, use BLOSUM matrices to calculate similarity levels.

---

tide backbone in a similar manner (Fig 8.1; Box 8.1). However, the level of degeneracy tolerated in terms of converting a primary amino acid sequence into a three-dimensional structure is much greater than this. Indeed, only a small number of residues in a globular protein are actually associated with protein function, and these may vary in their position within a sequence provided their spatial arrangement is conserved (Kassua & Thornton 1999). Thus, even sequences with very few conserved residues can generate similar structures (Finkelstein & Ptytsin 1987). It has been estimated that 20–30% of orphan genes could be annotated immediately by determining the structures of the encoded proteins and comparing these to known structures (Bork & Koonin 1998; Blundell & Mizuguchi 2000).

A hypothetical example of the power or structural genomics was discussed in a recent review by Shapiro & Harris (2000) using the proteins

haemoglobin and myoglobin. The solved structures of these proteins are very similar, reflecting the fact that they are both oxygen carriers (Aronson *et al*. 1994) (Fig. 8.2). However, the sequences are so divergent that a BLAST search with myoglobin would not elicit a hit on α- or β-globin, the components of haemoglobin (Fig 8.2). If the myoglobin sequence was to be obtained for the first time as part of a genome project, how would it be annotated? Sequence comparison would not be appropriate, but structural comparison would clearly show similarities to haemoglobin, allowing myoglobin to be classified as an oxygen carrier.

Conserved protein structure may imply conserved function but how strong is the likelihood? The frequency with which similar structure corresponds to similar function has been calculated by Koppensteiner *et al.* (2000). These investigators compared the structures and functions of a large

(a)

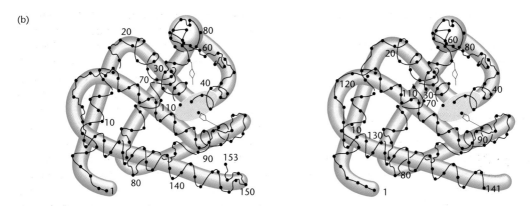

```
  1    V L S P A D K T N V K A A W G K V G A H A G E Y G A E A L E R M F L S F P T T K T Y F P H F – – – –     4 6
       . | | . . : . . . | . . . | | | | . | . . . . : | . | . | . | . : | . . . | . | . . . | . . |
  1    G L S D G E W Q L V L N V W G K V E A D I P G H G Q E V L I R L F K G H P E T L E K F D K F K H L K     5 0


 4 7   – – D L S H G S A Q V K G H G K K V A D A L T N A V A H V D D M P N A L S A L S D L H A H K L R V D     9 4
       | . . . . | . . : | . | | . . | . . | | . . . . : . . . . . . . | : . . | | . | . : :
 5 1   S E D E M K A S E D L K K H G A T V L T A L G G I L K K K G H H E A E I K P L A Q S H A T K H K I –     9 9


 9 5   P V N F – K L L S H C L L V T L A A H L P A E F T P A V H A S L D K F L A S V S T V L T S K Y R       1 4 1
       | | . : : . : | . | : : . . | . : . . | . : | . . . . . . : : : | . | . . . . . . . : . | . | :
1 0 0   P V K Y L E F I S E C I I Q V L Q S K H P G D F G A D A Q G A M N K A L E L F R K D M A S N Y K E L G F Q G   1 5 3
```

(b)

Fig. 8.2 (a) Alignment of human α-globin and human myoglobin using the EBI EMBOSS-Align program (http://www.ebi.ac.uk/emboss/align/). Note the sequences are very different (26% identity, 39% similarity). (b) Despite divergence at the sequence level, the three-dimensional structures are remarkably conserved.

number of protein folds and found correlation between structure and function in 66% of cases. This means that, if a hypothetical protein is structurally similar to a characterized protein, there is a 66% chance that the functions are also related. On the other hand, there is a 33% chance that they are not, and the relationship in structure is coincidental. This is partly because a number of protein folds are known to serve a variety of functions. These common structures are known as *superfolds* (Orengo *et al.* 1994). For example, the triose phosphate isomerase (TIM) barrel, which consists of eight α-helices and eight β-sheets, is associated with no less than 16 different functions (Hegyi & Gernstein 1999). It is found in most enzymes and if this struc-

ture is revealed in a new protein, it is generally of no practical help in functional annotation.

Approaches to protein structural genomics

The ability to annotate orphan genes systematically on the basis of protein structure is highly desirable. However, there are four points to consider in the development of a structural genomics initiative.

1 How to categorize protein structures systematically.

2 How to determine the structures of hypothetical proteins in a high throughput manner.

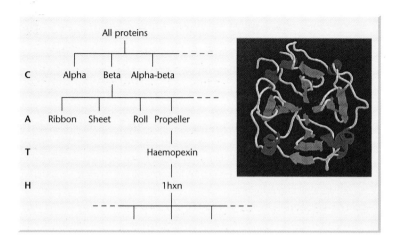

Fig. 8.3 The CATH (Class, Architecture, Topology, Homologous superfamily) hierarchical protein structure classification scheme. This example classification is featured on the CATH homepage (http://www.biochem.ucl.ac.uk/bsm/cath/lex/cathinfo.html), and shows the single domain haemopexin protein 1hxn (Baker *et al.* 1993). In CATH this protein is classified as C (mainly β) A (propeller) T (haemopexin) H (1hxn). Note that the classification system can branch further after level H.

3 How to use structural information for functional annotation.
4 How to choose appropriate targets for a structural genomics programme.
We consider these issues in turn below.

Classification of protein structures

The functional unit of a protein is known as a *domain*. Some proteins feature a single domain, whereas others have multiple domains which can function independently (see Chapter 7). Before attempting to use structural information as a means to assign functions to hypothetical proteins, it is necessary to have a systematic way of classifying protein structure. One way to do this would be to identify particular structural features, which are known as *folds*, and group protein domains according to which folds are present. This is not as easy as it sounds because there may be a continuous range of intermediate structures between two particular fold types, especially for those folds that appear in many proteins; this is known as the *Russian doll effect*. It is estimated that there are about 1000 superfamilies of protein folds in total (Brenner *et al.* 1997; Zhang & DeLisi 1998; Wolf *et al.* 2000b) although much higher numbers have been suggested (e.g. see Swindells *et al.* 1998). Classification schemes are generally hierarchical: at each level, proteins are classified according to the particular folds they possess, allowing new proteins to be assigned to the appropriate category (Fig. 8.3). There are various different classification schemes based on pairwise

structural comparisons which involve arbitrary thresholds of similarity. Consequently, different schemes can place the same protein in different families or subfamilies according to these cutoff levels. In a recent study, three structural classification schemes have been compared (Hadley & Jones 1999). These are Structural Classification of Proteins (SCOP; Murzin *et al.* 1995), Class, Architecture, Topology and Homologous superfamily (CATH; Orengo *et al.* 1997) and Fold classification based on Structure–Structure alignment of Proteins (FSSP; Holm & Sander 1996). SCOP uses a number of parameters, including human knowledge of evolutionary relationships. CATH and FSSP rely entirely on sequence comparisons and geometric criteria. Perhaps because of these differences, only about 60% of the proteins investigated showed the same results with all three classification schemes. Over one-third were placed in different groupings by the three schemes (Hadley & Jones 1999).

High throughput determination of protein structure

Solving protein structures experimentally

The most direct way to determine the three-dimensional structure of a hypothetical protein is to solve it using atomic resolution techniques, the principal methods being X-ray crystallography and nuclear magnetic resonance (NMR) spectroscopy (Box 8.2). However, both these procedures are notoriously

Box 8.2 Methods for solving protein structures

X-ray crystallography

The basis of X-ray crystallography is the ability of a precisely orientated protein crystal to scatter incident X-rays onto a detector in a predictable manner. The scattered X-rays can positively or negatively interfere with each other, generating signals called *reflections* (see Fig. B8.1). The nature of the scattering depends on the number of electrons in an atom, and electron density maps can thus be reconstructed using a mathematical function called the *Fourier transform*. The more data used in the Fourier transform, the greater the resolution of the technique, and the more accurate the resulting structural model. Structural determination depends on both the amplitude and phase of the scattering, but only the amplitude can be calculated from the intensities of the diffraction pattern. For a long time, the only way in which this phase problem could be addressed and macromolecular structures completely solved was *multiple isomorphous replacement*, i.e. the use of heavy atom derivatives of the protein crystal. The incorporation of heavy atoms into a protein crystal

by soaking alters the diffraction intensity allowing the phase to be calculated.

More recently, the laborious process of finding heavy atom derivatives has been superceded by techniques such as *multiwavelength anomalous diffraction* (MAD). MAD takes advantage of the presence of anomalously scattering atoms in the protein structure. The magnitude of anomalous scattering varies with the wavelength of the incident X-rays, and thus requires precisely tuneable incident X-radiation, which can be found only at *synchrotron radiation sources*. Isomorphous crystals can be produced rapidly by incorporating the amino acid derivative selenomethionine into the expressed protein (Hendrickson *et al*. 1990). Anomalous diffraction data collected at several wavelengths can then be used to determine both the phase and amplitude of the scattering, allowing high-throughput determination of protein structures. Recently, it has been reported that under optimal conditions a protein crystal structure can be solved using MAD analysis in about 30 min (Walsh *et al*. 1999).

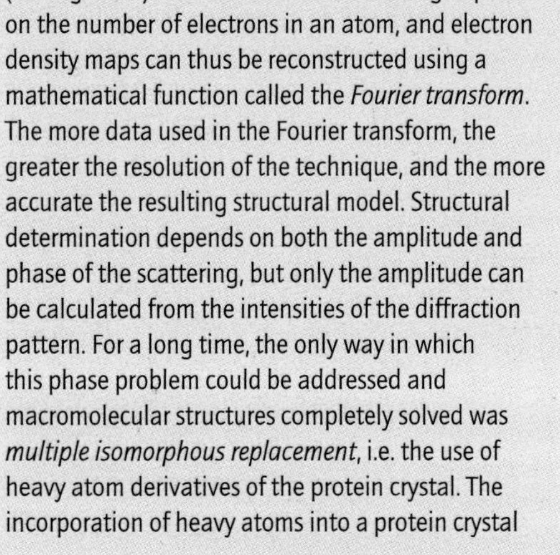

Nuclear magnetic resonance spectroscopy

NMR spectroscopy is used to determine the structure of proteins in solution. The basis of NMR is that some atoms, including natural isotopes of nitrogen, phosphorus and hydrogen, are intrinsically magnetic and can switch between magnetic spin states in an applied magnetic field. This is achieved by the absorbance of electromagnetic radiation, generating magnetic resonance spectra (see Fig. 2). While the resonance frequency for different atoms is unique, it is also influenced by the surrounding electron density. This means that magnetic resonance frequency is shifted in the context of different chemical groups (a so-called *chemical shift*) allowing the discrimination between e.g. methyl and aromatic groups. The manner in which NMR decays depends on the structure and spatial configuration of the molecule. For example, the *nuclear Overhauser effect* (NOE) results from the transfer of magnetic energy through space if interacting nuclei are < 0.5 nm

Fig. B8.1 X-ray diffraction image of a protein phosphatase. (Courtesy of Daniela Stock, MRC Laboratory of Molecular Biology, Cambridge.)

continued

Box 8.2 *continued*

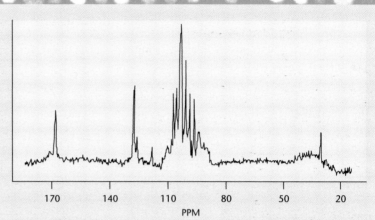

Fig. B8.2 Typical appearance of an NMR spectrum. This data can be built into a series of structural models based on atomic distance constraints.

apart. NOE spectroscopy (NOESY) shows proximal atoms as symmetrical peaks superimposed over the typical NMR spectrum. In this way, the analysis of NMR spectra can determine the three-dimensional spatial arrangement of atoms, and can allow a set of distance constraints to be built up to determine protein structure. The technique is generally applicable only to small proteins (< 25 kDa) although recent innovations have stretched this limitation to about 100 kDa (see main text).

Circular dichroism spectroscopy

Circular dichroism (CD) describes the optical activity of asymmetric molecules as shown by their differing absorption spectra in left and right circularly polarized light. CD spectrophotometry between 160 and 240 nm allows the characterization of protein secondary structure, because α-helices and β-sheets generate distinct spectra. With the recent application of synchrotron radiation CD (SRCD) to protein structural studies, this technique has emerged as a useful complement to X-ray crystallography and NMR spectroscopy for the analysis of protein structure. The use of the technique in structural genomics has been recently reviewed (Wallace & Janes 2001).

slow, involving many laborious preparative steps that have to be optimized for each protein analysed. A second bottleneck is data acquisition, especially for X-ray crystallography, where synchrotron radiation sources are required. A third bottleneck is data processing, i.e. converting the X-ray diffraction patterns and NMR spectra into structures that can be submitted to the protein structural databases. Recent advances in all these areas have made high-throughput structural determination a realistic goal.

Both X-ray crystallography and NMR spectroscopy require milligram amounts of very pure protein. The first challenge in structural genomics is therefore to find suitable expression systems and purification procedures which are widely applicable and produce proteins that fold properly to achieve their native state. Advantages of heterologous expression systems include the ease with which amino acid derivatives can be incorporated, as these are required for procedures such as multiwavelength anomalous diffraction phasing (Box 8.2), and the ability to include an affinity tag, which greatly simplifies protein purification (reviewed by Stevens 2000a). Otherwise, optimal chromatography procedures must be established for every protein, a significant bottleneck in sample preparation (Heinemann *et al.* 2001; Mittl & Grutter 2001). Various systems have been tested, including *Escherichia coli*, yeast, insect cells and cell-free systems (reviewed by Edwards *et al.* 2000). *E. coli* has the advantage of low cost, which is an important consideration in a large-scale structural genomics programme. However, eukaryotic systems are advantageous where protein folding requires chaperons and where post-translational modification is required to achieve the native structure.

Successful X-ray crystallography is dependent on the production of high quality protein crystals. Defining conditions to achieve crystallization has long been regarded as an art, and typically involves the laborious trial and error testing of hundreds of different conditions. A major recent breakthrough is the development of robotic workstations that can process hundreds of thousands of crystallization experiments in parallel, allowing rapid determination of optimum crystallization conditions (Stevens 2000b; Terwilliger 2000; Mueller *et al.* 2001). Automated handling of crystals can maximize the use of synchrotron radiation sources therefore increasing the number of crystals that can be analysed in a particular session (Abola *et al.* 2000). The final bottleneck in X-ray crystallography, that of data processing and structure determination itself, is being addressed with the development of powerful new software packages (reviewed by Lamzin & Perrakis 2000).

NMR spectroscopy is carried out in solution and is therefore suitable for the analysis of proteins that are recalcitrant to crystallization. Until recently, the technique was applicable only to proteins with a low molecular weight (< 25 kDa) but a number of advances in instrumentation have helped to increase this threshold significantly (see review by Riek *et al.* 2000). Following sample preparation, the major bottlenecks in NMR structure determination are data recording and analysis. Data acquisition is becoming quicker with improvements in instrument design, particularly novel probes (Medek *et al.* 2000) and higher frequency magnets (Lin *et al.* 2000). These give higher signal to noise ratios and greater resolution. Data analysis times, typically ranging from weeks to months, are being cut down through the use of integrated software packages for spectral analysis and model building (e.g. Koradi *et al.* 1998; Xu *et al.* 2001).

An important consideration is that increasing the speed at which protein structures are solved should not reduce the quality of the structures generated.

With this in mind, a number of computer programs have been developed for structural validation and checking the consistency of structural data, e.g. PROCHECK (Laskowski *et al.* 1993) and SFCHECK (Vaguine *et al.* 1999).

Prediction of protein structures

An alternative to solving protein structures directly is to predict how a given amino acid sequence will fold. If the sequence with unknown structure (the *target sequence*) shows > 25% identity to another sequence whose structure is known (the *template sequence*), then the structure of the target sequence can be predicted by *comparative modelling* (also called homology modelling). In this method, multiple sequence alignment is carried out with the target protein and one or more templates. Then a computer program such as SWISS-PDBVIEWER is used to align the target structurally on the template. For closely related proteins this is an easy process, but for distantly related proteins with some insertions and deletions, it can be more difficult. The compact core of the protein is always easier to model than the more variable loops, and algorithms are sometimes used that match the loops to databases of loops from other proteins. This is called a 'spare parts algorithm'.

Where the level of sequence identity is too low for comparative modelling, two broad classes of predictive methods can be used: those in which sequences are compared to known structures to test 'goodness of fit' (*fold recognition methods*) and those in which structural information is derived directly from primary sequence data without reference to known structures (ab initio *prediction methods*).

Fold recognition is also known as *threading*, and can be defined as a technique that attempts to determine whether the sequence of an uncharacterized protein is consistent with a known structure by systematically matching its compatibility to a collection of known protein folds (Jones *et al.* 1992, Bryant & Lawrence 1993). Recent algorithms such as GenTHREADER (Jones 1999a) incorporate aspects of both sequence profiling (as used in comparative modelling) and threading, to increase the confidence with which structures are assigned. Fold recognition algorithms are 'trained' on fold libraries, such as those found in structural databases like

Pfam and BLOCKS (Chapter 6), to establish parameters of recognition. This method is widely used to assign tentative functions to the large sets of anonymous cDNA sequences arising from genome projects (e.g. Fischer & Eisenberg 1997, 1999b; Rychlewski *et al.* 1998, 1999). *Ab initio* prediction of tertiary protein structure is still a distant goal, and the algorithms in current use are suitable only for the prediction of secondary structures. However, secondary structure prediction is achieved with reasonable accuracy. For example, the PSIPRED method devised by Jones (1999b) matches solved protein structures in more than 75% of cases. For the prediction of tertiary structures, a mini-threading algorithm (i.e. an *ab initio* method incorporating fold data in the form of fragments from resolved protein structures) has been used to predict successfully the structure of NK-lysin (Jones 1997, reviewed in Jones 2000).

The usefulness of structural modelling depends on the reliability with which the models can accurately predict protein structures. This has been assessed by rigorously comparing several thousand structural predictions with actual solved structures, in a series of Critical Assessment of Structural Prediction (CASP) experiments (reviewed by Moult *et al.* 1999; Murzin 2001). *Ab initio* methods are the least reliable and often fail to predict folds that are identified by fold recognition methods. However, fold recognition is limited by the accuracy with which the sequence of the hypothetical protein can be aligned over that of the template for comparison (reviewed by Moult & Melamud 2000). Programs such as Dali (Holm & Sander 1995) and VAST (Gibrat *et al.* 1996) can be used for structural alignment, as recently reviewed by Szustakowski & Weng (2000).

Using protein structure to predict function

Functional information from structural features of proteins

Once the solved or predicted structure of a protein is available, it is deposited in the Protein Databank (PDB), a universal database of three-dimensional structures (http://www.rcsb.org). By searching through these structures and looking for matches, it is often possible to assign a putative function to the new protein.

Where sequence and structural comparisons both fail to provide functional information, structural analysis can often reveal functional characteristics of proteins at a simpler level. For example, scanning the surface of a protein can reveal clefts that are likely to represent ligand-binding sites (Laskowski *et al.* 1996) or domains that probably interact with other proteins (Jones & Thornton 1995, 1997). This allows the low-resolution classification of proteins, e.g. a protein with a large cleft could be tentatively assigned the designation of 'enzyme'. Higher resolution characterization may be possible by matching the shape of the cleft to a library of small molecular shapes. This can be achieved with the help of drug design software such as DOCK (Briem & Kuntz 1996) and HOOK (Eisen *et al.* 1994) which match potential ligands to binding sites on the protein surface. A number of methods have been published recently that allow the prediction of active sites in enzymes based on the spatial distribution of critical residues in different fold families (Wallace *et al.* 1997; Russell *et al.* 1998).

Other routes to functional annotation

The Rosetta stone

A number of diverse approaches to functional annotation have received interest recently, because they do not, to a large degree, depend on homology between the hypothetical protein and one that has been previously annotated. One such approach is the *domain fusion method*, which searches for functionally related proteins that are separate in some organisms but fused into a single protein in others. The assumption is that two domains in a multidomain protein are likely to be involved in a common cellular function and in those organisms where they are encoded by separate genes, the individual products probably interact. For example, the ligand-binding and kinase domains of a receptor tyrosine kinase are both involved in signal transduction, and they may either form part of the same protein or an oligomeric complex. The 'fused' protein has been dubbed the Rosetta stone because it can reveal relationships between its components. As shown in Fig. 8.4, the Rosetta stone method works as follows. The sequence of protein X is used to screen another genome, revealing homologues of X and perhaps a

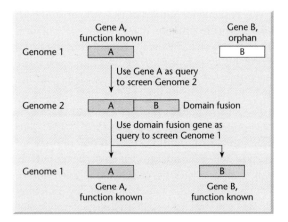

Fig. 8.4 Principle of the domain fusion (Rosetta stone) method of functional annotation. The sequence of gene A, of known function from genome 1, is used as a search query to identify orthologues in genome 2. The search may reveal single-domain orthologues of gene A, but may also reveal domain fusion genes such as AB. As part of the same protein, domains A and B are likely to be functionally related. The sequence of domain B can then be used to identify single-domain orthologues in genome 1. Thus gene B, formally an orphan with no known function, becomes annotated because of its association with gene A.

Rosetta stone protein X–Y. The sequence of Y can then be used to screen the original genome, from where X was obtained. Even in the absence of homology or any structural information, the function of Y can be linked to that of X. Several experiments involving functional annotation in bacteria and eukaryotes by this method have been reported (e.g. Marcotte *et al.* 1999a; Shirasu 1999) and it has proven possible to construct protein interaction maps on this basis (Enright *et al.* 1999). Other methods for protein interaction mapping are considered in Chapter 11.

Coevolution

Another 'guilt by association' method for functional annotation is the conservation of genes in the context of the genome as a whole. One way in which this can be exploited, at least in bacteria, is to look for conservation of neighbouring genes. In bacteria, many genes are found as functionally related (but structurally diverse) operons, the best-characterized

example of which is the lactose operon in *E. coli* (Jacob & Monod 1961). The three products of the lactose operon are enzymes involved in lactose metabolism. If two or more non-annotated genes tend to be found as neighbours in the genomes of several bacteria, it suggests they are linked in an operon and are not only functionally related but that their products probably also interact (Dandekar *et al.* 1998; Overbeek *et al.* 1999).

Another way in which evolution can be exploited in functional genomics is mapping *conserved phylogenetic profiles*. If a gene is found in several aerobic organisms but never in an anaerobic organism, it is likely to be involved in aerobic respiration. If a large number of genomes (e.g. > 10) is subject to phylogenetic analysis, genes that are coinherited are likely to be functionally related (Pellegrini *et al.* 1999). This reflects the fact that most proteins operate as part of a complex, and the loss of one component (e.g. by mutation) results in the failure of the entire complex. If this logic is applied on an evolutionary scale, the loss of one protein would render the others useless and their subsequent loss would be inconsequential. Such differential genome analysis has been used by Huynen *et al.* (1998a) to identify species-specific functions of *Helicobacter pylori*.

Structural genomics programmes

As discussed above, the determination of protein structure provides various routes to the functional annotation of orphan genes. Since 1997, a number of genome-wide initiatives have been launched in which protein fold assignment and homology modelling have been used to predict the structures and functions of genes in a variety of microbial genomes. The genome of *Mycoplasma genitalium* has been a popular target because of the small number of genes (Fischer & Eisenberg 1997; Huynen *et al.* 1998b; Rychlewski *et al.* 1998; Teichmann *et al.* 1998; Wolf *et al.* 1999) and the *Saccharomyces cerevisiae* genome has also been studied in this manner (Sanchez & Sali 1998). However, sequence comparison and structural modelling is not as reliable as solving protein structures directly by X-ray crystallography and/or NMR spectroscopy, and the technology now exists to apply such techniques in a high-throughput manner. Therefore a number of pilot structural

Table 8.1 A selection of structural genomics research consortia and their target organisms.

Programme	URL	Target organisms
Berkeley Structural Genomics Center	http://www.cchem.berkeley.edu/~shkgrp	*Methanococcus jannaschii* *Pyrococcus horikoshii* *Mycoplasma pneumoniae*
Joint Center for Structural Genomics	http://www.jcsg.org/scripts/prod/home.html	*Thermotoga maritima* *Caenorhabditis elegans*
Structural Proteomics (Ontario Clinical Genomics Center)	http://www.uhnres.utoronto.ca/proteomics/	*Methanobacterium thermoautotrophicum*
Structure 2 Function	http://s2f.carb.nist.gov	*Haemophilus influenzae*
Fold Diversity Project	http://proteome.bnl.gov/	Human disease and pathogen proteins
Mycobacterium tuberculosis Structural Genomics Consortium	http://www.doe-mbi.ucla.edu/TB/	*Mycobacterium tuberculosis*
Northeast Structural Genomics Consortium	http://www.nesg.org/	*Saccharomyces cerevisiae* *Caenorhabditis elegans* *Drosophila melanogaster* (and human homologues)
RIKEN Structural Genomics/Proteomics Initiative	http://www.rsgi.riken.go.jp/	*Thermus thermophilus* HB8 *Arabidopsis thaliana* *Mus musculis*
Southeast Collaboratory for Structural Genomics	http://secsg.org/	*Caenorhabditis elegans* *Pyrococcus furiosus* Human

genomics projects have been launched worldwide in which X-ray crystallography and NMR spectroscopy are applied to the systematic determination of protein structures (Table 8.1). Internet resources for structural genomics are listed in Box 8.3.

Perhaps the most fundamental issue regarding these projects is *target selection* (Brenner 2000; Linial & Yona 2000). In the broadest sense, structural genomics aims to provide total coverage of 'fold space' by solving representative structures from every protein superfamily. Once this has been achieved, structural databases will provide a comprehensive resource for the annotation of all future anonymous cDNAs. The problem is that the PDB is highly redundant in that most of the structures that have been added in the last 5 years match existing structures. This suggests that some structures

are very much under-represented. By deliberately choosing to solve the structures of hypothetical proteins, it is hoped to provide complete coverage of protein space, i.e. discover all protein folds in existence. Because of the degeneracy of the 'protein folding code' many different sequences can fold in a similar manner, and the total number of fold types that exist is likely to be relatively small, in the region of 1000 (Brenner *et al.* 1997; Zhang & DeLisi 1998; Wolf *et al.* 2000). It is estimated that up to 10 000 protein structures will have to be determined to meet this goal, because it is likely that many solved structures will emerge to be similar and there is no way to establish this prior to carrying out the experiments (Burley *et al.* 1999; Linial & Yona 2000). Research has focused on microbes, which have smaller genomes (and thus smaller proteomes) than higher

Box 8.3 Internet resources for structural genomics

Protein structure databases

http://www.rcsb.org Protein databank (PDB)

Protein structure classification schemes

http://scop.mrc-lmb.cam.ac.uk/scop/ SCOP
http://www.biochem.ucl.ac.uk/bsm/cath/
    lex/cathinfo.html CATH
http://www.ebi.ac.uk/dali/fssp/ FSSP

Technology development for structural genomics

http://www.nysgrc.org/ New York Structural
Genomics Research Consortium

Structure prediction resources

http://bioinf.cs.ucl.ac.uk/psipred *Ab initio* structure
    prediction tool
http://trantor.bioc.columbia.edu/cgi-bin/SPIN/
    A searchable database of Surface Properties of
    protein–protein Interfaces (SPIN)

---

eukaryotes, but a fundamentally similar basic set of protein structures. Several groups have chosen thermophilic bacteria such as *Methanococcus jannaschii* for their pilot studies, on the basis that proteins from these organisms should be easy to express in *E. coli* in a form suitable for X-ray crystallography and NMR spectroscopy (e.g. see Kim *et al.* 1998). In contrast, higher eukaryotes have a much larger proteome, much of which consists of redundant structures. A favourable strategy in model eukaryotes is to focus on proteins that are implicated in disease processes. This would include both human proteins encoded by 'disease genes' and proteins of pathogens that are responsible for disease. The rationale is that such research is more likely to receive generous funding from pharmaceutical companies looking for novel drug targets (see e.g. Burley *et al.* 1999; Heinemann *et al.* 2000; Schwartz 2000).

The progress of current structural genomics initiatives in Europe, North America and Japan has been reviewed recently in special issues of the journals *Nature Structural Biology* (supplement to vol. 7, 2000) and *Progress in Biophysics and Molecular Biology* (vol. 73, issue 2, 2000). Further information and updated results can be found on the websites related to the various projects, which are listed in Table 8.1. A common theme emerging from these projects is a 'funnel effect' in terms of the number of solved structures compared to the number of proteins chosen for analysis. This is because of the failure of a proportion of the target proteins at each stage of the analysis procedure (e.g. in the case of X-ray crystallography, the essential stages are: cloning, expression, solubilization, purification, crystallization and structural determination). Recent results from the *Methanobacterium thermoautotrophicum* project have been published, including structures and function predictions of the first 10 protein structures to be successfully solved (Christendat *et al.* 2000). Only three of these proteins contained a fold that was not already represented in the protein database, supporting the theory that the total number of protein folds in existence is actually quite limited (Table 8.2, and references therein).

Suggested reading

Brenner S.E. (2000) Target selection for structural genomics. *Nature Structural Biology* **7**, 967–969.

Brenner S.E. (2001) A tour of structural genomics. *Nature Reviews Genetics* **2**, 801–809.

Burley S.K. (2000) An overview of structural genomics. *Nature Structural Biology* **7**, 932–934.

Heinemann U., Illing G. & Oschkinat H. (2001) High-throughput three-dimensional protein structure determination. *Current Opinion in Biotechnology* **12**, 348–354.

Mittl P.R.E. & Grutter M.G. (2001) Structural genomics: opportunities and challenges. *Current Opinion in Chemical Biology* **5**, 402–408.

Norin M. & Sundstrom M. (2002) Structural proteomics: developments in structure-to-function predictions. *Trends in Biotechnology* **20**, 79–84.

Table 8.2 Functional annotation of the first 10 proteins to be structurally solved in the *Methanobacterium thermoautotrophicum* structural genomics programme (see http://www.uhnres.utoronto.ca/proteomics/ for details and updates). Note that only one-third of the proteins contained a novel fold, and two of these could be functionally annotated by other methods (in the case of MTH0152 by the presence of a bound co-factor, and in the case of MTH1048 because of similarity with a functional orthologue). In total, eight of the 10 proteins could be annotated based on structural data alone.

Protein ID	Recognized fold	Predicted function	Reference*
MTH0040	Three helix bundle	Zinc-binding RNA polII subunit	Mackereth *et al.* (2000)
MTH0129	TIM barrel	Orotidine monophosphate decarboxylase	Wu *et al.* (2000)
MTH0150	Nucleotide binding	NAD$^+$ binding protein	
MTH0152	(novel)	Nickel binding	
MTH0538	Rossmann fold	ATPase	Cort *et al.* (2000)
MTH1048	(novel)	RNA polII subunit	Yee *et al.* (2000)
MTH1175	Ribonuclease H	(unknown)	
MTH1184	(novel)	(unknown)	
MTH1615	Armadillo repeat	Transcription factor	
MTH1699	Ferredoxin-like	Transcriptional elongation	Kozlov *et al.* (2000)

* Proteins with no explicit references are discussed in Christendat *et al.* (2000).

A collection of excellent and comprehensive reviews that provide a starting point for the reader interested in the basis, procedures and practicality of functional genomics.

Christendat D., Yee A., Dharansi A., Kluger Y. *et al.* (2000) Structural proteomics of an archaeon. *Nature Structural Biology* **7**, 903–909. *A seminal publication, the first published results of a structural genomics project, containing details of the structural determination and functional annotation of 10 proteins from* Methanobacterium thermoautotrophicum.

Useful website

http://www.rcsb.org
The Protein Databank (PDB), a universal repository for all solved protein structures.

CHAPTER 9

Global expression profiling

Introduction

Important insights into gene function can be gained by *expression profiling*, i.e. determining where and when particular genes are expressed. For example, some genes are switched on (induced) or switched off (repressed) by external chemical signals reaching the cell surface. In multicellular organisms, many genes are expressed in particular cell types or at certain developmental stages. Furthermore, mutating one gene can alter the expression of others. All this information helps to link genes into functional networks, and genes can be used as *markers* to define particular cellular states.[1]

In the past, genes and their expression profiles have been studied one at a time. Therefore, defining functional networks in the cell has been rather like completing a large and complex jigsaw puzzle. More recently, technological advances have made it possible to study the expression profiles of thousands of genes simultaneously, culminating in *global expression profiling* where every single gene in the genome is monitored in one experiment. This can be carried out at the RNA level (by direct sequence sampling or through the use of DNA arrays) or at the protein level (by two-dimensional electrophoresis followed by mass spectrometry, or through the use of protein arrays). Global expression profiling produces a holistic view of the activity of the cell. Complex aspects of biological change, including differentiation, response to stress and the onset of disease, can thus be studied at the genomic level. Instead of defining cell states using single markers, it is now possible to use clustering algorithms to group data obtained over many different experiments and identify groups of co-regulated genes. This provides a new way to define cellular phenotypes, which can help to reveal novel drug targets and develop more effective pharmaceuticals. Furthermore, anonymous genes can be functionally annotated on the basis of their expression profiles, because two or more genes that are co-expressed over a range of experimental conditions are likely to be involved in the same general function.

Traditional approaches to expression profiling

Since the 1970s techniques have been available to monitor gene expression on an individual basis, and some of these are discussed briefly in Box 9.1. Analysis at the RNA level invariably relies on the specificity of nucleic acid hybridization. The target RNA is either directly recognized by a labelled complementary nucleic acid probe, or is first converted into cDNA and then hybridized to a specific pair of primers which facilitate amplification by the polymerase chain reaction (PCR). Analysis at the protein level also relies on specific molecular recognition, usually involving antibodies.

Unfortunately, none of these traditional techniques is particularly suited to global expression profiling. This is because the experimental design is optimized for single gene analysis, i.e. each experiment works on the principle that a single nucleic acid probe (or primer combination or antibody) is used to identify a single target. Although it is possible to modify at least some of the techniques for the parallel analysis of multiple genes (*multiplexing*), the procedure becomes increasingly technically demanding and laborious as more genes are assayed simultaneously. For global analysis, it has been necessary to develop novel technologies with a high degree of automation, which allow thousands or tens of thousands of genes to be assayed simultaneously with minimal labour. We discuss the

---

[1] In this context, a marker is a gene whose expression defines a particular cellular phenotype. For example, a neuronal marker is a gene expressed only in neurons and a cancer marker is a gene expressed only in tumours. The term *marker* is also used in a variety of alternative ways, e.g. to describe landmarks on physical and genetic maps, and to describe genes that confer selectable or scorable phenotypes on transformed cells and transgenic animals and plants. Standard sized proteins or nucleic acids used as references in electrophoresis experiments are also known as markers.

Box 9.1 Gene-by-gene techniques for expression analysis

The techniques described below are some of the most widely applied in molecular biology. For a more detailed discussion of these procedures and an extensive list of original references, the interested reader should consult Sambrook & Russell (2001).

Northern blot and RNA dot blot

In these similar, hybridization-based techniques, RNA from a complex source is transferred to a membrane and immobilized, either without prior fractionation (dot blot) or after fractionation by electrophoresis (northern blot). A labelled probe (DNA, antisense RNA or an oligonucleotide) is then hybridized to the immobilized RNA. The dot blot can indicate the presence or absence of a particular transcript and allows rough quantification of the amount of RNA if several samples are compared. In addition, the northern blot allows size determination and can reveal the presence of homologous transcripts of different sizes, such as alternative splice variants. In both cases, the probe is applied to the membrane in great excess to the target, and hybridization is carried out to saturation, so that the signal intensity reflects the abundance of the immobilized target. Disadvantages of these techniques include their low sensitivity and the large amount of input RNA required. It is difficult to detect rare transcripts using these methods.

Reverse northern blot

In this technique, individual target cDNAs or genomic DNA fragments are immobilized on a membrane and hybridized with a complex probe, i.e. a probe prepared from a heterogeneous RNA source. When carried out to saturation, this technique is often used simply to identify or confirm the presence of genes in large genomic clones (see Chapter 6). However, if non-saturating hybridization is carried out (i.e. with the immobilized target in great excess) the signal intensity reflects the abundance of hybridizing molecules in the probe. DNA array hybridization is based on this principle (see main text).

Nuclease protection

A labelled antisense RNA probe is hybridized in solution with a complex RNA population. The probe is present in excess and the mixture is then treated with a selective nuclease such as RNaseA, which digests single- but not double-stranded RNA. Hybridization between the probe and target RNA protects the probe from degradation and simultaneously allows the signal to be visualized on a sequencing gel. This technique is sensitive and, because probe-target binding is stoichiometric, allows quantification of the target molecule.

Reverse transcriptase polymerase chain reaction

In this technique a population of mRNA molecules is reverse transcribed to generate an equivalent population of cDNAs. These are then amplified by PCR using primers specific for a particular gene or genes. With appropriate controls, the reverse transcriptase polymerase chain reaction (RT-PCR) can be semi- or fully quantitative and the amplification is such that very rare target molecules can be detected and quantified. Many experiments can be carried out in parallel meaning that RT-PCR is easily adaptable for multiplex analysis.

Western blot

Western blot analysis is the equivalent, for proteins, to northern blot analysis of RNA. Proteins are denatured, separated by sodium dodecylsulphate polyacrylamide gel electrophoresis (SDS-PAGE), transferred to a membrane and immobilized. The immobilized proteins are then probed with antibodies, which are generally detected using secondary antibodies covalently attached to enzymatic labels. Like northern blots, western blots

continued on p. 148

Box 9.1 *continued*

are not particularly sensitive but allow rough quantification of the target protein and reveal its approximate molecular mass.

Enzyme-linked immunosorbent sandwich assay (ELISA)

This technique allows accurate determination of protein levels. Antibodies specific for a particular protein are used to coat a microtitre dish, which is then inoculated with the sample (cell lysate, etc.) The amount of bound protein is then determined using a second antibody joined to a colorimetric or fluorescent/chemiluminescent label.

In situ hybridization/immunohistochemistry

These techniques are similar in principle to the northern and western blots, but hybridization/ antibody binding occurs *in situ*, i.e. in sectioned material or whole biological specimens, allowing the expression pattern of particular genes to be determined. The material must be fixed to immobilize the target RNA/protein but not to such an extent that the probe cannot penetrate, an important consideration particularly in wholemount specimens. An alternative to *in situ* hybridization is *in situ* PCR (primed *in situ*, PRINS), a technically difficult procedure but one with very high sensitivity.

---

development and application of such technologies in the following sections, but it should be borne in mind that the principles of molecular recognition, which underlie the simpler methods such as northern and western blots, remain largely unchanged.

Global analysis of RNA expression

The transcriptome

The full complement of RNA molecules produced by the genome has been dubbed the *transcriptome* (Veculescu *et al.* 1997). However, it is important to realize that the transcriptome is potentially much more complex than the transcribed portion of the genome. This is because one gene can produce different mature transcripts, by alternative splicing or RNA editing, and each transcript may then yield a protein with a distinct function. If every gene in the genome produced two alternative transcripts, the complexity of the transcriptome would be double that of the genome. However, in reality, the level of complexity may be much higher. In extreme cases, where a gene has many introns and undergoes extensive differential processing, one gene may potentially produce thousands or even millions of distinct transcripts. An example is the *Drosophila* gene *Dscam* (the homologue of the human Down's syndrome cell adhesion molecule), which can be alternatively spliced to generate nearly 40 000 different mature transcripts (twice the number of genes in the *Drosophila* genome). Each of these transcripts potentially encodes a distinct receptor that may have a unique role in axon guidance (Schmucker *et al.* 2000). Other examples of this phenomenon are discussed in a recent review (Graveley 2001).

Complex as the transcriptome is, it is never seen as a complete system *in vivo*. This is because of gene regulation: there is no situation in which all the genes in the genome are simultaneously expressed. Cells transcribe a basic set of *housekeeping genes* whose activity is required at all times for elementary functions, but other *luxury genes* are expressed in a regulated manner, e.g. as part of the developmental programme or in response to an external stimulus. Similarly, post-transcriptional events such as splicing are also regulated processes. Researchers use phrases such as 'human brain transcriptome' or 'yeast meiotic transcriptome' to emphasize this. A typical human cell is thought to express, on average, about 15 000–20 000 different mRNAs, some of which have housekeeping functions and some of which are more specialized. A proportion of these will be splice variants of the same primary transcript. Some of the mRNAs will be very abundant, some moderately so and others very rare. For a truly global perspective of RNA expression in the cell, all of these transcripts must be quantified at the same time. This requires a highly parallel assay format which is both sensitive and selective.

There are two main types of strategy currently used for global RNA expression analysis.

1 Hybridization analysis with comprehensive non-redundant collections of DNA sequences immobilized on a solid support. These are known as *DNA arrays*.

2 *Direct sampling of sequences* from source RNA populations or cDNA libraries, or from sequence databases derived therefrom.

Although such analysis is often called *transcriptional profiling* it is important to emphasise that one is not really looking at the level of transcription, but at the steady state RNA level, which also takes into account the rate of RNA turnover. Furthermore, most of the transcriptional profiling techniques discussed below do not measure absolute RNA levels, but rather compare relative levels within and/or between samples.

DNA array technology

DNA array hybridization is the method of choice for high-throughput RNA expression analysis (Bowtell 1999; Granjeaud *et al.* 1999). Arrays comprise a series of DNA elements arranged as spots (*features*) in a grid pattern on a solid support. These arrayed targets are hybridized (*interrogated*) with a complex probe, i.e. a probe comprising many different sequences, which is prepared from an RNA population from a particular cell type or tissue.[2] The composition of the probe reflects the abundances of individual transcripts in the source RNA population. If an excess of target DNA is provided and hybridization occurs when the kinetics are linear, the intensity of the hybridization signal for each feature represents the relative level of the corresponding transcript in the probe. These conditions are generally met because, depending on the type of array, each individual feature comprises 10^6–10^9 molecules, only a small proportion of which will be

'occupied' during any hybridization reaction even in the case of abundant RNAs. The use of arrays allows simultaneous measurement of the relative levels of many transcripts, and the only intrinsic limitation of the technique is in the number and nature of sequences represented on the array. Since 1997, whole genome arrays have been available for a number of microbial species, thus providing a truly global perspective of gene expression.

Before discussing how DNA arrays are used for expression profiling, it is necessary to provide some background on the development of array technology and methodology. There are two major types of DNA array used in expression analysis: spotted DNA arrays and printed oligonucleotide chips. Their principal features are compared in Table 9.1.

Spotted DNA arrays

A spotted DNA array is made by transferring (*spotting*) actual DNA clones (or, more usually, PCR products derived therefrom) individually onto a solid support where they are immobilized. The technology arose directly from conventional hybridization analysis, and the first high-density cDNA arrays, now described as *macroarrays*, were essentially gridded reference libraries (see Primrose *et al.* 2001). Cloned cDNAs stored in a matrix format in microtitre plates were transferred to nitrocellulose or nylon membranes in a precise grid pattern, allowing rapid identification of the clones corresponding to positive hybridization signals. For expression analysis, complete libraries could be hybridized with complex probes, generating a 'fingerprint' specific to a particular RNA source (Gress *et al.* 1992; Zhao *et al.* 1995). Early examples of the use of macroarrays for expression analysis include studies of differential gene expression in the mouse thymus and human muscle (Nguyen *et al.* 1995; Pietu *et al.* 1996).

Technology development

Nylon macroarrays are generally about 10–20 cm^2 in size, and the feature density is low, with typically 1–2 mm between targets (10–100 targets per cm^2). This has some advantages: the arrays are easy to manufacture (and are therefore relatively inexpensive), and they are also simple to use because standard hybridization procedures are applicable.

---

[2] In order to maintain continuity in this book, we define a 'probe' as a labelled population of nucleic acid molecules in solution, and a 'target' as an unlabelled population of nucleic acid molecules usually immobilized on a solid support. These definitions are generally followed by researchers using spotted arrays. However, care should be exercised when reading literature concerning the use of Affymetrix GeneChips, because exactly the opposite convention is followed. Each feature on an Affymetrix GeneChip is termed a *probe cell* or simply a *probe*, and the labelled nucleic acids in solution, which hybridize to the features, are described as *targets*. We deliberately ignore this nomenclature for the sake of clarity.

Table 9.1 Properties of different types of DNA array for expression analysis.

Property	Spotted nylon macroarrays	Spotted glass microarrays	Affymetrix GeneChips
Target composition	dsDNA fragments (genomic or cDNA clones, or PCR products derived from them)		Single-stranded oligonucleotides
Target source	Maintained clone sets, either annotated or anonymous. Must be derived from source RNA or purchased from licensed vendors		Sequences derived from public and/or private databases. Chemically synthesized
Target size	Typically 100–300 bp		Typically 20–25 nt
Array format	Individual features represent non-redundant clones; hybridization sensitivity high		Single clones represented by sets of ~20 non-overlapping oligos to reduce false positives
Density (features per cm²)	1–10*	> 5000	64 000 for available chips, but experimental versions up to 1 000 000
Manufacture	Robotoized or manual spotting	Robotoized spotting	On-chip photolithographic synthesis
Substrate	Nylon	Glass	Glass or silicon
Probe labelling	Radioactive or enzymatic	Dual fluorescent	Fluorescent
Hybridization	High volume (up to 50 ml*), ~65°C	Very low volume (10 µl), ~65°C	Low volume (200 µl), 40°C
Data acquisition	Autoradiography or phosphorimager for isotopic probes, flatbed scanner for enzymatic probes	Confocal scanning	Confocal scanning
Cost of prefabricated arrays	Low	Moderate	High
In-house manufacture	Inexpensive	Expensive, but prices falling	Not currently available

* Note that nylon microarrays are also available: these have a density of up to 5000 features per cm², and require only 100–200 µl of hybridization solution.

For this reason, macroarrays are still manufactured by a number of commercial suppliers[3] and the technology for in-house array production is readily available, involving simple robotic devices or even handheld arrayers. The principal disadvantages of macroarrays are:

1 the low feature density limits the number of sequences that can be interrogated simultaneously; and

2 hybridization must be carried out in a large volume using a radioactive probe, the results being obtained by autoradiography or, preferably, using a phosphorimager.

Although radioactive probes are sensitive, comparative gene expression analysis (e.g. mutant vs. wild-type or stimulated vs. non-stimulated tissue) requires the preparation of duplicate arrays, or the sequential probing, stripping and reprobing of the same array with two different probes. Both these strategies can generate interexperimental variation that can give misleading results. Also, the large volume of solution required to cover the membrane lim-

---

[3] Although nylon macroarrays are sold by many biotechnology companies, they tend to be called microarrays in the accompanying literature. The distinction between a macroarray and a microarray is not clear-cut. The term *microarray* was initially coined to describe the high-density arrays printed on small glass chips, which contrasted sharply with the original macroarrays printed on large nylon membranes. Confusion arises now that nylon arrays can be manufactured with a size and feature density similar to that of the glass arrays. A convenient cutoff point for a microarray might be an overall size of 1–2 cm² and a spacing between spots of 0.5 mm, but this is purely arbitrary.

its the probe concentration, reducing the efficiency of the hybridization reaction. However, extensive miniaturization of nylon arrays has been difficult because the resolution of the signal provided by radioactive probes is poor. Fluorescent probes have a higher resolution but cannot be used on nylon membranes because the substrate has a high level of autofluorescence, generating a low signal to noise ratio. It has been possible to produce nylon microarrays with about 300 µm between features (up to 5000 targets per cm^2) but their analysis requires expensive high-resolution imaging devices (Bertucci *et al.* 1999). An alternative system, which uses enzymatic rather than radioactive probes, gives a high-resolution signal that can be detected with a low-cost scanning apparatus, but with some loss of sensitivity (Chen *et al.* 1998).

A breakthrough in spotted array technology came with the development of microarrays on glass chips (Schena *et al.* 1995, 1996). Glass is an inert substrate and must be coated with an agent such as poly-L-lysine or aminosilane before DNA will adhere; the DNA must then be crosslinked to the surface. Because glass is non-porous and has very little autofluorescence, fluorescent probes can be used and they can be applied in very small hybridization volumes. The greater resolution afforded by fluorescent probes allows feature density to be increased significantly compared to nylon macroarrays, and the small hybridization volume improves the kinetics of the reaction. Together, these advantages mean that more features can be assayed simultaneously with the same amount of probe without loss of sensitivity. Thus, glass arrays can routinely be manufactured with up to 5000 features per cm^2. However, the major advantage of fluorescent probes is that different fluorophores can be used to label different RNA populations. These can be simultaneously hybridized to the same array, allowing differential gene expression between samples to be monitored directly (Shalon *et al.* 1996). The most common strategy is to use Cy3 and Cy5, which have different emission wavelengths, to label different probes. If a particular cDNA is present only in the Cy3-labelled population, the spot on the array appears on one channel. If another cDNA is present only in the Cy5-labelled population, the spot on the array appears on the other. cDNAs that are equally represented in both populations contain equivalent proportions of each label and appear on both. In this way, it is easy to identify potentially interesting differentially expressed genes (Fig. 9.1).

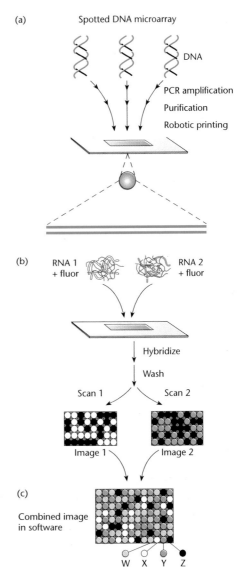

Fig. 9.1 Spotted DNA arrays. (a) Principle of array manufacture. Each robotically-printed feature corresponds to one gene or cDNA. (b) Hybridization of glass arrays with differentially labelled RNA probes followed by scanning on separate channels to detect fluorescence. (c) Combined image showing four types of signal: W – genes represented equally in both RNA populations; X – genes represented only in RNA 1; Y – genes represented only in RNA 2; Z – genes represented in neither RNA population. ((a) and (b) redrawn with permission from Harrington *et al.* 2000 by permission of Elsevier Science.)

Box 9.2 Selected sources of clone sets for the manufacture of spotted arrays

Company	Website	Resources
Research Genetics	http://www.resgen.com	Human UniGene collection Mouse UniGene collection Rat cDNA clone collection Genome-wide *Caenorhabditis elegans* partial ORF primers *Drosophila* cDNA collection Genome wide *Saccharomyces cerevisiae* ORF primers Genome wide *S. cerevisiae* intergenic primers
Incyte Genomics	http://www.incyte.com	Human UniGene collection Mouse UniGene collection 8000 *Arabidopsis thaliana* cDNA clones *Candida albicans* complete ORF collection
Genosys Biotech	http://www.genosys.com	*Escherichia coli* complete ORF collection *Bacillus subtilis* complete ORF collection Partial clone collections for several other bacteria

Practical considerations in the manufacture of glass microarrays

Because of the many advantages of glass microarrays, including their amenability for automated spotting, this format has emerged as the most popular type of spotted array for expression profiling. However, until very recently, a major disadvantage was the cost of production, with the result that the technology was beyond the reach of all but the best-funded laboratories. Researchers were faced with the initial choice of purchasing ready-made arrays from a commercial source or investing in the resources required for in-house array manufacture. Prefabricated glass arrays can cost up to £500 each, and are designed for single use. Therefore, a simple series of experiments with an appropriate number of replicates can carry a hefty price tag. Unfortunately, the cost of a commercially available precision robot for array manufacture is even greater: £50 000 or more for those with even the simplest specifications. Additionally, clone sets usually need to be purchased to provide the features for a home-made array (Box 9.2).

Genomic clones tend to be used to derive features for bacterial arrays because the lack of introns makes them essentially equivalent to cDNAs. This is also true in the case of the yeast *Saccharomyces cere-*visiae, where introns are small and few in number (p. 14). In higher eukaryotes, where introns are larger and more common, it is much more convenient to use cDNAs instead of genomic clones. However, full-length cDNA clones are neither necessary nor particularly desirable because of the prevalence of large gene families with conserved sequences. The use of partial cDNA sequences that exploit the differences between related clones avoids cross-hybridization. Most of the cDNA sequence information that exists in databases is in the form of expressed sequence tags (ESTs; p. 51) and these are a valuable resource for the manufacture of spotted arrays.

It is beyond the scope of most laboratories to prepare comprehensive clone sets *de novo* for array manufacture. Only large-scale sequencing projects can provide the materials and data required to make comprehensive arrays, and this is the domain of biotechnology and pharmaceutical companies, consortia of academic laboratories, and collaborations between academic institutes and industry. Typically, such organizations make their clone sets available commercially through licensed vendors (for a selected list, see Box 9.2). An example is the UniGene collection of human (and mouse) clustered sequence-verified ESTs. This collection is available from Incyte Genomics and Research Genetics.

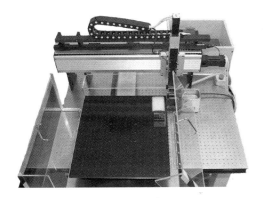

Fig. 9.2 A home-made microarraying robot, which can be constructed for approximately £18 000. (Reprinted from Thompson *et al.* 2001 by permission of Elsevier Science.)

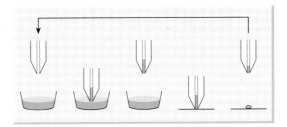

Fig. 9.3 Principle of array manufacture by capillary contact printing.

Fig. 9.4 A non-contact printing method for array manufacture. This is the 'pin and ring' system developed by Genetics Systems and currently marketed by Affymetrix. (Courtesy of Affymetrix.)

Over the last year or so, the financial barriers to the general use of microarrays have begun to fall. This reflects a number of factors, including competition between companies producing prefabricated arrays, increasing numbers of universities investing in microarray core facilities, the availability of protocols that allow robots for array manufacture to be built in the laboratory for under £20 000 (Fig. 9.2), and the development of novel printing technologies (see below). Instructions for building arraying robots using simple and readily available components are available on the Internet (Box 9.3) and have been discussed in several recent articles (e.g. see Bowtell 1999; Cheung *et al.* 1999; Duggan *et al.* 1999; Thompson *et al.* 2001).

Printing technology for spotted arrays

Two general types of printing method have been developed for making spotted arrays (Duggan *et al.* 1999; Xiang & Chen 2000). The original contact printing method was pioneered at Stanford University, in the laboratory of Pat Brown, and involves the use of a capillary spotting pin (or quill) which draws up a defined amount of liquid from wells in a microtitre plate (Fig. 9.3). The pin is then placed in contact with the array surface, and this causes some liquid to be deposited. The pin is thoroughly washed and dried in an automated cycle before returning to the microtitre dish for the next sample. The speed at which arrays can be produced is increased by using multiplex print heads which deposit samples in a block.

A number of alternative 'non-contact' printing methods are also available. The *pin and ring system*, devised at Genetic Microsystems and currently marketed by Affymetrix, is popular (Fig. 9.4). The 'ring' is inserted into the well of a microtitre plate and draws up a certain amount of liquid. The 'pin' then extends through the ring and carries a smaller droplet of solution down onto the array surface. Emerging non-contact printing technologies include piezoelectric devices similar to those found in inkjet printers, and bubblejet print heads which deposit DNA samples on the substrate as a bubble extended from the nozzle (Okamoto *et al.* 2000). These methods

Box 9.3 Internet resources for microarrays and oligo chips

Selected commercial sources of prefabricated nylon and glass arrays

http://www.clontech.com for Atlas cDNA arrays (human, mouse, rat cDNA)

http://www.incyte.com for Membrane Expression arrays (human and *Arabidopsis* cDNA)

http://lifesciences.perkinelmer.com for Micromax arrays containing human transcription factor, kinase and phosphatase cDNAs

http://www.resgen.com for GeneFilters microarrays (human and rodent cDNA, yeast open-reading frames)

http://www.stratagene.com for various arrays

http://www.hyseq.com for various arrays

Selected commercial sources of robots for in-house array manufacture

http://www.affymetrix.com pin and ring device

http://www.biorobotics.co.uk contact printing

http://www.cartesiantech.com contact printing

http://www.genetix.com contact printing

http://www.genomicsolutions.com contact printing

Sites describing the construction of home-made robots for array manufacture

http://cmgm.stanford.edu/pbrown/mguide/

Printing technology development

http://bt.swmed.edu/DOC.htm digital optical chemistry

Companies that produce prefabricated or custom oligonucleotide chips

http://www.affymetrix.com high density GeneChips produced by on-chip photolithographic synthesis

http://www.operon.com oligo arrays produced by conventional contact printing

Selected outlets for array scanning systems and image analysis software

http://www.axon.com for GenePix 400 scanning system and GenePix Pro analysis software

http://www.genomicsolutions.com for GeneTAC 1000 scanning system

http://www.affymetrix.com for HP GeneArray scanner

http://www.tigr.org/softlab/ for TIGR Spotfinder software

http://www.scanalytics.com for MicroArray Suite software

Sites providing data from transcriptional profiling experiments (fragment arrays and chips)

http://cellcycle-www.stanford.edu (Spellman *et al.* 1998)

http://cmgm.stanford.edu/pbrown/sporulation (Chu *et al.* 1998)

http://genomics.stanford.edu (Wodika *et al.* 1997; Cho *et al.* 1998)

---

provide a more uniform spot size reducing the variation between features.

Printed oligonucleotide chips

The alternative to a spotted DNA array is a high-density prefabricated oligonucleotide chip (Lockhardt *et al.* 1996; Lipschutz *et al.* 1999). These are similar to DNA arrays in that they consist of gridded DNA targets which are interrogated by hybridization. However, while DNA arrays consist of double-stranded clones or PCR products that may be up to several hundred base pairs in length, oligo chips contain single-stranded targets ranging from 25–70 nucleotides. Oligo chips can be made in the same way as spotted DNA arrays, by robotically transferring chemically synthesized oligonucleotides from microtitre dishes to a solid support, where they are

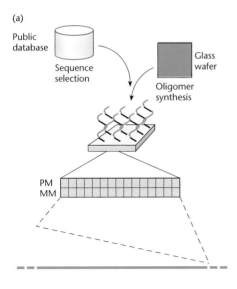

(a)

Public database

Sequence selection

Glass wafer

Oligomer synthesis

PM
MM

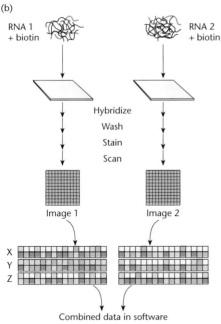

(b)

RNA 1 + biotin

RNA 2 + biotin

Hybridize
Wash
Stain
Scan

Image 1

Image 2

X
Y
Z

Combined data in software

Fig. 9.5 Oligonucleide chips. (a) Principle of chip manufacture. Note that sequences are obtained from public or private databases and sythesized on the chip. Each gene is represented by 20 non-overlapping oligonucleides, each with a perfect match (PM) and mismatch (MM) feature. (b) Principle of chip hybridization using biotin-labelled cRNA probes. (Redrawn with permission from Harrington *et al.* 2000.)

immobilized (e.g. see Yershov *et al.* 1996). However, the maximum array density is increased almost 10-fold if the oligos are printed directly onto the glass surface. This can be achieved using a process developed by the US company Affymetrix Inc, which is the major player in DNA chip manufacture (see below). Dual hybridization is not used for expression profiling on oligo chips. Instead, probes for chip hybridization are made from cleaved biotinylated cRNA (RNA that has been transcribed *in vitro* from cDNA). Comparative expression analysis is carried out by hybridization of alternative cRNA samples to identical chips, followed by comparison of signal intensities (Fig. 9.5).

On-chip printing technology

Direct 'on-chip' synthesis of high-density oligonucleotide arrays was developed by Steve Fodor *et al.* at Affymetrix, using a light-directed printing technology known as photolithography (Fodor *et al.* 1991, 1993; Pease *et al.* 1994). The procedure is complex but essentially involves the use of a glass slide coated in such a way that DNA can be covalently attached to the surface in a simple chemical reaction. However, the covalent binding sites are blocked by a photolabile protecting group. A mask is then applied to the surface of the chip which determines which areas are exposed to light. Under illumination, the protecting group in these areas is removed, allowing the addition of a single nucleotide, which is also blocked with a photolabile protecting group. If this process is repeated using a series of different masks and different nucleotides, a precise grid can be generated containing millions of defined and precisely arrayed oligos. The process is highly accurate and allows the production of the densest arrays currently available. Commercially available Affymetrix GeneChips have a density of 64 000 features over an area slightly larger than 1 cm^2, but experimental chips with a density of $> 10^6$ targets per cm^2 have been produced. GeneChips can also be used for other types of sequence analysis, such as genotyping and sequencing by hybridization. These applications are discussed in Chapter 5.

The sequential use of different masks makes photolithographic technology much more expensive than the production of oligonucleotide or cDNA arrays by robotic spotting. Consequently, each

1 cm^2 chip carries a price tag of about £3000. For the routine use of DNA chips, the hybridization apparatus, scanning equipment and the necessary software for data analysis cost in the region of £100 000. Therefore, despite the obvious advantages of increased feature density, the expense of high-density oligo chips prevents their use in many laboratories. As with spotted arrays, several companies offer to carry out hybridization experiments with Affymetrix GeneChips and provide the expression data in a form that the customer can subject to further analysis. The costs can be upward of £7000 per experiment. Novel printing techniques may eventually begin to bring the prices down. For example, Singh-Gasson *et al.* (1999) describe a novel *in situ* synthesis method that obviates the need for photo-lithographic masks. Instead, a digital mask is created using selectively focused ultraviolet light, allowing the rapid generation of unique oligo chips in a machine no larger than a standard photocopier, at much reduced expense. Inkjet technology has also been developed to print oligonucleotides onto glass chips (Blanchard *et al.* 1996).

Comparison of spotted arrays and oligo chips

Although subject to some debate, it is believed that spotted microarrays and high density oligo chips perform equally well in terms of sensitivity (see discussion by Granjeaud *et al.* 1999). However, an important difference between the two types of array is that the manufacture of oligo chips relies entirely on pre-existing sequence information. Conversely, spotted DNA arrays can be generated using anonymous (i.e. non-annotated) clones from uncharacterized cDNA libraries, and can therefore be used for *de novo* gene discovery. Oligonucleotide chips can be designed on the basis of genome sequence data or a collection of cDNA or EST sequences. The advantage of this is that chips can be devised *in silico*, i.e. using computer databases as a source of information, with no need to maintain physical DNA clone sets. Oligonucleotide arrays are therefore highly advantageous for expression profiling in those organisms with complete or near-complete genome sequences, or comprehensive EST collections, but less useful for other organisms. In contrast, it is quite possible to generate cDNA arrays for largely uncharacterized organisms, although positive hybridization signals

then have to be further characterized by sequencing the corresponding clone.

The specificity of hybridization to spotted DNA arrays is relatively high, because of the length of each target. Distinction between members of gene families can be achieved by selecting the least conserved region of the cDNA. Hybridization specificity is more of a problem when the target size is smaller, and this has been addressed in the case of Affymetrix GeneChips by the use of hybridization controls and redundant targets (Lipschutz *et al.* 1999). Each mRNA is represented by up to 20 non-overlapping oligonucleotides, such that the likelihood of obtaining a false positive result is greatly reduced. The chips are designed to contain both 'perfect match' (PM) and 'mismatch' (MM) oligos for each specific target (Fig. 9.5). The PM is expected to hybridize along its whole length, while the MM contains at mismatching base at a central site, thus acting as a control for cross-hybridization. The signal obtained from the MM control is subtracted from that of the corresponding PM to reveal the actual level of specific hybridization. Distinction between members of gene families is achieved by designing oligos matching the least conserved regions of the gene.

Data acquisition and analysis

The raw data from DNA array experiments are monochrome images of hybridized arrays. Where dual fluorescence has been used, images are obtained on two channels, rendered in false colours and combined. These visual data must be quantified, and the software for this is often provided with the image recording apparatus. The signal intensity for each feature has to be corrected for background, which may be generated by non-specific hybridization, autofluorescence, dust and other contaminants or poor hybridization technique. The background may not be constant over the entire array, so local background values must be obtained.

The result of data collection is a *gene expression matrix*, which shows the normalized signal intensities for each feature over a range of experimental conditions. With dual fluorescence, two measurements are taken, one from each channel. In other cases, readings may be taken from a number of identical arrays representing a series of developmental time points or a cell culture exposed to different concentrations of a drug. It is important to have control features on each array so that the data can be normalized for variation across arrays. The data can then be grouped according to similar expression profiles using a clustering algorithm (reviewed by Eisen *et al.* 1998; Bittner *et al.* 1999; Brazma & Vilo 2000; Hess *et al.* 2001; Raychaudhuri *et al.* 2001; Noordewier & Warren 2001). This involves converting the gene expression matrix into a distance matrix showing the pairwise differences between the expression levels of each possible combination of genes. The data are then clustered to generate a tree-like graph called a dendrogram (see Fig. 6.6). In *hierarchical clustering* methods, the two most similar genes are clustered first and these define a new merged data point. The analysis is repeated until all the genes are clustered together. Other popular methods include *k-means clustering*, in which the expected number of clusters is specified at the outset, and the generation of *Kohonen self-organizing maps*, a similar process refined by the use of neural nets. These algorithms can take a long time to run if the data set is very large. Run times can be limited by employing *feature reduction* strategies, such as the elimination or merging of redundant and uninformative genes or expression profiles. Several bioinformatics tools are available over the Internet to carry out clustering analysis of microarray expression data, such as the EPCLUST program, which is part of the Expression Profiler suite (http://ep.ebi.ac.uk/EP/).

Despite the many advances in array-based technology, it is notable that functional annotation on the basis of expression data alone is still only about 30% reliable. For this reason, a number of integrated functional genomics resources have been developed which include expression data and information from other sources, such as the biomedical literature, structural genomics programmes, mutation analysis, proteomics (see below) and protein interaction databases. Details of these resources can be found by consulting Marcotte *et al.* (1999b), Brown *et al.* (2000), Shatkay *et al.* (2000) and Wilson *et al.* (2000).

Expression profiling with DNA arrays

Because DNA array technology is still relatively new, much of the early literature concerned methodology development and proof-of-principle studies.

Indeed, between 1995 and 1999 the number of papers describing the theory and practice of array hybridization far outweighed the number of papers reporting actual experiments! In the last 2 years, there has been an exponential increase in the number of array-based experiments and the applications are extremely diverse, covering many different organisms and ranging from basic studies of biological processes to clinical applications and pharmacology. Comprehensive coverage of this burgeoning field would require an entire book so we therefore restrict the discussion below to two important applications with relevance to core functional genomics: genome-wide expression profiling in microbes, and disease profiling in humans. Further applications of expression profiling – in agriculture, biotechnology and developmental biology – are considered in Chapter 12.

Global profiling of microbial gene expression

In 1997, the first genome-wide expression profiling experiments were reported. Spotted arrays were manufactured containing PCR-amplified open-reading frames (ORFs) representing most of the 6200 genes in the *Saccharomyces cerevisiae* genome (De Risi *et al.* 1997; Lashkari *et al.* 1997). These investigators analysed the transcriptional profile of yeast cells shifted from fermentation (anaerobic) to aerobic metabolism, and as they were subjected to a variety of environmental manipulations, including heat shock. In each case, about 5% of the interrogated genes showed highly significant changes in expression induced by the experimental conditions, when unstimulated yeast cells were used as a source of control RNA. Genome-wide expression profiling with arrays has also been carried out for a number of complex biological processes in yeast, such as sporulation (Chu *et al.* 1998) and the cell cycle (Spellman *et al.* 1998). These studies have allowed tentative functions to be assigned to a number of previously uncharacterized genes, based on their informative expression patterns. In the study by Spellman *et al.* (1998) 800 cell cycle-regulated genes were identified, about 400 of which were inducible by cyclins. Genes have also been identified whose expression is dependent on the ploidy (number of chromosome sets) of the cell (Galitski *et al.* 1999). Furthermore, transcriptional profiling of yeast cells exposed to drugs has allowed novel drug targets to be identified (e.g. see Lockhart 1998; Marton *et al.* 1998).

Affymetrix GeneChips have also been manufactured representing all the ORFs in the yeast genome. Wodicka *et al.* (1997) reported the first use of GeneChips for transcriptional profiling in yeast, when they compared yeast grown on minimal and rich media. The Affymetrix yeast GeneChip has also been used to profile yeast cells exposed to alkylating agents (chemicals that cause damage to DNA). Forty-two genes were found to be induced by DNA damage, and for almost all of these genes, the results of the chip experiment were confirmed by traditional gene-by-gene northern blot hybridization (Jelinsky & Samson 1999). Genome-wide transcriptional analysis of the mitotic cell cycle has also been carried out (Cho *et al.* 1998) as well as a comprehensive analysis of the meiotic transcriptome (Primig *et al.* 2000).

Bacteria have smaller genomes than yeast, which should make transcriptional profiling using DNA arrays a simpler process. An array containing all 4290 genes of the *Escherichia coli* genome was produced by Tao *et al.* (1999) and interrogated using RNA from bacteria growing on glucose-rich and minimal medium. Over 200 genes were shown to be induced on minimal medium, including a number of previously identified stress-response genes. About 120 genes were induced by growth on rich medium, many of these involved in protein synthesis. Note that array hybridization in bacteria is complicated by the difficulty in selectively labelling mRNA. In eukaryotes, mRNA has a polyadenylated tail which can be used to selectively prime first-strand cDNA synthesis, generating a labelled probe devoid of rRNA and tRNA. Bacterial mRNA generally lacks a polyadenylated tail, so hybridization is carried out with total RNA. Nevertheless, the presence of rRNA does not appear to interfere with the sensitivity of the hybridization (Richmond *et al.* 1999). Richmond *et al.* looked at the genome-wide transcriptional profile of *E. coli* after exposure to heat shock and the lactose analogue isopropylthio-β-D-galactopyranoside (IPTG). They used both glass microarrays with a fluorescent probe and nylon arrays with a radioactive probe, finding that the former produced more reliable and consistent results. Expression profiling in bacteria has also been used to identify potential new drug targets. Wilson *et al.* (1999) exposed *Mycobacterium tuberculosis* to isoniazid, a drug commonly used to treat tuberculosis. RNA extracted from treated and untreated bacteria was used to interrogate a genome-wide DNA array. As well as identifying genes involved in the biochemical pathway representing the known mode of action of the drug, a number of other genes were induced which could be exploited in the development of novel therapeutics. DNA microarrays have also been used to study differences between Bacille–Calmette–Guerin (BCG) vaccines (which are live attenuated strains of *Mycobacterium bovis*) and *Mycobacterium tuberculosis*, the organism against which these vaccines are used (Behr *et al.* 1999).

Over the last couple of years, there has been a shift in emphasis in genome-wide array hybridization experiments. The aim of the expression profiling studies discussed above has been, essentially, to assemble a list of genes that are specifically induced or repressed under particular conditions. With the development of more sophisticated data analysis tools (see below), more recent experiments have tracked transcriptional changes over tens or hundreds of different conditions. Clustering these data allows subtle changes in gene expression patterns to be revealed. As an example we consider a series of experiments carried out by Hughes *et al.* (2000) in which yeast cells were exposed to drugs such as itraconazole, which inhibits sterol biosynthesis. This treatment resulted in significant changes in the expression of hundreds of genes, suggesting that the drug had many specific targets in the yeast cell. However, by looking at the data generated in this experiment in concert with the expression profiles revealed under 300 other conditions (including various mutants, chemical treatments and physiological parameters) it became apparent that most of the effects were non-specific, and that the only genes specifically affected by itraconazole were those involved in the sterol biosynthetic pathway. The large-scale use of microarrays in this series of experiments resulted in a compendium expression database, which allowed expression profiles over multiple conditions to be compared. In this way, it was not only possible to compile lists of co-regulated genes, but also assign functions to orphan reading frames and identify drug targets. Thus an anonymous transcript known as *YER044c* was shown to be co-regulated with other sterol biosynthetic genes, strongly indicating a role in sterol metabolism, and expression of the *erg2p* gene was shown to be influenced by the anaesthetic drug dyclonine, suggesting a candidate homologous target in humans (Hughes *et al.* 2000).

Applications of expression profiling in human disease

Genome-wide arrays are not yet available for higher eukaryotes, although this is likely to change as more genome projects are completed. However, cDNA arrays containing from a few hundred to several thousand features have been generated for a variety of organisms, ranging from the fruit fly *Drosophila melanogaster*, various species of plants, to rodents and humans.

Arrays have been widely used to investigate transcriptional profiles associated with human disease, to identify novel disease markers and potential new drug targets. Many investigators have used arrays to profile transcriptional changes associated with cancer and this area of research has been reviewed recently (Lockhart & Winzeler 2000; Marx 2000; Young 2000). In one of the earliest studies, De Risi *et al.* (1996) used cDNA arrays to investigate the ability of human chromosome 6 to suppress the tumorigenic phenotype of the melanoma cell line UACC-903. A number of novel tumour-suppressor genes were identified. Spotted cDNA arrays have also been used to investigate global gene expression in rheumatoid arthritis and inflammatory bowel disease (Heller *et al.* 1997), insulin resistance (Aitman *et al.* 1999) and asthma (Syed *et al.* 1999). In some cases, the investigation of global gene expression profiles has led to the discovery of novel links between biological processes. Iyer *et al.* (1999) investigated the transcriptional profile of serum-starved cells following the addition of fresh serum, using a human cDNA array containing approximately 8600 genes. While many of the genes induced at early time points were well-characterized proliferation-response genes, the investigators also found that a large number of genes induced at later time points were known to be involved in the wound response, an example being *fgf-7*. A functional link between serum starvation and wounding had not previously been identified. Affymetrix produce a range of different chips for human, mouse and plant genomes in addition to yeast and bacteria, thus

human GeneChips have also been used for disease profiling. The HUGeneF1 chip contains features representing nearly 7000 human genes. This was interrogated using RNA from human foreskin fibroblasts at several time points after infection with human cytomegalovirus (Zhu *et al.* 1998). One day post-infection, 364 transcripts were shown to have undergone significant changes in expression level and it is likely that some of these may strongly influence the progress of the infection, and could represent useful drug targets.

As array technology has matured, one emerging application with an important impact on medicine is the use of expression profiling for the classification of tumours. Unlike the clustering approaches discussed above, which are unsupervised (i.e. there are no pre-defined groups), tumour classification is a supervised type of analysis (i.e. the data are placed into categories that have already been defined). Different forms of cancer are generally identified by a histological phenotype, which is subject to visual interpretation and human error. Recently, a number of studies have shown that gene expression profiles can be a useful way to classify tumours, and that such profiles can be defined more rigorously. For example, a systematic study of 60 diverse cancer cell lines held at the National Cancer Institute using an array containing about 10 000 cDNAs showed that each line could be distinguished clearly on the basis of its expression profile (Ross *et al.* 2000). The same cell lines have recently been profiled to determine relationships between RNA levels and drug responses (Scherf *et al.* 2000).

Expression profiling has also been useful for distinguishing very similar types of cancer, an approach called *class prediction* where subtypes of a disease are known, and *class discovery* where they are not. Perou *et al.* (1999, 2000) used expression profiling to distinguish different classes of breast cancer, while Golub *et al.* (1999) used the *self-organizing maps* algorithm to analyse the transcriptional profiles of a number of leukaemia samples, correctly placing them into the two known categories: acute myeloid (AML) and acute lymphoblastic leukaemia (ALL). An array containing 7000 cDNAs was used in this analysis, and about 50 genes were shown to be differentially expressed. Similarly, Alizadeh *et al.* (2000) used the 18 000 gene 'lymphochip' array available from Research Genetics to study non-

Hodgkin's lymphoma. Interestingly, this experiment revealed two previously unknown subclasses of the disease with different clinical characteristics. Bittner *et al.* (2000) have used expression profiling for class discovery in cutaneous melanoma, and were also able to distinguish aggressive metastatic melanomas through the analysis of microarray hybridization results. In each of these studies, prediction of the correct type of cancer helps to ensure appropriate treatment is carried out, and the microarray experiments themselves may even reveal novel drug targets.

Direct quantification of RNA levels by sequence sampling

The usefulness of a DNA array is limited by the targets available for hybridization. For example, a human cDNA array containing 8000 features can provide expression data only on the 8000 genes represented on the array. There will be no information on any other genes, even though the transcripts may be represented in the probe. The array can be described as a 'closed system' because it accepts interrogation on a restricted basis. The direct sampling of sequences from the RNA population used to prepare the probe, however, could be described as an 'open system' because there are no restrictions to the data that can be obtained.

Expression profiling by direct sequence sampling involves the large-scale random sampling of sequences representing a given RNA source (Okubo *et al.* 1992). For example, a cDNA library that has not been 'normalized' is representative of the mRNAs in the source population used to prepare it. Some mRNAs (and corresponding cDNAs) are likely to be highly abundant, and some extremely rare. If 5000 clones are picked randomly from the library and partial sequences obtained, abundant transcripts would be more frequently represented among the sequences than rare transcripts. Statistical analysis of these results would allow relative expression levels to be determined, and comparisons of libraries prepared from different sources (disease vs. normal, induced vs. uninduced) should facilitate the identification of differentially expressed genes. The limitation of this approach is the expense involved in producing cDNA libraries and carrying out the large-scale sequencing projects required to make the data statistically significant (Audic & Claverie

1997). However, as the amount of EST data continues to increase for model organisms, it is now becoming a viable approach to interrogate these data to see how often particular genes are represented, giving a digital representation of gene expression. This approach is not particularly sensitive and depends on the availability of ESTs from appropriate sources, but it demonstrates the principle that direct sampling of sequences *in silico* can be used to derive expression data. Indeed, there have been several reports of differentially expressed genes identified using EST sequence sampling (e.g. see Claudio *et al.* 1998; Vasmatzis *et al.* 1998).

Serial analysis of gene expression

The idea of expression analysis by sequence sampling has been taken to the extreme by Veculescu *et al.* (1995) in a technique called serial analysis of gene expression (SAGE). Essentially, this involves the generation of very short ESTs (9–14 nt), known as SAGE tags, which are joined into long concatamers that are cloned and sequenced. The size of the SAGE tag approaches the lower limit for the unambiguous identification of specific genes (Adams 1996). If we consider a random sequence of 9 nt, there are 4^9 possible combinations of the four bases, or 262 144 sequences. This is approximately fivefold the estimated number of genes in the human genome. However, an 8-nt SAGE tag would allow only about 65 000 variations, which is in the same order of magnitude as some estimates of the human gene number. The concatamerized tags are sequenced and the sequence is analysed to resolve the individual tags; the representation of each tag provides a guide to the relative abundances of the different mRNAs. Compared to the random sequencing of cDNA libraries, SAGE is up to 50 times more efficient, because each concatamer represents the presence of many cDNAs. However, the major advantage of SAGE is that the data obtained are digital representations of absolute expression levels, which allows direct comparison between new experiments and existing databases. Thus, as the amount of SAGE data grows it becomes increasingly possible to carry out computer searches to identify differentially expressed genes (e.g. see Veculescu *et al.* 1999).

The SAGE method as originally described is shown in Fig. 9.6. Poly(A)$^+$ RNA is reverse transcribed

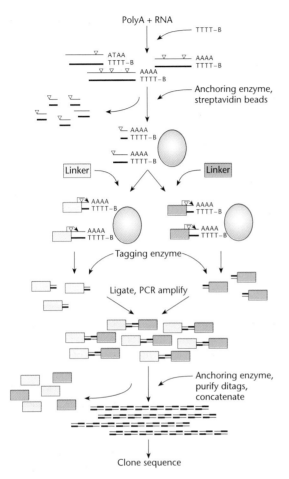

PolyA + RNA

TTTT–B

ATAA
TTTT–B

AAAA
TTTT–B

AAAA
TTTT–B

Anchoring enzyme,
streptavidin beads

AAAA
TTTT–B

AAAA
TTTT–B

Linker   Linker

AAAA
TTTT–B

AAAA
TTTT–B

AAAA
TTTT–B

AAAA
TTTT–B

Tagging enzyme

Ligate, PCR amplify

Anchoring enzyme,
purify ditags,
concatenate

Clone sequence

Fig. 9.6 Principle of serial analysis of gene expression (SAGE).
(Adapted from Veculescu *et al.* 1997.) ▽ Anchoring enzyme
(*Nla* III); ⌐► Tagging enzyme (*Fok* I)

using a biotinylated oligo-dT primer and the cDNA is
digested with a restriction enzyme that cleaves very
frequently, such as *Nla*III (this recognizes the 4-bp
sequence CATG and would be expected to cut, on
average, every 250 bp). The 3′ end of each cDNA is
then captured by affinity to streptavidin, resulting
in a representative pool of cDNA ends which can be
used to generate the SAGE tags. Note that, by selecting 3′ ends, there is generally more sequence
diversity. Although many genes are members of
multigene families with conserved sequences, these
tend to diverge in the 3′ untranslated region even
when the coding sequences are very similar. The
pool is then split into two groups, each of which
is ligated to a different linker. The linkers contain
recognition sites for a type IIs restriction enzyme

such as *Fok*I, which has the unusual property of
cutting outside the recognition site, but a specific
number of base pairs downstream. Cleavage with
such an enzyme therefore generates the SAGE tag
attached to part of the linker. The tags are ligated
'tail-to-tail' to generate dimers called 'ditags', and
then amplified by PCR using the linkers as primer
annealing sites. The amplified products are then
cleaved with *Nla*III to remove the linkers, and the
ditags are concatamerized. Concatamers are then
cloned by standard methods, and the resulting
plasmids are sequenced to reveal the composition
of the concatamer. Accurate sequencing is essential
in SAGE because even single nucleotide errors (*mis-
calls*) could result in the incorrect identification of a
tag and false expression data for a particular gene. A
similar error rate in standard cDNA sequencing
would be irrelevant, because of the length of the
sequence. More recently, the SAGE method has been
adapted so it can be used with small amounts of
starting material (Datson *et al.* 1999; Peters *et al.*
1999; Bosch *et al.* 2000; Neilson *et al.* 2000; Ye *et al.*
2000).

The SAGE method of expression profiling, with
various modifications to increase the likelihood of
identifying genes unambiguously, has been applied
to many different systems, including the yeast
transcriptome (Veculescu *et al.* 1997), the analysis
of changes in mRNA levels associated with cancer
(Polyak *et al.* 1997; Zhang *et al.* 1997) and other
human diseases (e.g. de Waard *et al.* 1999; Ryo *et al.*
1999). In 2000, over 50 research articles were
published featuring SAGE analysis, many characterizing the transcriptional profiles associated with
human diseases either in human tissues or rodent
models. Much of these data are available over the
Internet (Box 9.4). A recent study has used SAGE to
characterize genes associated with developmental
arrest and longevity in *Caenorhabditis elegans*: over
150 genes were identified in this investigation,
including genes encoding histones and a novel
telomere-associated protein (Jones *et al.* 2001).

Massively parallel signature sequencing

In a novel approach to global expression analysis,
Brenner *et al.* (2000) have described massively parallel signature sequencing (MPSS). This is a hybrid
of microarray technology and sequence sampling,

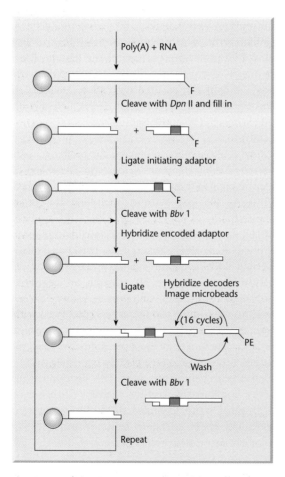

Fig. 9.7 Principle of massively parallel signature sequencing (MPSS) technique. PE = fluorescent label. (Adapted from Brenner 2000.)

in which millions of DNA-tagged microbeads are aligned in a flow cell and analysed by fluorescence-based sequencing. The principle of the technique is that cDNA clones can be sequenced by sequential rounds of cleavage with a type IIs restriction enzyme, followed by adapter annealing, with each adapter able to 'decode' the sequence of bases left in the overhang of the restriction cleavage site. Decoding is achieved through the use of conjugated labels, which can be analysed by flow cytometry. The highly parallel nature of the technique results from the ability to simultaneously analyse the fluorescent signal of thousands of microbeads in a flow cell, and provides the same degree of through-put as large-scale cDNA sequencing.

The method is complex and is shown in Fig. 9.7. Initially, a comprehensive population of 32mer oligonucleotides is synthesized, which are used as tags. At the same time, an equivalent population of complementary 'anti-tags' is synthesized, and co-valently attached to plastic microbeads. The 32mer tags are then mixed as a 100-fold excess with a population of cDNAs, and ligated to form conjugates. In the original method, about 10^7 different 32mers were mixed with a population of about 5×10^4 cDNAs to generate approximately 5×10^{11} different conjugates. One per cent of this mixture was taken, ensuring that each cDNA was likely to be represented and attached to a different tag, i.e. only 1% of the available tags were used. In the next stage of the procedure, amplified tagged cDNAs were end-labelled with a fluorescent probe and attached to microbeads bearing complementary anti-tags. Because only 1% of the tags were used, only 1% of the anti-tags were recognized and 99% of the beads were discarded; this was achieved by fluorescence-activated cell sorting (FACS). Next, the cDNA was cleaved by the restriction enzyme *Dpn*II to remove the fluorescent label and generate a cohesive site to which an initiating primer could anneal. The initiating primer recognizes the *Dpn*II overhang, and carries the recognition site for a type IIs enzyme, *Bbv*1, which cleaves a specific number of bases downstream of the recognition site, therefore chewing a small fragment from the cDNA. The resulting 4-base overhang is dependent on the cDNA sequence not the restriction enzyme. In the next stage, the cleaved cDNA is annealed to a set of 16 encoded adapters. Each adapter recognizes a specific 4-nt overhang and carries a unique sequence at the other end which is recognized by a fluorescent-labelled decoding oligonucleotide. Scanning of the bead after each round of hybridization therefore reveals the 4-bp overhang. Importantly, the encoding adapter also carries a *Bbv*I site, allowing repetition of the process. Therefore, a series of 4-bp calls can be followed on the same microbead in a flow cell, generating a sequence signature for each cDNA.

The accuracy of the method was determined by carrying out MPSS analysis on early and late log phase yeast cells and comparing the signatures obtained with the sequences in public databases; over 90% were represented. Expression analysis was carried out on human THP-1 cells. Over 1.5 million MPSS signatures were obtained from induced THP-1 cells, while nearly 1850 cDNA clones were conventionally sequenced. For most of the genes analysed, the expression levels revealed by cDNA sequence sampling and MPSS analysis were very similar, although there were discrepancies for a small number of genes which remain unexplained. Developments of this technique which allow the detection of unlabelled cDNA sequences (Steemers *et al.* 2000) and which increase sensitivity by using gold nanoparticles rather than fluorescent tags (Taton *et al.* 2000) have been reported.

Expression profiling: standardization of data presentation

Large-scale expression profiling by DNA array hybridization and sequence sampling methods produces a great deal of data. Like DNA sequences and protein structures, enormous benefits are gained if this information is freely shared among the scientific community (Chapter 6). However, unlike DNA sequences and protein structures, gene expression profiles are not easy to present in a standardized manner. In the case of DNA arrays, expression data are in the form of signal intensities from hundreds or thousands of features – for glass arrays, dual fluorescent probes are used to determine differential expression profiles directly, whereas for nylon arrays and Affymetrix GeneChips, pairwise or group comparisons across multiple experiments are required. In the case of SAGE and direct sequence sampling, the expression data are in the form of frequencies of different sequence signatures or concatamerized tags.

Recently, there has been an international effort to develop a set of rules and conventions for the standardization of microarray data presentation. The conventions have been devised by the microarray and gene expression (MAGE) group and are discussed in detail on the Microarray Gene Expression Database (MGED) (http://www.mged.org). One convention is known as minimum information about a microarray experiment (MIAME). It comprises six properties of microarray experiments that can be used as descriptors to ensure that experiments can be repeated accurately. These properties are: overall experimental design, array layout, probe source and labelling method, hybridization procedures and parameters, measurement and normalization procedure and details of any controls. Standard formats for data transfer have also been proposed, including a microarray and gene expression object model and mark-up language (MAGE-OM, MAGE-ML). There are now a number of microarray databases on the Internet (Box 9.3) and many of these are beginning to adopt the MAGE conventions.

Global analysis of protein expression

The proteome

The entire complement of proteins synthesized by a given cell or organism has been termed the *proteome* (Wasinger *et al.* 1995). The field of *proteomics* has arisen around this concept, and encompasses such aims as the identification and cataloguing of all the proteins produced in a cell (the protein equivalent of whole genome annotation), the functional annotation of genes based on protein structure (Chapter 8) and the global analysis of protein–protein interactions using the yeast two-hybrid system and related technologies (Chapter 11). It also includes the analysis of protein expression on a global scale, the protein equivalent of transcriptional profiling.

Considering the spectacular success of the transcriptional profiling techniques discussed above, why is it also necessary to look at protein expression? The main reason is that the ultimate product of most genes is a protein, not an RNA molecule, and it is therefore proteins that carry out the actual function of the gene by interacting with the cellular and extracellular environment. For example, while transcriptional profiling can help to reveal novel drug targets, it is the proteins in the cell that actually interact with drugs and mediate their effects. If there was one-to-one correspondence between mRNAs and proteins this would not be an important issue, but RNA levels do not necessarily reflect either the abundance or diversity of proteins in the cell. Many genes are regulated post-transcriptionally, so protein levels may not be related to RNA levels (e.g. see Gygi *et al.* 1999). Furthermore, extra diversity can be introduced at the level of protein synthesis if a single mRNA gives rise to alternative proteins, or proteins undergo differential post-translational modifications. Thus, just as the transcriptome is more complex than the expressed portion of the genome, the proteome can be more complex still. Many types of post-translational modification are intimately associated with protein function, e.g. phosphorylation. Therefore, direct analysis at the protein level is both beneficial and desirable.

The global analysis of protein expression presents some technical hurdles that are not problematical in transcriptional profiling, and these reflect the concepts of target abundance and molecular recognition. Transcriptional profiling using arrays depends on two processes: the ability to selectively amplify particular DNA molecules as targets for hybridization (this may be achieved by molecular cloning, PCR amplification or direct synthesis in the case of oligonucleotides), and the ability of the probe to hybridize to all targets under similar conditions. Because all DNA and RNA molecules, regardless of

sequence, obey similar hybridization kinetics, thousands of DNA sequences arrayed on a solid substrate can be hybridized simultaneously with a heterogeneous probe. For proteins, there is no 'amplification procedure' as can be applied to nucleic acids, and molecular recognition is mediated by antibodies, other proteins or other ligands, and for each type of protein the reaction kinetics are distinct. Therefore, while there has been considerable interest in the development of 'protein arrays' as a direct parallel to DNA arrays, the difficulty with determining a standard set of conditions for molecular recognition has pushed proteomic research in alternative directions. Currently, the most widely used technology for proteome analysis is protein separation by two-dimensional gel electrophoresis (2DE), followed by annotation using high-throughput mass spectrometry.

Protein separation technology by two-dimensional gel electrophoresis

Technology for the global analysis of proteins was available as long ago as 1975, following the development of a two-dimensional polyacrylamide gel electrophoresis procedure that could be applied to complex protein mixtures such as those extracted from whole cells and tissues (Klose 1975; Scheele 1975). The method involves first dimension isoelectric focusing (i.e. separation in a pH gradient on the basis of charge) followed by second dimension fractionation according to molecular mass. Size fractionation is achieved by first equilibrating the isoelectric focusing gel in a 2% solution of the detergent sodium dodecylsulphate (SDS) which binds non-specifically to all proteins and confers a uniform negative charge. After fractionation, the protein gel is stained. Silver staining and traditional dyes such as Coomassie brilliant blue have been shown to lack compatibility. Recently, a new series of reagents has been developed with proteomics very much in mind, particularly the highly sensitive non-covalent fluorescent stains known as SYPRO dyes (Lim *et al.* 1997). The outcome is a unique pattern of dots, each dot representing a protein, providing a fingerprint of the proteins in the cell (Fig. 9.8). There remain problems of sensitivity. Even moderately abundant proteins may not be seen on a two-dimensional gel if

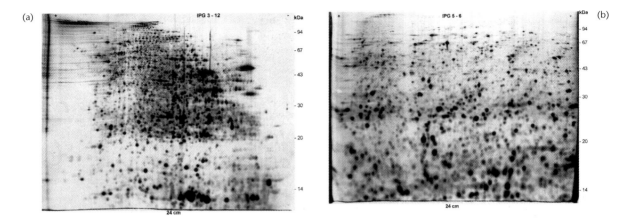

Fig. 9.8 Two-dimensional gel electrophoresis of mouse liver proteins followed by silver staining. (a) Wide immobilized pH gradient (3–12). (b) Narrow immobilized pH gradient (5–6) showing much greater resolution of proteins. (Reproduced with permission from Angelika Görg and *Electrophoresis*.)

the signal is obscured by a nearby spot representing a highly abundant protein, and scarce proteins – which are thought to represent up to 50% of the proteome – may not be detected at all. This can be improved by fractionating the total protein sample prior to electrophoresis, and by running gels that are selective for particular types of protein, e.g. membrane proteins or proteins with a particular pH range (Rabilloud *et al.* 1997; Sanchez *et al.* 1997). Differentially modified variants of the same protein tend to have different mass : charge ratios and therefore occupy different positions on the gel. Up to 25% of proteins on a typical two-dimensional gel may be modified versions of proteins represented elsewhere on the same gel (Celis *et al.* 1995).

The first 'global' 2DE experiment was reported by O'Farrell (1975), who was able to resolve more than 1000 proteins from the bacterium *E. coli*. More recent developments in 2DE have increased the resolution of the technique, allowing up to 10 000 proteins to be resolved in one experiment. Many large-format gels can now be run in parallel, increasing the reproducibility and reliability of the results (reviewed by Herbert *et al.* 2001). In this way, 2DE can be used simply to catalogue proteins, by systematic analysis of every spot. However, by comparing gels prepared using different samples and looking for spots present on one gel but absent or under-represented on another, differentially expressed proteins can be identified. This was first demonstrated in *E. coli*,

where 2DE was used to reveal changes in protein expression patterns between differentially treated samples (O'Farrell & Ivarie 1979). The same principle can be applied to higher organisms. For example, one could compare two-dimensional gels prepared from the tissue of a healthy individual and from another with a particular disease. A protein that is present only in the disease sample could represent a valuable drug target or, if that was not the case, at least as a diagnostic disease marker. The large number and complex distribution of spots on a typical two-dimensional gel means that comparisons across gels must be carried out using image analysis software. One problem with this is that, because of minor variations in the chemical and physical properties of electrophoretic gels, it is impossible to reproduce exactly the conditions from one experiment to another. Gel matching software, such as the freely available program MELANIE II (http://www.expasy.ch/ch2d/melanie), typically works by establishing the positions of unambiguous landmark spots and then stretching, skewing and rotating parts of the image to match other spots. This may involve calculating values for spot intensities and interspot distances between neighbouring spots as variables in the algorithm (reviewed by Pleissner *et al.* 2001). Another useful program is the Java applet CAROL (http://gelmatching.inf.fu-berlin.de/Carol.html) which can be used to compare any two gel images over the Internet.

Alternatives to two-dimensional gel electrophoresis

2DE is the workhorse of proteomics but it has some limitations. This technique has recently been combined with reverse phase microcapillary liquid chromatography (RPMLC) to enhance the separation of proteins. Multidimensional chromatography (serial chromatographic separation according to different properties) has also become a popular alternative to 2DE. Indeed, a procedure involving cation-exchange liquid chromatography followed by RPMLC has been combined with tandem mass spectroscopy to characterize the yeast proteome. This *multidimensional protein identification technology* (shortened to MudPIT) facilitated the identification of nearly 1500 proteins in the yeast proteome (Washburn *et al.* 2001).

High-throughput protein annotation

As early as the beginning of the 1980s it was suggested that a 2DE database should be established to catalogue human proteins and identify proteins whose presence or absence was associated with disease (Anderson & Anderson 1982). However, because there was little information available in sequence databases and no high-throughput techniques for protein annotation, such differentially expressed proteins often remained uncharacterized. In the mid 1980s, the first convenient methods for protein sequencing became available. This began with the development of automated Edman degradation 'sequenators', and a few years later a method was devised for blotting two-dimensional gels onto polyvinylidine difluoride (PVDF) membranes, which allowed direct protein sequencing. However, these techniques require relatively large amounts of starting material and are insufficient to cope with the vast amounts of data arising from today's 2DE experiments.

Mass spectrometry for correlative annotation

A breakthrough in high-throughput protein annotation came with improvements in mass spectrometry techniques coupled with the development of algorithms allowing protein databases to be searched on the basis of molecular mass (Anderson & Mann 2000; Lahm & Langen 2000; Yates 2000; Mann & Pandey 2001). Mass spectrometry involves the ionization of target molecules in a vacuum and accurate measurement of the mass of the resulting ions. A mass spectrometer has three component parts:

1 an ionizer, which converts the analyte into gas phase ions,

2 a mass analyser, which separates the ions according to their mass : charge ratio (m : z); and

3 an ion detector.

Generally, large molecules such as proteins and nucleic acids are broken up and degraded by the ionization procedure but, more recently, sensitive instruments that are capable of *soft ionization*, i.e. the ionization of large molecules without significant degradation, have been developed. This allows accurate mass measurements of whole proteins and peptide fragments, data that can be used to search protein databases to identify particular proteins (Fig. 9.9).

Two major strategies are used to annotate proteins. The first is *peptide mass fingerprinting*. This is generally carried out using a mass spectrometer with a matrix-assisted laser desorption ionization (MALDI) source coupled with a time of flight (TOF) analyser. A glossary of these terms is provided in Box 9.5. Briefly, protein spots are excised from a two-dimensional gel and digested with a specific endopeptidase, such as trypsin, to generate peptide fragments. These are then analysed by mass spectrometry to determine their molecular masses, and these data are used to search protein databases. Computer algorithms have been developed by a number of groups for correlating mass spectrometry determined peptide masses with virtual peptide masses derived from protein databases (Henzel *et al.* 1993; James *et al.* 1993; Mann *et al.* 1993; Pappin *et al.* 1993; Yates *et al.* 1993). The searches are useful only if a significant amount of sequence data exist, i.e. in the case of organisms with complete or well-advanced genome projects. In yeast, for example, it has been possible to calculate the masses of the tryptic peptide fragments from every protein, based on the translation of all known ORFs, thus allowing rapid and precise protein annotation. Unfortunately, EST sequences are not useful for peptide mass fingerprinting because the algorithms require correlation between the masses of several peptide fragments from the same protein and a given database entry.

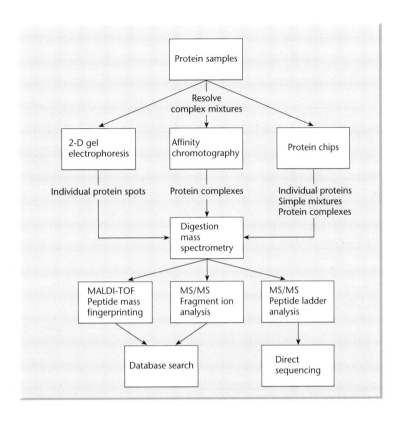

Fig. 9.9 Current routes to protein annotation in proteomics.

ESTs are generally too short to represent a significant number of peptide fragments. The problem is compounded by the propensity for mass spectrometry data not to agree with stored sequences. This may reflect genuine sequence errors in some cases, but often occurs because of unrecognized post-translational modifications to the protein, non-specific proteolysis, the presence of contaminating proteins, or the absence of a particular sequence from the database. The accuracy of the mass spectrometry measurement may also vary, although this has been addressed by the development of improved MALDI-MS instruments with ion focusing mirrors and delayed extraction devices to help reduce their spread of kinetic energy and thus increase accuracy and resolution.

The limitations of peptide mass fingerprinting can be overcome in a second strategy, called fragment ion searching. Fragmentation is generally achieved by tandem mass spectroscopy (MS/MS) either with quadrupole or TOF analysers, or a mixture, separated by collision cell to induce fragmentation. Searching databases with both a peptide mass and the mass of

its fragment ions is very useful, because fragment ions can be searched against ESTs (see Lamond & Mann 1997, reviewed by Choudhary *et al.* 2001).

Mass spectrometry for de novo sequencing

The utility of peptide mass fingerprinting and fragment ion searching is limited to those species for which sufficient sequence data exist. However, MS/MS can also be used for *de novo* protein sequencing. In a similar theme to chain terminator DNA sequencing, the technique relies on the production of comprehensive peptide ladders, i.e. nested collections of peptides from a single protein, whose length differs by a single amino acid. The masses of the peptides are ranked, and the differences between them compared to a standard table of amino acids. In this way, the sequence can be unambiguously determined, although leucine and isoleucine cannot be distinguished because they have the same mass. Note that the method is complicated by the fact there are actually two series of fragments generated at the same time: a *B series* (N-terminal fragments)

Box 9.5 Glossary of terms used in mass spectrometry, and useful Internet resources

Mass spectrometry

A technique for accurately determining molecular masses by calculating the mass : charge ratio of ions in a vacuum. A mass spectrometer is an instrument combining a source of ions, a mass analyser that can separate ions according to their mass : charge ratio, and an ion detector.

Soft ionization

The ionization of large molecules, such as proteins and nucleic acids, without causing significant amounts of fragmentation.

Matrix-assisted laser desorption ionization (MALDI)

A soft ionization method used for peptide mass fingerprinting. The analyte, a mixture of peptide fragments resulting from tryptic digestion of a particular protein, is first mixed with a light-absorbing 'matrix compound' such as dihydroxybenzoic acid, in an organic solvent. The solvent is then evaporated to form crystals and these are transferred to a vacuum. The dry crystals are targeted with a laser beam. The laser energy is absorbed and then emitted (desorbed) as heat, resulting in expansion of the matrix and anylate into the gas phase. A high voltage is applied across the sample to ionize it, and the ions are accelerated towards the detector.

Electrospray ionization (ESI)

A soft ionization method used for fragment ion searching. The analyte is dissolved in an appropriate solvent and pushed through a narrow capillary. A potential difference is applied across the capillary such that charged droplets emerge and form a fine spray. Under a stream of heated inert gas, each droplet rapidly evaporates so that the solvent is removed as the analyte enters the mass analyser and the ions are accelerated towards the detector.

Quadrupole

A mass analyser that determines the mass : charge ratio of an ion by varying the potential difference applied across the ion stream, allowing ions of different mass : charge ratios to be directed towards the detector. A quadrupole comprises four metal rods, pairs of which are electrically connected and carry opposing voltages which can be controlled by the operator. More than one quadrupole may be connected in series, as in triple quadrupole mass spectrometry. Varying the voltage steadily over time allows a mass spectrum to be obtained.

Time of flight (TOF)

A mass analyser that determines the mass : charge ratio of an ion by measuring the time taken by ions to travel down a flight tube to the detector (Karas & Hillenkamp 1988).

Tandem mass spectrometry (MS/MS)

Mass spectrometry using an instrument with two mass analysers, either of the same type or a mixture. A number of hybrid quadrupole/TOF instruments have been described (e.g. Krutchinsky *et al.* 2000; Shevchenko *et al.* 2000). The mass analysers may be separated by a collision cell that contains inert gas and causes ions to dissociate. MS/MS is generally used for fragment ion analysis, because particular peptide fragments can be selected using the first mass analyser, fragmented in the collision cell and the fragments can be separated in the second analyser.

Collision-induced dissociation (CID)

The use of a collision cell between mass analysers to excite ions and make them dissociate into fragments.

Mass spectrometry resources on the Internet (peptide mass and fragment fingerprinting tools)

http://www.expasy.ch/tools/#proteome
http://www.seqnet.dl.ac.uk/Bioinformatics/
http://www.narrador.embl-heidelberg.de/
  GroupPages/PageLink/peptidesearchpage.html
http://prospector.ucsf.edu

and a *Y series* (C-terminal fragments). Unambiguous sets of ladders can be obtained by labelling the C-terminus of the protein for example, so that each Y-peptide has an additional known mass attached to it and can be identified on this basis. The short sequences thus obtained can be used directly in homology searches with dedicated algorithms such as MS-BLAST.

Two-dimensional gel electrophoresis databases

A number of databases have been established to store annotated digital images of 2DE gels. There has been some success in standardizing the presentation of 2DE data, including the use of grid references representing pI values and molecular masses, and the annotation of known proteins with links to database entries in SWISS-PROT (Appel *et al.* 1996). Conveniently, a list of current 2DE databases is maintained on the WORLD-2DPAGE (http://www.expasy.ch/ch2d/2d-index.html) and this site includes links to further resources that provide laboratory services, training and data analysis software relevant to 2DE.

Protein arrays and protein chips

Two-dimensional gel electrophoresis is an open system for proteome analysis, rather in the same way as direct sequence sampling is an open system for transcriptome analysis (p. 148). The advantage of an open system is that potentially all proteins can be detected, but the disadvantage is that they also have to be characterized, which relies on downstream annotation by mass spectrometry (similarly, directly sampled cDNAs have to be characterized by downstream sequencing). DNA arrays are closed systems in transcriptome analysis, i.e. the data obtained are constrained by the number and nature of sequences immobilized on the array. However, it is not necessary to characterize any of the features on the array by sequencing because the sequences are already known. Similarly, *protein arrays* are emerging as a useful closed system for proteome analysis. These are miniature devices in which proteins, or molecules that recognize proteins, are arrayed on the surface (Cahill 2000; Walter *et al.* 2000; Wilson & Nock 2001; Zhu & Snyder 2001; Schweitzer & Kingsmore 2002; Templin *et al.* 2002)

In concept, protein arrays are no different to DNA arrays, but they suffer from several practical limitations. First, the manufacture of DNA arrays is simplified by the availability of methods, such as PCR, for amplifying any nucleic acid sequence. No amplification procedure exists for proteins. Secondly, all DNA sequences are made of the same four nucleotides and hence behave similarly in terms of their chemical properties. The principles of molecular recognition (hybridization between complementary base pairs) apply to all sequences. For this reason, hybridization reactions can be carried out in highly parallel formats using a single complex probe. Conversely, proteins are made of 22 amino acids with diverse chemical properties; e.g. some proteins are soluble in water while others are lipophilic. Recognition parameters vary widely so the same reaction conditions could never be used for all proteins. Thirdly, the homogeneity of DNA molecules means that labels are incorporated evenly and labelling does not interfere with hybridization. Binding to solid substrates such as nylon and glass does not interfere with hybridization either. However, the labelling of proteins is much more variable and both labelling and attachment to a substrate could interfere with protein binding, either by affecting the way the protein folds or by blocking the binding site. Despite these differences, protein arrays have been manufactured in many of the ways discussed above for DNA arrays, including variations on standard contact printing (Lueking *et al.* 1999; Mendoza *et al.* 1999), inkjetting (Ekins 1998) and photolithography (Jones *et al.* 1998; Mooney *et al.* 1998).

The concept of the protein array, protein microarray or protein chip covers a wide range of different applications. Some of these are considered below.

Antibody arrays (microimmunoassays)

On these devices, antibodies are attached to the array surface, so the protein array or chip can be thought of as a miniaturized solid state immunoassay. Antibodies interact with specific proteins and are highly discriminatory, so they are suited to the detailed analysis of protein profiles and expression levels. The feasibility of this approach has been demonstrated, e.g. using a recombinant Staphylococcal protein A covalently attached to a gold surface. The recombinant protein A has five immunoglobulin G binding domains, allowing antibodies to be attached by the Fc region, therefore exposing the antigen-binding domain (Kanno *et al.* 2000). Antibody arrays have also been generated by using banks of bacterial strains expressing recombinant antibody molecules (de Wildt *et al.* 2000). There are three different formats for this type of assay.

1 A standard immunoassay in which the antibodies are immobilized and are used to capture labelled proteins from solution. Protein expression levels are quantified by measuring the signal (usually fluorescent) which has been incorporated into the proteins. A recent report (Haab *et al.* 2001) describes the use of such an array comprising 115 different antibodies. Another ground-breaking aspect of this report was that two protein samples, each labelled with a different fluorophore, were exposed to the array simultaneously, and differential protein expression could be monitored.

2 A miniature sandwich assay, in which unlabelled proteins are captured from solution and detected with a second, labelled antibody. Although this format requires two antibodies recognizing distinct epitopes for each protein, it is not necessary to label the target population of proteins, a process which is inefficient and variable. An example of this approach for the detection of human cytokines is provided by Moody *et al.* (2001).

3 The third format involves a tertiary detection system and therefore offers even greater sensitivity. One example is the *immunoRCA technique* which involves rolling circle amplification (Schweitzer *et al.* 2000). The principle of this technique is that a protein, captured by an immobilized antibody, is recognized by a second antibody in a sandwich assay as above, but the second antibody has an oligonucleotide covalently attached to it. In the presence of a circular DNA template, DNA polymerase and the four dNTPs, rolling circle amplification of the template occurs resulting in a long concatamer comprising hundreds of copies of the circle, which can be detected using a fluorescent-labelled oligonucleotide probe.

Antigen arrays (reverse microimmunoassays)

On these devices, protein antigens are attached to the array surface. They are used for reverse immunoassays, i.e. detecting antibodies in solution (see e.g. Joos *et al.* 2000; Paweletez *et al.* 2001).

General protein arrays

This type of device can contain any type of protein, and is used to assay protein–protein interactions and protein interactions with other molecules. A range of detection strategies may be used, including labelling of the interacting molecules, or label-independent methods such as surface plasmon resonance (see below). Ge *et al.* (2000) developed a system for studying molecular interactions using a universal protein array (UPA) system where protein samples are transferred from 96-well microtitre plates to nylon membranes. The technology has also been applied to the arraying of cDNA expression libraries, such that screening can be carried out not only with nucleic acid probes but also with antibodies or other ligands directed at the recombinant proteins (Bussow *et al.* 1998). One of the most impressive demonstrations of the power of protein array technology was provided by MacBeath & Schreiber (2000). They used an array of proteins on a glass slide to screen for ligands, enzyme substrates and protein–protein interactions.

Functional assays can also be carried out on such arrays. Zhu *et al.* (2000) produced an array containing nearly all the protein kinases of the yeast proteome and carried out kinase assays on 17 different substrates. More notably, the same group also produced a glass microarray containing nearly all the proteins in the yeast proteome (5800 spots) and used this to screen for various functions such as phospholipid binding and interactions with calmodulin (Zhu *et al.* 2001).

Arrays of specific capture agents

These are not protein arrays in the strict sense because they do not consist of arrayed proteins. However, they are considered here because they are analogous to antibody arrays, i.e. they contain specific capture agents that interact with proteins. DNA arrays fall within this class if they are used to analyse DNA–protein interactions (see e.g. Buluyk *et al.* 2001; Iyer *et al.* 2001). Aptamers, single-stranded nucleotides that interact specifically with proteins, could also be used in this manner (reviewed by Mirzabekov & Kolchinsky 2001).

Molecular imprints

We move on now to devices which can be called protein chips but do not really fit under the umbrella of arrays because nothing is actually arrayed on the surface. An area of active current research is the development of chips containing artificial recognition sites for proteins. The concept of *molecularly imprinted polymers* (MIPs) depends on the ability to emboss a polymeric substrate with recognizable molecular imprints that mimic the actual recognition molecules. Shi *et al.* (1999) have described a procedure for coating recognition molecules in sugar, which is then overlain by a hexafluoropropylene polymer. They coated a mica surface with streptavidin and then covered it with a disaccharaide which moulded the shape of the protein. After applying the polymer, the protein and mica were removed leaving a streptavidin MIP. This was subsequently shown to bind biotin preferentially. It is not yet clear whether this technology has the sensitivity or specificity to be applied to proteome-wide expression analysis.

Broad-capture agent biochips

Instead of specific molecular interactions, these devices use broad-specificity capture agents. As above, they can be termed protein chips but not protein arrays. Ciphergen Biosystems Inc. market a range of ProteinChips with various surface chemistries to bind different classes of proteins. Although relatively non-specific compared to antibodies, complex mixtures of proteins can be simplified and then analysed by mass spectrometry. An advantageous feature of this system is the ease with which it is integrated with downstream MS analysis, because the ProteinChip itself doubles as a modified MALDI plate (Fung *et al.* 2001; Weinberger *et al.* 2001). After the chip has been washed to remove unbound proteins, it is coated with a matrix solution and analysed by TOF-MS. This allows surface-enhanced laser desorption and ionization (SELDI) which provides more uniform mass spectra than MALDI and allows protein quantification.

Other protein chip platforms use surface plasmon resonance to detect and quantify protein binding. This involves measuring changes in the refractive index of the chip surface caused by increases in mass (Malmqvist & Karlsson 1997). Protein chips produced by the US company BIAcore are based on this concept, and other chips combine surface plasmon resonance measurements with MS (Nelson *et al.* 2000).

Solution arrays

Recent developments in microfluidic devices indicate that the next generation of miniature protein assay platforms may be *solution arrays*. These could provide increased sensitivity (as a result of the kinetics of binding in solution) and higher throughput (because the arrays are constructed in three dimensions). For a rewarding review of this emerging technology, including details of arrays based on fluorescent beads and barcoded gold nanoparticles, see Zhou *et al.* (2001).

Suggested reading

Schena M., Shalon D., Davis R.W. & Brown O.P. (1995) Quantitative monitoring of gene expression patterns with a complementary DNA microarray. *Science* **270**, 467–470.

Veculescu V.E., Zhang L., Vogelstein B. & Kinzler K.W. (1995) Serial analysis of gene expression. *Science* **270**, 484–488.

Two seminal papers in the development of high-throughput expression profiling at the mRNA level.

Altman R.B. & Raychaudhuri S. (2001) Whole-genome expression analysis: challenges beyond clustering. *Current Opinion in Structural Biology* **11**, 340–347.

Blohm D.H. & Guiseppi-Elie A. (2001) New developments in microarray technology. *Current Opinion in Biotechnology* **12**, 41–47.

Bowtel D.D.L. (1999) Options available – from start to finish – for obtaining expression data by microarray. *Nature Genetics* **21** (Suppl), 25–32.

Lander E.S. (1999) Array of hope. *Nature Genetics* **21** (Suppl), 3–4.

Lipshutz R.J., Fodor S.P.A., Gingeras T.R. & Lockhart D.J. (1999) High density synthetic oligonucleotide arrays. *Nature Genetics* **21** (Suppl), 20–24.

Quackenbush J. (2001) Computational analysis of microarray data. *Nature Reviews Genetics* **2**, 418–427.

A selection of excellent reviews charting the development of array technology for expression profiling, and recent advances in manufacture, methodology and data analysis.

Andersen J.S. & Mann M. (2000) Functional genomics by mass spectrometry. *FEBS Letters* **480**, 25–31.

Lee K.H. (2001) Proteomics: a technology-driven and technology-limited discovery science. *Trends in Biotechnology* **19**, 217–222.

Schweitzer B. & Kingsmore S.F. (2002) Measuring proteins on microarrays. *Current Opinion in Biotechnology* **13**, 14–19.

Templin M.F., Stoll D., Schrenk M., Traub P.C., Vohringer C.F. & Joos T.O. (2002) Protein microarray technology. *Trends in Biotechnology* **20**, 160–166.

Zhou H., Roy S., Schulman H. & Natan M.J. (2001) Solution and chip arrays in protein profiling. In: *A Trends Guide to Proteomics II*, supplement to *Trends in Biotechnology* **19**, S34–S39.

A selection of reviews discussing 2DE and mass spectrometry in proteome research, and alternative technologies based on protein chips.

Useful websites

http://www.ebi.ac.uk/microarray
Information on cDNA microarrays for transcriptional profiling.

http://www.affymetrix.com
Affymetrix, developers of the GeneChip oligonucleotide arrays.

http://www.expasy.ch/ch2d/2d-index.html
Worldwide 2DE homepage.

CHAPTER 10

Comprehensive mutant libraries

Introduction

One of the most powerful ways to determine the function of a gene is to mutate it and study the resulting phenotype. In this respect, the link between gene and function can be approached from two directions. Traditional phenotype-driven 'forward genetics' involves random mutagenesis followed by screening to recover mutants showing impairment for a particular biological process. The essence of forward genetics is that one starts by identifying a mutant phenotype and then works towards the gene by mapping and cloning. 'Reverse genetics' is the opposite approach, where one starts with a cloned gene whose function is unknown. This gene is mutated deliberately and reintroduced into the host organism to study its effect. In the post-genome era, gene-driven reverse genetics is assuming more and more importance as thousands of gene sequences with unknown functions accumulate in databases. In order to assign functions to these anonymous sequences, researchers have sought high-throughput strategies to mutate every gene in the genome, producing comprehensive genome-wide mutant libraries. Such libraries have been generated in a number of model organisms, ranging from bacteria and yeast to plants and mammals (Coelho *et al.* 2000; Hamer *et al.* 2001; Ramachandran & Sundaresan 2001).

There are three basic types of mutational genomics strategy. First, the systematic approach of deliberately mutating every single gene in the genome, one at a time, and generating banks of specific mutant strains. This can only yield a comprehensive mutant library if the entire genome sequence is available. The second is a random approach in which genes are mutated indiscriminately. Individual mutations are then catalogued by obtaining flanking sequence tags, and genes are annotated by matching the tags to entries in sequence databases. This method can be applied to any species, even if there is little or no existing sequence information. Moreover, in species with complete or well-advanced genome projects, random mutagenesis may uncover genes that have been missed by other annotation methods. Each of these strategies has further advantages and disadvantages. Although the systematic approach provides exhaustive genome coverage, it is a labour-intensive process and depends on pre-existing sequence information. The random mutagenesis approach is rapid and relatively inexpensive, but there is no control over the distribution of mutations so saturation may be difficult to achieve. The third approach encompasses a group of techniques which generate functional *phenocopies* of mutant alleles, i.e. the likeness of a mutation without actually altering the DNA sequence of the organism. Examples include the use of genome-wide RNA interference in the nematode *Caenorhabditis elegans*, and virus-induced gene silencing in plants.

High-throughput systematic gene knockout

Mutations can be introduced into pre-defined genes *in vivo* through a process termed *gene targeting* which involves homologous recombination. Where the aim is to completely inactivate the target gene and generate a null allele, the term *gene knockout* is often used. Homologous recombination occurs to a greater or lesser degree in all organisms, but the efficiency varies considerably. In bacteria and yeast, and also in certain mosses, the process is highly efficient and gene transfer with a suitable targeting vector results in homologous recombination more than 90% of the time. Because microbial genomes also contain fewer genes than those of higher eukaryotes, these species are ideally suited to a functional genomics strategy based on systematic gene knockouts.

Homologous recombination in higher eukaryotes occurs at a much lower efficiency. Even if a homologous target is available in the genome, DNA introduced into most animal and plant cells is 100 000 times more likely to integrate randomly than recombine with its target. Until very recently, the only higher eukaryote species that was amenable to gene targeting technology was the mouse (reviewed by Muller 1999), and this is because of the special properties of embryonic stem (ES) cells. ES cells can be cultured like any established cell line but they are derived from the very early mouse embryo and are therefore *pluripotent*. This means that if the cells are injected into a mouse blastocyst they can colonize the embryo and contribute to all its tissues, including the germ line. The other important property of ES cells is that they have an unusual propensity for homologous recombination. Although random integration still occurs 1000-fold more frequently, PCR-based screening or appropriate selection strategies can be use to identify correctly targeted cells. These can be injected into mouse blastocysts to give rise to genetic chimaeras. If colonization of the germ line has occurred in these animals, their offspring will carry the targeted mutation in every cell.

Very recently, homologous recombination has also been achieved in *Drosophila* (Rong & Golic 2000) and in sheep fibroblasts subsequently used for nuclear transfer (McCreath *et al.* 2000). Gene targeting has also been achieved in plants (e.g. Kempin *et al.* 1997) but the efficiency is extremely low.

Functional genomics by systematic gene knockout in yeast

The genome of *Saccharomyces cerevisiae* contains about 6200 open-reading frames, which by comparison to higher eukaryotes is a small number. Since the yeast genome has been completely sequenced, several systematic gene knockout projects have been initiated. One is being carried out by a consortium of European research organizations named the European Functional Analysis Network (EUROFAN; Dujon 1998). This project involves the use of PCR-generated targeting cassettes in which a selectable marker is placed between ~50 bp sequences corresponding to the flanking sequences of each yeast gene. Targeting with such constructs

Fig. 10.1 Barcoding strategy for yeast deletion strains. Dark green boxes represent yeast homology regions, which recombine with the endogenous gene (crosses). Grey boxes represent unique oligonucleotide barcodes for unambiguous strain identification. Pale green represents selectable marker gene. See main text for details.

results in the replacement of the entire endogenous coding region with the marker, thus generating a null allele (Baudin *et al.* 1993). In a similar strategy, the *Saccharomyces* Gene Deletion Project (SGDP) consortium, which comprises a number of US and European research institutions, has generated targeting cassettes corresponding to about 85% of yeast genes (Winzeler *et al.* 1999). Each contains, in addition to the selectable marker and yeast flanking sequences, two unique 20-bp 'barcodes' placed just inside the yeast homology region at each end (Fig. 10.1). These provide a means to rapidly detect the presence of specific strains in a population by hybridization of PCR-products to an oligonucleotide chip, known as a barcode chip (Shoemaker *et al.* 1996). Thus, the growth properties of potentially all the targeted yeast strains can be assayed in parallel (Giaever *et al.* 1999).

The advantage of these systematic approaches is that strains can be maintained as a central resource and then distributed to laboratories worldwide for functional analysis. Typically, each mutant strain will be subjected to many different assays in parallel to rapidly determine the function of the missing gene product. Such functions may be described in terms of normal metabolism or cellular activity (e.g. failure to synthesize a particular metabolite, inability to carry out meiosis, etc.) or specialized screening may be carried out (e.g. to assay for sensitivity or resistance to particular drugs; Bianchi *et al.* 1999). Both EUROFAN and SGDP have websites where further information on the yeast knockout programmes can be found, and specific strains can be obtained (Box 10.1).

Prospects for comprehensive gene targeting projects in other organisms

Despite the benefits of gene targeting, in particular

Box 10.1 Internet resources for gene targeting

Websites providing information on systematic knockout projects in yeast

http://mips.gsf.de/proj/eurofan/index.html
The EUROFAN website (Dujon 1998)
http://sequence-www.stanford.edu/group/
yeast_deletion_project The *Saccharomyces* Gene
Deletion Project website (Winzeler *et al.* 1999)

Websites containing information and resources for transgenic and gene targeted mice

http://jaxmice.jax.org/index.shtml The Jackson
Laboratory website, describing over 2500 strains
of targeted mutant mice
http://tbase.jax.org TBASE, a comprehensive
transgenic and targeted mutant database run by
the Jackson Laboratory
http://www.bioscience.org/knockout/knochome.htm
Frontiers in Science gene knockout database
http://biomednet.com/db/mkmd BioMedNet
Mouse Knockout Database

the precision with which specific mutant alleles can be designed, there has been no co-ordinated programme for systematic gene knockouts in the mouse. However, as a result of the efforts of individual researchers, it is likely that up to 20 000 independently derived targeted mouse strains will have been produced by the time the mouse genome project is completed (Capecchi 2000). Depending on which of the current estimates of the total gene number is most accurate, this could represent 20–50% of all genes. Several excellent and comprehensive databases are available on the Internet which list these mouse strains (Box 10.1). It is therefore possible that a mouse knockout project could evolve over the next 5 years with the aim of completing the knockout catalogue and providing a comprehensive genome-wide data set.

In the near future, the only other higher eukaryote for which a systematic gene knockout project seems likely is the fruit fly, *Drosophila melanogaster*. The current gene disruption programme, which involves random mutagenesis with P-elements (see below), is restricted by the sequence specificity of P-element insertions. For this reason, only one-third of the genes have been tagged. Also, a significant number of insertions fail to generate mutant phenotypes (Spradling *et al.* 1995, 1999). Gene targeting could address both these issues, and help to functionally annotate the estimated 7000–10 000 genes that remain to be analysed.

Genome-wide random mutagenesis

Saturation mutagenesis has been used for many years to identify mutations affecting specific biological processes. Essentially, the idea behind such an experiment is to mutagenize a population of whatever species is under study and recover enough mutants to stand a reasonable chance that each gene in the genome has been 'hit' at least once. This population can then be screened to identify mutants in a particular function. Large-scale screening has been carried out in the past to look for replication mutants in bacteria, cell cycle mutants in yeast and, more recently, developmental mutants in *Drosophila* and the zebrafish. The difference between these traditional studies and the new science of functional genomics is that, in the former, most of the mutants were discarded. Researchers focused on a particular area and ignored mutants affecting other processes because they were not interested in them. In functional genomics, *all* mutations are interesting, and the idea is to catalogue them, generate a sequence signature from each affected gene, and use these signatures to annotate full-length genomic and cDNA sequences housed in databases.

Insertional mutagenesis

By far the most popular mutagenesis strategy in functional genomics is *insertional mutagenesis*, where

Box 10.2 Gene traps and other advanced gene tagging vectors

Gene tagging is the use of an insertional mutagen to mark interrupted genes with a unique DNA sequence. This DNA sequence subsequently can be used as a target for hybridization or as an annealing site for PCR primers, allowing flanking sequences to be isolated. By careful design, however, simple gene tagging vectors can be modified in a number of ways to expedite cloning and provide more information about the interrupted genes. Some of these refinements are considered below.

Plasmid rescue vector

The insertion element in this type of vector contains the origin of replication and antibiotic resistance marker from a bacterial plasmid (Perucho *et al.* 1980). Genomic DNA from a tagged organism is digested with a restriction enzyme that does not cut in the insert, and the resulting linear fragments are self-ligated to form circles. The complex mixture of circles is then used to transform bacteria, which are grown under antibiotic selection. The circle containing the origin of replication and resistance gene is propagated as a plasmid while all the other circles are lost. In this way, the genomic sequences flanking the insert can be isolated and selectively amplified in a single step. Although more time-consuming than the direct amplification of flanking sequences by PCR, 'rescued' plasmids can be maintained as a permanent resource library.

Gene trap vector

In this vector, the insertion element contains a visible marker gene such as *lacZ* (encoding β-galactosidase) or *gusA* (encoding β-glucuronidase) downstream of a splice acceptor site (Gossler *et al.* 1989; Friedrich & Soriano 1991; Skarnes *et al.* 1992; Wurst *et al.* 1995). The marker gene is therefore activated only if the element inserts within the transcription unit of a gene and generates a transcriptional fusion. This strategy selects for insertions into genes and is very useful in animals and plants with large amounts of

non-genic DNA (Evans *et al.* 1997; Springer 2000). Early gene trap vectors depended on in-frame insertion, so up to two-thirds of all 'hits' on genes were not recognized. Furthermore, expression of the marker relied on the transcriptional activity of the surrounding gene, so inserts into non-expressed genes were not detected. The use of internal ribosome entry sites has obviated the need for in-frame insertion and has greatly increased the hit rate of gene traps. The incorporation of a second marker, which is driven by its own promoter but carries a downstream splice donor making it dependent on the surrounding gene for polyadenylation, has facilitated the detection of non-expressed genes (Zambrowicz *et al.* 1998).

Enhancer trap vector

This construct comprises a visible marker gene downstream of a minimal promoter. Under normal circumstances the promoter is too weak to activate the marker gene and it is not expressed. However, if the construct integrates in the vicinity of an endogenous enhancer, the marker is activated and reports the expression profile driven by the enhancer (O'Kane & Gehring 1987). The enhancer is often a long way from the gene so enhancer trapping is not a convenient method for cloning novel genes. However, it can be exploited in other ways, e.g. to drive the expression of a toxin and thus ablate a specific group of cells. This technique has been widely used in *Drosophila* (O'Kane & Moffat 1992).

Activation tagging

In this technique, the insertion element carries a strong outward-facing promoter. If the element integrates adjacent to an endogenous gene, that gene will be activated by the promoter. Unlike other insertion vectors, which cause loss of function by interrupting a gene, an activation tag causes gain of function through overexpression or ectopic expression (Kakimoto 1996).

a piece of DNA is randomly inserted into the genome causing gene disruption and loss of function. The DNA may constitute a *transposable element*, i.e. a sequence that can jump from site to site in the genome when supplied with the necessary enzyme (*transposase*), or it may be a foreign DNA sequence which is introduced into the cell. The main advantage of this strategy over traditional forms of mutagenesis is that the interrupted gene becomes 'tagged' with the insertion element, hence the strategy is sometimes termed *signature-tagged mutagenesis* (STM). Simple hybridization- or PCR-based techniques can be used to obtain the flanking DNA. The sequence of this flanking DNA can then be used to interrogate sequence databases, allowing the tagged gene to be associated with its 'parent' genomic clone or cDNA. If the insertion also generates a mutant phenotype, the gene in the database can then be ascribed a tentative function. Another advantage of insertional mutagenesis is that the insertion element can be modified into a gene trap vector to provide information about the gene it interrupts. The principle of the gene trap is outlined in Box 10.2.

Genome-wide insertional mutagenesis in yeast

In yeast, both endogenous transposons and heterologous (bacterial) transposons have been used for saturation mutagenesis. The endogenous retrotrans-poson *Ty* has been used as an insertional mutagen, and libraries of mutants have been generated carrying the *Ty* element as a 'genetic footprint' (Smith *et al.* 1995, 1996). Several copies of the *Ty* element are normally present in the yeast genome, so an element modified to carry a unique DNA signature was used as a mutagen to enable unambiguous identification of interrupted genes. A PCR-based strategy was developed in which a primer annealing to the modified element was used in combination with a number of gene-specific primers to identify insertions at particular loci (Fig. 10.2). This strategy, which has been widely adopted in other functional genomics programmes (see below), allows highly parallel analysis of large populations of yeast cells, increasing the likelihood that an insertion will be detected in a given gene. A disadvantage of *Ty*-based functional genomics is the tendency for the element to insert at the 5′ end of genes transcribed by RNA polymerase III, i.e. mainly tRNA genes (Ji *et al.* 1993, p. 28). Protein-encoding genes, which represent the vast majority of the transcriptome, are transcribed by RNA polymerase II.

Genome-wide insertional mutagenesis in yeast has also been performed with an *Escherichia coli* Tn*3* transposon modified to make a very sophisticated gene trap vector (Ross-Macdonald *et al.* 1997, 1999). Transposons from *Escherichia coli* are not mobilized in yeast, but mutagenesis can be carried

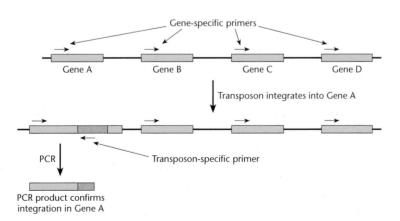

Fig. 10.2 Principle of target-selected random mutagenesis. Random insertional mutagenesis is carried out on a population. Mutants for any particular gene can then be identified by PCR amplification using one gene-specific primer and one primer specific for the insertional mutagen. For organisms with completely sequenced genomes, inserts can be identified in any gene. For organisms with incomplete genome sequences, the target gene has to be isolated and sequenced first so that gene-specific primers can be designed. In large populations, pools of cells/seeds/embryos, etc. can be screened and then deconvoluted to identify individual mutant strains.

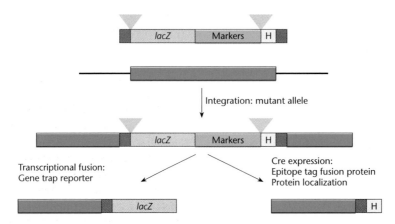

Fig. 10.3 Multifunctional *Escherichia coli* Tn 3 cassette used for random mutagenesis in yeast. The cassette comprises Tn 3 components (black), *lacZ* (grey), selectable markers (light green) and an epitope tag (very light green, H). The *lacZ* gene and markers are flanked by *loxP* sites (triangles). Integration generates a mutant allele which may or may not reveal a mutant phenotype. The presence of the *lacZ* gene at the 5′ end of the construct allows transcriptional fusions to be generated, so the insert can be used as a reporter construct to reveal the normal expression profile of the interrupted gene. If Cre recombinase is provided, the *lacZ* gene and markers are deleted, leaving the endogenous gene joined to the epitope tag, allowing protein localization to be studied.

out in the surrogate environment of a bacterial cell transformed with a yeast genomic construct. From there, the genomic DNA is reintroduced into yeast cells, where it recombines with the endogenous genome, replacing the existing sequence. Ross-Macdonald *et al.* have generated a bank of *E. coli* strains housing genomic DNA clones covering the entire yeast genome. Mutant clones were then isolated from these strains and reintroduced into diploid yeast cells, generating a library of 12 000 mutant yeast strains. There are several advantages to this strategy, including the fact that stable insertions are generated, allowing the production of mutant lines as a permanent resource, and that *E. coli* transposons are not subject to the insertional bias of yeast *Ty* elements, and therefore provide a better random coverage of the genome. Even so, there is still a certain degree of site selection, resulting in the preferential recovery of mutant alleles for some genes and the omission of others. The modified Tn 3 transposon comprises a reporter gene downstream of a splice acceptor bracketed by subterminal *loxP* sites, allowing excision of most of the insert following expression of Cre recombinase. Outside the *loxP* sites, the transposon also carries an in-frame epitope tag. Thus, following Cre-mediated excision of the transposon, the endogenous gene becomes

fused to the epitope tag, allowing the characterization and localization of proteins (Fig. 10.3).

Genome-wide insertional mutagenesis in vertebrates

Two of the key 'functional' objectives of the mouse genome project are the development of a systematic genome-wide series of deletion strains and the assembly of a comprehensive insertional mutagenesis library in ES cells (Denny & Justice 2000). The first objective is being addressed in the form of the DELBank project, overseen by John Schimenti at the Jackson Laboratory, Bar Harbor, Maine, USA. Deletion strains are generated by introducing a negative selectable marker into ES cells, irradiating them to induce deletions, and then selecting cells where the deletion has eliminated the marker. Mapping the marker gene prior to irradiation therefore provides a route to generating deletions at specific chromosomal regions. The initial aim of the project is to assemble a panel of 500 or more ES cell lines with the marker distributed at 5–10 cM intervals, such that deletions can be induced in any region of the genome.

The procedure for generating deletions was described by You *et al.* (1997) and is shown in Fig. 10.4.

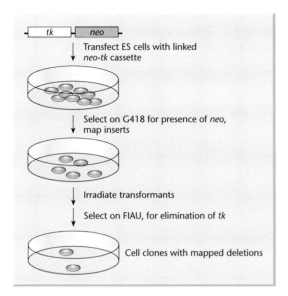

Fig. 10.4 Procedure for generating defined chromosomal deletions in mouse embryonic stem (ES) cells.

First, a construct carrying the positive marker *neo* and the negative marker *tk* is introduced into ES cells by transfection. The *neo* marker confers antibiotic resistance, allowing the cells to grow in the presence of G418, while the *tk* marker encodes the enzyme thymidine kinase, which makes cells sensitive to toxic thymidine analogues such as fialuridine (FIAU). Successfully targeted cells are propagated under antibiotic selection and then irradiated to induce chromosome breaks. The breaks occur randomly and deletions arise as the DNA is repaired. In a few of the cells the deletion will actually eliminate the *tk* marker and only these cells will survive in the presence of FIAU.

Two techniques have been used for mapping the inserts. Initially, plasmid vectors were used for transfection and flanking sequences were obtained by plasmid rescue (Box 10.2). Polymorphisms detected in these flanking sequences were then mapped against an interspecific backcross panel. Later, retroviral gene trap vectors were used for transfection and inserts into genes were identified. The flanking sequences could therefore be rescued by a technique called 3′ rapid amplification of cDNA ends (RACE) (Zambrowicz *et al.* 1998) and used to search cDNA and EST databases for matches. In many cases, mouse or human cDNAs and ESTs have

already been mapped onto the genome, so further mapping of the inserts was unnecessary.

Gene trapping in ES cells is being carried out by several groups to generate comprehensive ES cell-based insertional mutagenesis libraries. Tagging by insertional mutagenesis has a long history in mouse genetics. It has been demonstrated serendipitously in transgenic mice in cases where the transgene has by chance integrated into an endogenous gene and inactivated it (see e.g. Rijkers *et al.* 1994). Also, it is notable that approximately 5% of all naturally occurring recessive mutations in the mouse are caused by a particular family of retroviruses when one of them integrates into or adjacent to an endogenous gene. Indeed, the cloning of several mouse genes has been facilitated by virtue of their linkage to a proviral sequence (e.g. Bowes *et al.* 1993).

The concept of ES gene trap libraries was first put into practice by Hicks *et al.* (1997), who analysed 400 inserts and identified 63 genes and anonymous cDNAs on the basis of flanking sequence tags. The German Gene Trap Consortium (GGTC) is aiming to produce and characterize 20 000 gene trap ES lines using a retroviral gene trap vector. So far, more than 12 000 lines have been produced and flanking sequences have been determined to generate a database of gene trap sequence tags (GSTs; Wiles *et al.* 2000). A similar project is underway at the US company Lexicon Genetics. The result is a library of gene trap ES cell lines called OmniBank, represented by over 50 000 'OmniBank sequence tags' corresponding to both expressed and non-expressed genes. Both these organizations have websites providing more information on the programmes and a searchable database to identify tagged genes (Box 10.3).

Murine retrovirus vectors have also been used for genome-wide mutagenesis projects in the zebrafish (Gaiano *et al.* 1996; Amsterdam *et al.* 1999). These projects were conceived for traditional forward genetics experiments, but over 36 000 lines were generated and these should also be suitable for functional genomics.

Insertional mutagenesis in plants

Insertional mutagenesis is the most appropriate option for functional genomics in plants because of the very low efficiency with which homologous

Box 10.3 Internet resources for genome-wide random mutagenesis projects

Insertional mutagenesis in yeast

http://ygac.med.yale.edu/mtn/insertion_libraries.stm

Insertional and ethylnitrosourea mutagenesis in the mouse

For a more extensive list of mouse resources see Brekers & Harabe de Angelis (2001) and Sanford *et al.* (2001).

http://lena.jax.org/~jcs/Delbank.html DELBank project information

http://www.lexgen.com/omnibank/omnibank.htm OmniBank gene trap ES cell library, information and resources

http://baygenomics.ucsf.edu/ BayGenomics gene trap project

http://socrates.berkeley.edu/~skarnes/resource.html A useful resource for mouse gene trap insertions

http://tikus.gsf.de German Gene Trap Consortium ES cell library, information and resources

http://www.gsf.de/ieg/groups/enu-mouse.html German Human Genome Project mouse ENU mutagenesis screen

http://www.mgu.har.mrc.ac.uk/mutabase UK/French consortium mouse ENU mutagenesis programme

Insertional mutagenesis in plants

http://www.biotech.wisc.edu/NewServicesAndResearch/Arabidopsis/default.htm *Arabidopsis*

Knockout Facility at the University of Wisconsin

http://mbclserver.rutgers.edu/~dooner/PGRPpage.html Maize '*Ac* gene engine' project to generate evenly distributed *Ac* inserts in the maize genome

http://www.zmdb.iastate.edu/zmdb/sitemap.html Maize Gene Discovery and *RescueMu* project

http://mtm.cshl.org Maize Targeted Mutagenesis database

http://nasc.nott.ac.uk/ima.html Information and database of *Ds* insertion lines and flanking sequences (Parinov *et al.* 1999) with instructions for obtaining seeds

http://www.jic.bbsrc.ac.uk/sainsbury-lab/jonathan-jones/SINS-database/sins.htm Information and database of sequenced insertion sites (SINS) from *dSpm* lines (Tissier *et al.* 1999) with instructions for obtaining seeds

http://signal.salk.edu/cgi-bin/tdnaexpress The Salk Institute Genome Analysis Laboratory (SIGnAL)

http://nasc.nott.ac.uk Nottingham *Arabidopsis* Stock Centre

http://arabidopsis.org/abrc *Arabidopsis* Biological Resource Center

Insertional mutagenesis in *Drosophila*

http://www.fruitfly.org/p_disrupt/index.html Berkeley *Drosophila* gene disruption project

recombination occurs in higher plant species. A number of different systems have been developed which essentially fall into two groups: those involving the use of T-DNA from *Agrobacterium tumefaciens* (Azpiroz-Leehan *et al.* 1997; Krysan *et al.* 1999), and those involving the use of plant transposons (Walbot 2000; Hamer *et al.* 2001; Ramachandran & Sundaresan 2001). In each case, if a large enough population of tagged lines is available, there is a good chance of finding a plant carrying an insert within any gene of interest. Mutations that are homozy-

gous lethal can be maintained in the population as heterozygous plants.

T-DNA mutagenesis

Agrobacterium tumefaciens is a soil bacterium that has been used extensively to generate transgenic plants. Plant transformation involves the transfer of a small segment of DNA, called the T-DNA, from a plasmid harboured by the bacterium into the plant genome. This T-DNA can act as an insertional mutagen, and has thus far been used for genome-wide mutagenesis programmes in *Arabidopsis thaliana* and rice. As T-DNA is not a transposon, it has no ability to 'jump' following integration, and so has the advantage of generating stable insertions in primary transformants. T-DNA integration is also believed to favour genes but within the 'gene-space' of the genome integration is essentially random. This is not the case for most plant transposons (see below). One disadvantage of T-DNA, however, is its tendency to generate complex multicopy integration patterns, sometimes involving the deletion or rearrangement of surrounding genomic DNA. This can complicate subsequent analysis, especially if the PCR is used to confirm integrations within specific genes.

A number of research groups have invested resources in generating banks of *Arabidopsis* T-DNA insertion lines (Feldmann & Marks 1987; Bouchez *et al.* 1993; Campisi *et al.* 1999; Krysan *et al.* 1999; Weigel *et al.* 2000). This reflects the availability of simple and convenient techniques for *Agrobacterium*-mediated transformation of *Arabidopsis*, facilitating rapid saturation of the genome (Bechtold & Pelletier 1998; Clough & Bent 1998). Furthermore, *Arabidopsis* is a gene-dense plant with small introns and little intergenic space; about 80% of the genome is thought to represent genes.

Currently, over 130 000 T-DNA tagged *Arabidopsis* lines are made available by the University of Wisconsin *Arabidopsis* Knockout Facility,[1] which maintains a searchable database of mutant lines (Box 10.3). These lines comprise two populations: one generated by the insertion of a simple T-DNA construct (Krysan *et al.* 1999), and one generated by

the insertion of an activation tag (Weigel *et al.* 2000). Such populations can be used for comprehensive reverse genetics screens (McKinney *et al.* 1995; Winkler *et al.* 1998; Krysan *et al.* 1999). DNA from the tagged lines is maintained as a series of hierarchical pools. These can be screened in several rounds of PCR using gene-specific primers supplied by the customer and T-DNA-specific primers supplied by the facility. Indeed, the system is very similar to that used in yeast (Fig. 10.2). If a 'hit' is achieved, corresponding seeds can be ordered and the customer can then grow plants with a particular gene disrupted.

Modified T-DNA vectors have been used in *Arabidopsis* not only as activation tags (Weigel *et al.* 2000) but also as gene and promoter traps (Feldmann 1991; Lindsey *et al.* 1993; Babiychuk *et al.* 1997; Campisi *et al.* 1999). A gene trap T-DNA vector has also been used by Jeon *et al.* (2000) in a genome-wide screen of rice. These investigators produced over 22 000 primary transformants carrying the T-DNA insertion, more than half of which contained multiple T-DNA copies. Over 5000 tagged lines were analysed for reporter gene expression in leaves and roots, 7000 lines were analysed for expression in flowers and 2000 for expression in seeds. Overall, about 2% of the lines showed marker gene activity, in some cases ubiquitous but in many cases restricted to highly specific cell types or tissues.

Transposon mutagenesis in plants

Transposons have been widely used for insertional mutagenesis in plants and this has led to the discovery of many new genes (reviewed by Gierl & Saedler 1992). Several transposons have been used for genome-scale mutagenesis projects, including *Activator* (*Ac*), *Suppressor–mutator* (*Spm*) and *Mutator* (*Mu*) from maize, *Tam 3* from *Antirrhinum majus* and *Tph1* from *Petunia*. Unlike T-DNA, transposons tend to generate simple single-copy insertions. However, as is the case for *Ty* elements in yeast, most plant transposons show pronounced 'target site preference', which can make it difficult to achieve whole-genome saturation. Also, while T-DNA generates stable inserts, additional crossing steps are required to stabilize those generated by plant transposons. This is because transposons have the intrinsic ability to mobilize unless their source of transposase is

---

[1] Some confusion in terminology can arise here. A *gene knockout* was originally defined as a null mutation produced by gene targeting (homologous recombination). More recently, the term has been used to describe any sort of induced null mutation, including those generated by (random) insertional mutagenesis. Hence, the *Arabidopsis* Knockout Facility maintains a collection of null mutants generated not by gene targeting but by random T-DNA insertion. Even worse, it is becoming common for mutations generated by random insertion to be called 'targeted mutations' even though this description is strikingly inaccurate. Hence, the Maize Targeted Mutagenesis Project concerns a population of random insertional mutants generated by the transposon *Mutator*. The term 'target-selected mutagenesis' is more accurate, referring to the fact that researchers can identify a mutation in a particular target gene, from a randomly mutagenized population, using a PCR-based assay on pooled DNA.

removed. Control is generally achieved by the use of 'two-component' transposon systems, comprising an autonomous (self-mobilizing) element and non-autonomous derivative. For example, the maize transposon *Activator* (*Ac*) is autonomous because it encodes its own transposase, but shorter derivatives of *Ac* called *Dissociation* (*Ds*) lack the transposase gene. However, *Ds* elements can transpose if transposase is provided by *Ac*. Thus, where *Ac* and *Ds* are present in the same genome, both elements can be mobile. However, if *Ac* is removed by crossing, progeny plants can be recovered with stable *Ds* insertions.

The properties of transposons vary, making different transposon families suitable for different applications. Although maize *Ac*/*Ds* has been widely used for genome-scale mutagenesis, it actually demonstrates a phenomenon called 'local transposition', i.e. it jumps preferentially to linked sites. This can make saturation difficult, particularly if there is a small number of founder lines, but it can also be an advantage for generating multiple mutant alleles in one gene, or for generating mutations in several clustered genes in a local genomic region. If necessary, selection systems can be devised to select against closely linked transpositions (Parinov *et al.* 1999; Tissier *et al.* 1999). This problem can also be circumvented by generating a population of maize plants with *Ac*/*Ds* elements spaced at regular 10–20 cM intervals throughout the genome. Such a project is indeed in progress at the Waksman Institute, Rutgers University, and can be seen at the following website: http://mbclserver.rutgers.edu/~dooner/PGRPpage.html.

Another maize transposon, *Mutator* (*Mu*), does not show preferential local transposition and is therefore a potentially better global mutagen than *Ac*. However, it does preferentially insert into transcription units, making it an excellent tool for gene disruption. Several genome-wide mutagenesis projects have therefore been established in maize using *Mu*, including the Trait Utility System for Corn (TUSC) developed by Pioneer Hi-Bred International, and the Maize Targeted Mutagenesis project (MTM). In each case, PCR primers facing away from the transposon are used in combination with a gene-specific primer to identify insertions into specific genes, with DNA pooled from maize plants in the field as the template. These resources have been used

successfully by a number of investigators (e.g. Bensen *et al.* 1995; Das & Martiensen 1995; Hu *et al.* 1998). The Maize Gene Discovery Project uses a modified *Mu* transposon called *RescueMu*, which can be used for plasmid rescue from whole genomic DNA. The rescued plasmids have been used to generate DNA libraries containing the *Mu* insertion sites. PCR is carried out on these plasmids to identify insertions in genes of interest. In all the facilities, seeds corresponding to each insertion can be supplied. Websites with further information on maize genome projects, search facilities and databases of tagged genes are listed in Box 10.3.

Some transposons, such as the *Drosophila* P-element (see below), can only function in their host species. Others, including *Ac* and another maize transposon called *Spm*, are more promiscuous and these can be used in a range of heterologous plants (Osborne & Baker 1995). The transposons must initially be introduced into the foreign genome either as a T-DNA or as a conventional transgene delivered by a method such as particle bombardment. Once integrated into the genome, however, normal transposition may then occur. Functional genomics programmes using *Ac* have been initiated in *Arabidopsis* (Ito *et al.* 1999; Seki *et al.* 1999) and tomato (Meissner *et al.* 2000). *Spm* has been used for several large-scale *Arabidopsis* mutagenesis projects (e.g. Speulman *et al.* 1999; Tissier *et al.* 1999). Where *Ac* has been used in *Arabidopsis*, the investigators have exploited local transposition to saturate genomic regions surrounding the original integration site. In both studies (Ito *et al.* 1999; Seki *et al.* 1999), a cDNA scanning strategy was used to isolate ESTs from this region, leading to rapid gene annotation.

The genome-wide transposon mutagenesis projects in *Arabidopsis* have produced a large number of mutant lines. As for the T-DNA insertion lines discussed above, many of the transposon lines are now maintained at a central resource, in this case the Nottingham *Arabidopsis* Stock Centre in the UK, and the *Arabidopsis* Biological Resource Center in the USA. This includes 960 DNA pools from *Spm* insertion lines corresponding to 48 000 inserts (Tissier *et al.* 1999) and about 2600 DNA pools from multiple *Spm* insertion lines, representing up to 65 000 insertions (Speulman *et al.* 1999). The flanking sequences have been determined and analysed by BLAST searches, and the results are available on searchable databases, allowing the rapid identification of interrupted genes (Seki *et al.* 1999; Parinov *et al.* 1999; Speulman *et al.* 1999, Tissier *et al.* 1999). Websites for these resources are listed in Box 10.3.

Insertional mutagenesis in Drosophila

Genome-wide insertional mutagenesis programmes have been initiated in several other animals. For example, the *Tc1* transposon has been used to generate a frozen bank of insertion mutants of the nematode *C. elegans*. This produced 5000 lines and 16 newly identified genes (Zwaal *et al.* 1993). P-elements are transposable elements that, under certain circumstances, can be highly mobile in the germ line of the fruit fly *Drosophila*. These transposons have been developed as insertional mutagens (Spradling & Rubin 1982) and have been used to clone and characterize many *Drosophila* genes (reviewed by Cooley *et al.* 1988). Currently, a genome-wide mutagenesis programme is ongoing at the Berkeley *Drosophila* Genome Project, with the aim of generating a comprehensive library of mutant fly strains (Spradling *et al.* 1995, 1999). In the initial phase of the programme, about 4000 mutagenized lines were examined for P-element insertions. Redundant strains (i.e. allelic mutations) were eliminated, leaving 1045 unique inserts identifying over 1000 genes with homozygous lethal phenotypes. This corresponds to approximately 25% of all 'essential' genes on the autosomal chromosomes. Saturation of the genome has proven difficult because P-elements, like many transposons, show a pronounced insertional bias (Liao *et al.* 2000). Also, many interrupted genes do not reveal a mutant phenotype, perhaps because the element has inserted into a non-essential region such as an intron. With the recent completion of the *Drosophila* genome sequence (Adams *et al.* 2000b), Spradling *et al.* have begun to collect sequence signatures from the flanks of each P-element insertion and identify genes by comparing these signatures to the genomic sequence. This removes the dependency on a mutant phenotype. At the same time, additional P-element vectors are being used which function as activation tags (Rorth 1996; Rorth *et al.* 1998). It is to be hoped that many of the genes which do not reveal informative loss of function phenotypes will show gain of function phenotypes in this

screen. The Berkeley *Drosophila* Genome Project maintains websites for both the P-element mutagenesis programme and the P-element gene misexpression programme (Box 10.3). Many of the disrupted lines are also available from *Drosophila* resources, such as the Bloomingdale Stock Centre: http://flybase.bio.indiana.edu/stocks.

Functional genomics using chemical mutagens

While insertional mutagenesis is likely to remain the most popular approach to functional genomics in many species, the use of chemicals to generate point mutations or deletions is also being explored by several groups. We consider two examples below.

Chemical mutagenesis in Caenorhabditis elegans

Liu *et al.* (1999b) used chemical mutagenesis in concert with a PCR-screening strategy to generate panels of deletion mutants in *C. elegans*. This high-throughput strategy relies on the parallel screening of genomic DNA samples pooled from groups of 96 mutagenized worm populations maintained individually in microtitre plates. Once a deletion is detected in a given pooled sample, a second round of PCR can be carried out on the 96 individual genomic DNA samples that contributed to the pool, and this leads to the identification of a particular worm population whose individuals can then be tested directly. This is a rapid approach for the production of mutant libraries. It can be applied to any organism that can be similarly maintained as arrays of frozen stocks and whose genome has been completely sequenced. Therefore it should be particularly useful for functional genomics in microbes. Internet resources for functional genomics in *C. elegans* are listed in Box 10.4.

Ethylnitrosourea mutagenesis in mice

The mouse is the model organism that most closely resembles humans, so a number of research groups and consortia have investigated the possibility of carrying out large-scale mutagenesis programmes in this species. Insertional mutagenesis projects in the mouse are discussed on p. 178. However, it is

Box 10.4 *Caenorhabditis elegans* RNA interference resources

Resources specific for RNAi

http://mpi-web.embl-heidelberg.de/dbScreen/
http://worm-srv1.mpi-cbg.de/dbScreen
http://www.rnai.org

General

http://www.wormbase.org

http://elegans.swmed.edu
http://nema.cap.ed.ac.uk/nematodeESTs/
nembase.html

Also of interest

http://nematode.lab.nig.ac.jp Expression pattern
database

notable that many human diseases are caused by point mutations and other small lesions, while fewer are caused by dramatic events such as insertional disruption. Insertional mutagenesis tends to generate null alleles (complete loss of gene function) which in many cases is lethal, while point mutations often have less severe effects. Chemical mutagenesis generates point mutations, so this strategy might be more useful in generating a range of informative mutants in mice.

The alkylating agent ethylnitrosourea (ENU) is the most powerful mutagen available for mice (Russell *et al.* 1979, reviewed by Justice *et al.* 1999). Large-scale mutagenesis screens have been carried out successfully in other species using this chemical (e.g. Mullins *et al.* 1994 used ENU in zebrafish) but only limited screens have been attempted in mice (e.g. Bode *et al.* 1988; Shedlovsky *et al.* 1988; Rinchik *et al.* 1990). Recently, two groups of researchers reported the first genome-wide ENU mutagenesis screens in mice. As part of the German Human Genome Project, Hrabe de Angelis & Balling (1998) and Hrabe de Angelis *et al.* (2000) have screened 14 000 ENU-mutagenized mouse lines for dominant and recessive mutations affecting a large number of clinically important phenotypes. Categories included allergy, immunology, clinical chemistry, nociception (response to pain) and dysmorphology (abnormal structure). In the initial study, 182 mutants were catalogued and many more are still undergoing analysis. Simultaneously, a consortium of UK and French researchers reported a similar large-scale experiment (Nolan *et al.* 2000). In this case, 26 000 mice were screened for dominant mutations and 500 were recovered. The mice were tested for visible developmental defects from birth to

weaning, and then subjected to a battery of functional, behavioural and biochemical tests over the next 8 weeks. In both programmes, mutations were mapped by interspecific backcrossing using a genome-wide panel of microsatellite polymorphisms. Websites for both projects provide information, resources and updates on newly discovered phenotypes (Box 10.3).

Functional genomics using phenocopy libraries

A phenocopy has the same appearance as a mutant phenotype, but there are no changes to the DNA sequence. While phenotypes are caused by mutations, phenocopies are generated by interfering with gene expression. For example, antisense RNA can be used to inactivate the messenger RNA corresponding to a particular gene, or antibodies can be used to inactivate the protein. In each case, there is a loss of gene function while the gene itself remains intact.

A number of strategies have been developed recently which would in principle be suitable for high-throughput functional inactivation. The phenomenon of virus-induced gene silencing (VIGS) in plants is one example (reviewed by Baulcombe 1999). VIGS is thought to result from a plant defence system which specifically recognizes viral nucleic acids and targets them for degradation. This only works if the virus is replication competent, suggesting that the double-stranded RNA (dsRNA) intermediate formed during viral replication is involved. The defence mechanism is sequence-specific, so only viral RNA is attacked. However, if the virus contains a transgene homologous to an endogenous gene, the defence system regards any

RNA with that sequence as a target, even if it represents the transcript of an endogenous gene. Therefore, RNA viruses such as tobacco mosaic virus and potato virus X can be engineered to carry host-derived transgene for the specific purpose of silencing the corresponding endogenous gene (e.g. see Kumagai *et al.* 1995; Kjemtrup *et al.* 1998; Ruiz *et al.* 1998). So far, the potential of VIGS has not been explored for a co-ordinated functional genomics programme, but its most attractive aspect is the rapidity with which silencing constructs could be produced, compared to the long time taken to generate mutagenized populations.

Genome-wide RNA interference in Caenorhabditis elegans

The genome sequence of the nematode worm *C. elegans* was published by the *C. elegans* Genome Sequencing Consortium (1998). Of the higher (multicellular) eukaryotes, *C. elegans* has the smallest genome and the most convenient biological properties for high-throughput handling: it is small, hermaphrodite and can be stored as frozen stocks. Therefore the worm is a very attractive target for functional genomic studies. Coincidentally, *C. elegans* is the model organism in which the phenomenon of RNA interference was first documented (Fire *et al.* 1998). RNA interference (usually abbreviated to RNAi) concerns the ability of dsRNA to induce potent and specific post-transcriptional silencing of a homologous endogenous gene or transgene; its mechanism is thought to be very similar to VIGS (reviewed by Hammond *et al.* 2001). Over the last few years, RNAi has become the routine procedure used to generate functional gene knockouts in the worm. In many cases the phenocopies of the RNAi functional knockouts are equivalent to genuine knockout phenotypes produced by traditional mutagenesis strategies.

Microinjection is currently the most consistently effective way to induce RNAi. Typically, *in vitro* synthesized dsRNA is injected into the germline of adult worms, and progeny are screened for RNAi-induced phenocopies. Because of the laborious nature of the microinjection procedure, carrying out individual injections corresponding to each of the 15 000 genes in the *C. elegans* genome would be a Herculean undertaking. Nevertheless, a recent report by Gonczy

et al. (2000) describes just such an approach. This group synthesized over 2200 individual dsRNA molecules (corresponding to over 95% of the genes on chromosome III) and then carried out systematic microinjections followed by screening for RNAi-induced phenocopies that affected the first two cleavage divisions of development. Remarkably, they obtained a hit rate of over 6% (133 genes were found to be involved in cleavage). This group of genes included all seven chromosome III genes that had previously been shown through traditional mutagenesis to affect cell division thereby providing important validation of the procedure.

RNAi is a systemic phenomenon, so microinjection is not the only way it can be achieved. Technically simpler ways to induce RNAi include adding dsRNA to the worms' liquid medium, or even feeding the worms on bacteria that have been engineered to express dsRNA. The former strategy was used by Maeda *et al.* (2001) to test the function of approximately 2500 genes represented in the *C. elegans* EST database. They found that nearly one-third of the genes revealed an RNAi phenocopy that could be easily scored under the dissecting microscope. The latter strategy has been developed on a genomic scale by Fraser *et al.* (2000) to identify systematically genes on chromosome I affecting morphological characteristics. A library of bacterial strains expressing over 2400 different dsRNAs was constructed, representing just under 90% of the genes on chromosome I. Worms then were fed on these bacterial strains and RNAi-induced phenocopies were sought in the progeny. Fourteen per cent of the genes tested revealed observable phenocopies. In 90% of the genes that had previously been characterized by mutagenesis, the mutant phenotypes and RNAi phenocopies were concordant. Interestingly, the success rate for neuronal genes was much lower. Only about 12% of neuronal genes were affected by RNAi, a phenomenon that has been observed in earlier single-gene studies. This problem can be overcome by expressing 'hairpin' dsRNA from a heritable transgene, although this procedure is too complex to be employed on a genomic scale (Tavernarakis *et al.* 2000).

Because of the viability of frozen stocks, the sharing and distribution of *C. elegans* mutant libraries is very simple. Furthermore, bacterial strains expressing dsRNA can be maintained at a central resource and distributed to other laboratories on request. Thus, it is not necessary to generate RNAi-inducing strains from first principles each time a particular study is undertaken. This resource can be expanded and refined, eventually providing a comprehensive coverage of the entire *C. elegans* genome.

Suggested reading

Anderson K.V. (2000) Finding the genes that direct mammalian development: ENU mutagenesis in the mouse. *Trends in Genetics* **16**, 99–102. *A commentary on the use of ENU for large-scale mutagenesis studies in the mouse, focusing on development.*

Barstead R. (2001) Genome-wide RNAi. *Current Opinion in Chemical Biology* **5**, 63–66.

Kim S.K. (2001) Http://C. elegans: mining the functional genomic landscape. *Nature Reviews Genetics* **2**, 681–689. *Current accounts of functional genomics in* Caenorhabditis elegans *using RNA interference and other methods.*

Beckers J. & Hrabe de Angelis M. (2001) Large-scale mutational analysis for the annotation of the mouse genome. *Current Opinion in Chemical Biology* **6**, 17–23. *An excellent summary of mutational analysis in the mouse, covering transgenics, gene-traps, knockouts and ENU mutagenesis.*

Cecconi F. & Meyer B.I. (2000) Gene trap: a way to identify novel genes and unravel their biological function. *FEBS Letters* **480**, 63–71.

Stanford W.L., Cohn J.B. & Cordes S.P. (2001) Gene-trap mutagenesis: past, present and beyond. *Nature Reviews Genetics* **2**, 756–768. *Two good summaries of insertional mutagenesis projects in the mouse, using gene trap vectors.*

Coelho P.S.R., Kumar A. & Snyder M. (2000) Genome-wide mutant collections: toolboxes for functional genomics.

Current Opinion in Microbiology **3**, 309–315. *An excellent and very accessible review concentrating on gene knockout and random mutagenesis projects in yeast and other microbes, with comparison to similar projects in higher eukaryotes.*

Hamer L., DeZwaan T.M., Montenegro-Chamorro M.V., Frank S.A. & Hamer J.E. (2001) Recent advances in large-scale transposon mutagenesis. *Current Opinion in Chemical Biology* **5**, 67–73. *A summary of recent transposon mutagenesis projects, concentrating on yeast and plants.*

Maes T., De Keukelerie P. & Gerats T. (1999) Plant tagnology. *Trends in Plant Sciences* **4**, 90–96.

Parinov S. & Sundaresan V. (2000) Functional genomics in *Arabidopsis*: large-scale insertional mutagenesis complements the genome sequencing project. *Current Opinion in Biotechnology* **11**, 157–161. *These reviews provide broad coverage of transposon and T-DNA mutagenesis studies in plants.*

Useful websites

These websites are particularly informative and user-friendly sites addressing mutagenesis projects in a variety of organisms.

http://mips.gsf.de/proj/eurofan/index.html
The EUROFAN website (Dujon 1998).

http://sequence-www.stanford.edu/group/yeast_deletion_project
The *Saccharomyces* Gene Deletion Project website (Winzeler *et al.* 1999).

http://www.biotech.wisc.edu/NewServicesAndResearch/Arabidopsis/default.htm
Arabidopsis Knockout Facility at the University of Wisconsin.

http://www.fruitfly.org/p_disrupt/index.html
Berkeley *Drosophila* gene disruption project.

CHAPTER 11

Mapping protein interactions

Introduction

We have learned in the last four chapters that information about the function of a gene can be gained from the analysis of DNA sequence, genome organization, protein structure, expression profile and mutant phenotype. However, this information rarely lets us see the whole picture. More often it provides suggestions or clues that need to be followed up by further experiments. This being the case, how can we rigorously define the function of a gene? At the most fundamental level, gene function reflects the behaviour of proteins. It is the proteins that actually carry out cellular activities and interact with the environment. Thus gene function can ultimately be broken down into a series of molecular interactions that take place among proteins and between proteins and other molecules. When things go wrong, through mutation or otherwise, it is ultimately the result of the failure of these normal interactions. Our efforts to treat diseases, through the use of drugs, ultimately depend on the ability of those drugs to modulate protein interactions in a beneficial manner.

In this final functional genomics chapter we discuss the techniques that are used to study protein interactions, and how these techniques have been adapted for high-throughput analysis on a proteomic scale. The function of an uncharacterized protein is often suggested by its spectrum of interactions. If protein X is uncharacterized but interacts with proteins Y and Z, both of which are part of the RNA splicing machinery, it is likely that protein X is involved in this process also. If this reasoning is applied on a global scale, every protein in the cell can eventually be linked into a functional network (Walhout & Vidal 2001). This functional network has been called the *interactome* (Sanchez *et al.* 1999).

Methods for protein interaction analysis

In the pre-genomic era, protein interaction studies – like all other genetic and biochemical analyses – were carried out largely on an individual basis. There are many different ways in which protein interactions can be inferred or demonstrated, some of which are discussed briefly in Box 11.1. A number of classical genetic approaches can be used, including screening for suppressor mutations, i.e. mutations in one gene that partially or fully compensate for a mutation in another (Hartman & Roth 1973). In many cases such mutations exist because the proteins encoded by the two genes interact. The primary mutation causes a change in protein structure that prevents the interaction, but the suppressor mutation introduces a complementary change in the second protein that restores the interaction (Fig. 11.1). Such genetic techniques have been widely employed in amenable organisms such as *Drosophila* and yeast.

The mainstay of protein interaction analysis, however, has been a core of biochemical methods providing direct evidence for interactions. Techniques such as co-immunoprecipitation, affinity chromatography and crosslinking have been employed for over 25 years to characterize the interactions of individual proteins (reviewed by Phizicky & Fields 1995). Methods such as X-ray crystallography and nuclear magnetic resonance spectroscopy can be used to characterize protein interactions at the atomic level. These techniques were discussed in detail in Chapter 8. There are also techniques applied at the level of cell biology, which provide correlative evidence for interaction between specific proteins. Studies in which two proteins are localized *in situ* with labelled antibodies can show they coexist in the same cellular compartment at the same time.

Box 11.1 Some traditional methods for detecting and characterizing candidate protein–protein interactions

Suppressor mutations

Interaction between protein X and protein Y is inferred where a mutation in gene Y compensates for a mutation in gene X. One explanation is that a conformational change in protein X prevents interaction, but this is compensated by a complementary change in protein Y that restores interaction (Hartman & Roth 1973). A variation on this theme is where overproduction of one protein leads to a mutant phenotype which can be compensated by the overproduction of a second (interacting) protein (see e.g. Rine 1991).

Synthetic effects

Interaction between protein X and protein Y is inferred where an XY double mutant generates a more severe phenotype than either single mutant alone. One explanation is that individual mutations in X or Y cause changes that nevertheless preserve protein interaction, but the combined effect of both mutations prevents that interaction (see e.g. Koshland *et al.* 1985; Huffaker *et al.* 1987).

Dominant negatives

Multimerization of protein X can be inferred where overproduction of a mutant form of the protein causes loss of function despite the presence of wild-type protein subunits. One explanation is that the mutant subunits sequester all the wild-type subunits into non-functional multimers (Herskowitz 1987).

Co-immunoprecipitation

Interaction between protein X and protein Y is demonstrated by the addition of (usually monoclonal) antibodies against protein X to a cell lysate. Precipitation of the antibody–protein X complex results in the coprecipitation of protein Y.

Affinity chromatography

Interaction between protein X and protein Y is demonstrated by the 'capture' of protein X on some kind of affinity matrix, e.g. a Sepharose column, when a cell lysate is passed through. Protein Y also remains attached to the column by virtue of its interaction with protein X, while non-interacting proteins are washed through. Affinity capture may be achieved using antibodies against protein X. Alternatively, protein X may be expressed as a fusion with an epitope tag or a molecule such as glutathione-*S*-transferase (GST) which binds to glutathione-coated Sepharose beads. In a related technique, protein X can be immobilized on a membrane in a manner similar to the western blot and used to screen for interacting proteins in a cell lysate. This has been called a far-western blot (Blackwood & Eisenman 1991).

Crosslinking

Interaction between protein X and protein Y is demonstrated where cells or cell lysates are exposed to a crosslinking agent, and immunoprecipitation of protein X results in the coprecipitation of protein Y. Protein Y can be released by cleavage of the crosslink.

Fluorescence resonance energy transfer

Interaction between protein X and protein Y is demonstrated where energy is transferred from an excited donor fluorophore to a nearby acceptor fluorophore, a phenomenon called fluorescence resonance energy transfer (FRET). FRET occurs only when the two fluorophores are up to 10 nm apart, and can be detected by the change in the emission wavelength of the acceptor fluorophore. FRET analysis can be carried out if protein X and protein Y are conjugated with fluorophores such as Cy3 and

continued

Box 11.1 *continued*

Cy5. Alternatively, they can be expressed as fusions with different fluorescent proteins, e.g. enhanced cyan fluorescent protein (donor) and enhanced yellow fluorescent protein (acceptor), in which case the technique may be called *bioluminescence resonance energy transfer* (BRET). The advantages of FRET/BRET analysis are that the normal physiological conditions inside the cell are maintained (analysis can be carried out *in vivo*) and that transient as well as stable interactions can be detected (Day 1998; Mahajan *et al.* 1998).

Surface plasmon resonance spectroscopy

Surface plasmon resonance (SPR) is an optical resonance phenomenon occurring when surface plasmon waves become excited at the interface between a metal surface and a liquid. Interaction between protein X (immobilized on the metal surface) and protein Y (free in solution) is demonstrated by a change in the refractive index of the surface layer. This is the technology exploited in protein chips marketed by the US company BIAcore. Information concerning the use of this technique for the detection of protein interactions, plus an extensive list of relevant publications, can be found at the following website: http://www.biacore.com

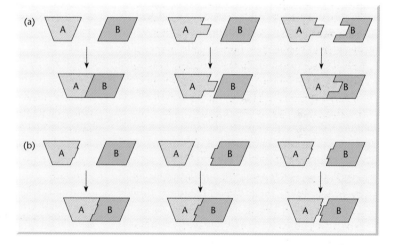

Fig. 11.1 Genetic tests for protein interactions. (a) Suppressor mutations. Two proteins, A and B, normally interact. A mutation in A prevents the interaction, causing a loss of function phenotype, but this can be suppressed by a complementary mutation in B which restores the interaction. (b) Synthetic lethal effect. The same two proteins can still interact if there is a mutation in either A or B which does not drastically affect the interaction between them. However, if the mutations are combined, protein interaction is abolished and a loss of function phenotype is generated.

While these traditional genetic, biochemical and physical techniques are highly informative on an individual basis, none is particularly suited to high-throughput analysis. Nor do they provide an easy way to link newly identified interacting proteins with their corresponding genes. Because the aim of functional genomics is to determine functions for the many anonymous genes and cDNAs amassing in databases, high-throughput strategies that link genes and proteins into functional networks are essential. This has been achieved by technology development in three areas.

1 Library-based interaction mapping.
2 High-throughput protein analysis and annotation.
3 Bioinformatics tools and databases of interacting proteins, which provide a platform for organizing and querying the increasing amount of interaction data.

Library-based methods for screening protein interactions

Library-based methods for protein interaction analysis allow hundreds or thousands of proteins to be screened in parallel and all experimentally identified proteins are linked to the genes or cDNAs that encode them. Therefore, once interacting proteins have been detected, the corresponding clones can be rapidly isolated and used to interrogate DNA sequence databases (Pelletier & Sidhu 2001). Essentially, there are two broad classes of library: those in which protein interactions are assayed *in vitro* and those where the interactions take place within the environment of a cell.

In vitro libraries

In principle, it is possible to use any standard expression library for protein interaction screening. Immunological screening (the use of antibodies as probes) was developed in the 1970s and is essentially a specialized form of interaction analysis (Broome & Gilbert 1978). There is no reason why other proteins should not be used as 'probes'. Indeed, a diverse range of proteins has been used in this manner with the aim of pulling out interacting partners. One example is provided by MacGregor *et al.* (1990) who sought interacting partners for the transcription factor c-Jun by screening a cDNA expression library with a biotin-labelled c-Jun probe.

The traditional clone-based library, however, is not an ideal platform for proteome-wide interaction screening. The studies discussed above involved labour-intensive and technically demanding screening procedures which would be unsuitable for high-throughput studies. Phage display is a more suitable alternative (Smith 1985). The principle of phage display is the expression of fusion proteins in such a way that a foreign peptide sequence is 'displayed' on the bacteriophage surface. Libraries of phage can be produced and screened to identify peptides that interact with a given probe (such as an antibody) which is immobilized on a membrane or in the well of a microtitre plate. Screening is basically a reiterative affinity purification process in which non-interacting phage are discarded and bound phage eluted and used to re-infect *Escherichia coli*. After several rounds of such 'panning', the remaining tightly bound phage are isolated and the inserts sequenced to identify the interacting peptides (for a recent review see Sidhu *et al.* 2000). The advantages of phage display over other *in vitro* library systems are that libraries of great complexity (up to

10^{12}) can be generated, and several rounds of highly selective screening can be carried out with intrinsic amplification at each step. Screening and panning can be carried out in an array format in microtitre dishes, making the technique amenable to high-throughput processing. One disadvantage of the system, however, is that only small foreign peptides can be incorporated into the coat protein. This may limit the number of interactions that can be detected. Thus, despite their potential for scale-up, phage display, and similar methods in which peptides are displayed on the surface of cells, have yet to be exploited for genome-scale interaction analysis.

In vivo libraries: the yeast two-hybrid system

The major disadvantage of all *in vitro* library systems is that interactions occur in an unnatural environment where the protein may be incorrectly folded or partially unfolded. The yeast two-hybrid (Y2H) system, initially described by Fields & Song (1989), is the prototype of a range of related techniques in which protein interactions are assayed *in vivo*. The principle of the system is that proteins often comprise several functionally independent domains. These can function not only when they are covalently linked in the same polypeptide chain, but also when they are brought together through non-covalent interactions. Transcription factors generally contain independent DNA-binding and transactivation domains and this means that a functional transcription factor can be created if separately expressed DNA-binding and transactivation domains can be persuaded to interact. On this basis, it is possible to use the two-hybrid system to confirm interactions between known proteins and to screen for unknown proteins that interact with a given protein of interest.

It is the latter approach that is most relevant in functional genomics. The general strategy is as follows: protein X is expressed as a fusion (a hybrid) with the DNA-binding domain of a transcription factor to generate a 'bait'. A library of 'prey' is then generated in which each clone is expressed as a fusion protein with the transactivation domain of the transcription factor. The final component of the system is a reporter gene which is activated specifically by the two-hybrid transcription factor. Mating between

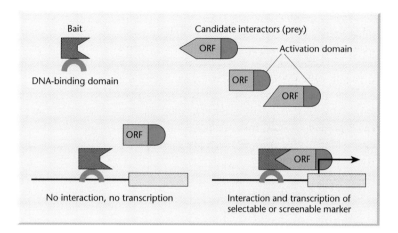

Fig. 11.2 Principle of the basic yeast two-hybrid system.

haploid yeast cells carrying the bait construct and those carrying the library of prey results in diploid cells carrying both components. In those cells where the bait interacts with the prey, the transcription factor is assembled and the reporter gene activated, allowing the cells to be isolated and the DNA sequence of the interactor identified. This principle is illustrated in Fig. 11.2.

To make the two-hybrid system suitable for genome-wide interaction mapping, comprehensive libraries of baits must be used to screen comprehensive libraries of prey, resulting in huge numbers of combinatorial interactions. Two general strategies have been devised: the matrix approach and the random library method. These are discussed in turn below.

Matrix approach

Matrix interaction screening involves panels of defined bait and prey, i.e. constructs derived from known open-reading frames (ORFs), which are mated systematically in an array format (Fig. 11.3). Because this approach depends on the availability of sequence data corresponding to each protein, it can only be used for predefined proteins. However, interacting proteins can be identified immediately on the basis of their array coordinates and the corresponding cDNAs or genomic clones can be retrieved. The advantage of the matrix approach is that it is fully comprehensive and can provide exhaustive proteome coverage. However, it is also laborious because each bait and prey construct must be prepared indi-

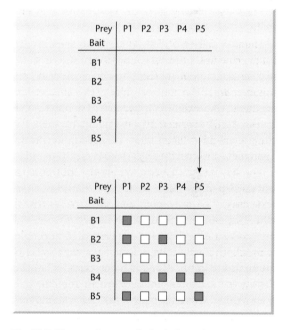

Fig. 11.3 The matrix system for high-throughput two-hybrid screening. Defined panels of bait and prey constructs maintained in haploid yeast strains are exhaustively tested for interactions by systematic mating. (Reprinted from Legrain *et al.* 2001 by permission of Elsevier Science.)

vidually by PCR followed by subcloning in the appropriate expression vector. Haploid yeast cells of opposite mating types are then transformed with the bait and prey constructs, respectively, and arrayed in microtitre plates. Specific pairwise combinations are generated by mating, and candidate interactions are assayed in the resulting diploid cells.

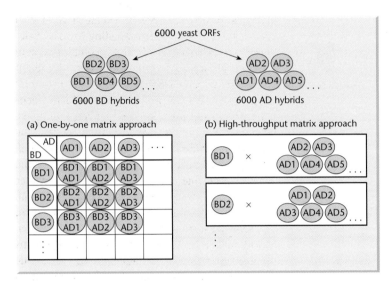

Fig. 11.4 The matrix and pooled matrix strategies. In the pooled matrix system, defined baits are tested against haploid yeast strains carrying pools of potential prey. If an interaction is detected, the pooled strain can be deconvoluted to identify individual interactors. The pooling strategy increases the throughput of the assay, depending on the number of constructs per pool. AD = activation domain; BD = binding domain. (Reprinted from Legrain & Selig 2000 by permission of Elsevier Science.)

The first matrix type study to be reported was a small-scale but systematic analysis of interactions among the proteins of the *Drosophila* cell cycle. In this investigation, Finley & Brent (1994) screened a panel of known cyclin-dependent kinases, and revealed a network of 19 protein interactions including many cyclins. More recently, the matrix approach has been used for several proteome-wide interaction screens in yeast. Uetz *et al.* (2000) used a standard matrix approach to screen a proteome-wide library of prey constructs (over 6000 ORFs) with 192 baits. To increase confidence in potential interactions, the screening was repeated and only interactions identified in both screens were selected for further analysis. Using this approach, 87 of the baits were shown to be involved in reproducible interactions with approximately three interactions per bait (a total of 281 interactions altogether). The same authors also described a modified matrix assay for high-throughput interaction screening (Fig. 11.4). Instead of generating cell lines expressing specific prey constructs, cells were transformed en masse with pools of prey. These were screened with 5300 ORF baits and 692 interacting protein pairs were identified. About half of these interactions were reproducible.

Pools of clones have also been used in another global study of protein interactions in yeast. Ito *et al.* (2000) used pools of 96 baits and 96 prey in 430 combinatorial assays such that over 4 million potential interactions were tested in parallel.

Approximately 850 positive colonies were obtained, and short regions of the bait and prey plasmids were sequenced to derive short sequence signatures termed *interaction sequence tags* (ISTs). This experiment identified 175 interacting protein pairs, only 12 of which were previously known. Four million interactions is approximately 10% of the total number of potential interactions within the yeast proteome assuming that every protein can interact with every other protein. Scaling this experimental format up to the proteome-wide level, Ito *et al.* (2001) identified 4549 interactions among 3278 proteins, 841 of which demonstrated three or more independent interactions.

A matrix format was also used to screen for interactions within the vaccinia virus proteome. All 266 ORFs were systematically tested against each other by McCraith *et al.* (2000), resulting in about 70 000 individual matings. Thirty-seven interactions were detected, only nine of which were previously known. Most recently, a matrix assay was used in a genome pilot study in the mouse, the first large-scale Y2H screen in a mammalian system (Suzuki *et al.* 2001).

Random library screening

The alternative to matrix format experiments is to generate prey libraries from random genomic fragments (Fig. 11.5). The prey can be screened using defined ORFs as baits or, for comprehensive

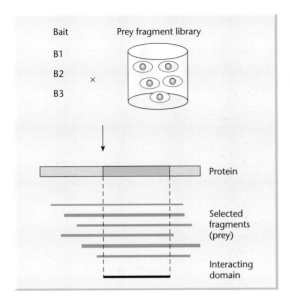

Fig. 11.5 The random library system for high-throughput two-hybrid screening. Defined panels of bait are tested for interactions against a panel of potential prey maintained as a high complexity random library. Each interactor may be defined by a number of overlapping constructs, therefore allowing specific interacting domains to be identified. (Reprinted from Legrain *et al.* 2001 by permission of Elsevier Science.)

proteome × proteome analysis, random libraries can be prepared for bait as well as prey. Unlike the matrix method, where all constructs are predefined and candidate interactors can be traced on the basis of their grid positions in the array, interacting clones in the random library must be characterized by sequencing and then compared to sequence databases for annotation.

In the first description of such an experiment, Bartel *et al.* (1996) generated random libraries of DNA-binding domain fusions and activation domain fusions from the genome of the *E. coli* bacteriophage T7. This encodes 55 proteins. The authors reported 25 interactions between separate proteins, only four of which had been previously described. In several cases, interactions were also found between different domains of the same protein. Interestingly, a significant number of interactions that had been previously demonstrated using biochemical and genetic techniques were not detected in this assay.

The use of defined ORFs to screen for interactions in a random library has been carried out predominantly in the study-specific protein complexes or

biological processes. Fromont-Racine *et al.* (1997) used as baits 15 ORFs corresponding to yeast splicing proteins. They screened a highly complex random genomic library containing more than 1 million clones and identified 145 potential prey in a total of 170 interactions. Approximately half of the identified prey were already known splicing proteins, whereas the other half were uncharacterized. The same genomic library was screened by Flores *et al.* (1999) using 15 of the 17 known components of the yeast RNA polymerase III complex, and in a further study of the spliceosome (Fromont-Racine *et al.* 2000) using 10 baits omitted from the initial screen. The same strategy was employed by Walhout *et al.* (2000) to investigate the interactions of 29 proteins involved in vulval development in the nematode worm *Caenorhabditis elegans*. Vulval development in this animal is an important model system in developmental biology and previous studies have reported direct and indirect evidence for at least 11 specific protein interactions within this group (reviewed by Kornfeld 1997). Therefore, the 29 proteins were first tested against each other in a conventional matrix format: six known interactions were confirmed and two novel interactions were revealed. Twenty-seven of the proteins were then used as baits in a random library screen and 17 of them were shown to take part in a total of 148 potential interactions.

To date the only genome-wide study involving arrayed baits screening a random library of prey was performed by Rain *et al.* (2001) to build a protein interaction map of the bacterium *Helicobacter pylori*. The Y2H assay was used to screen 261 *H. pylori* proteins against a highly complex library of genomic fragments. Over 1200 interactions were identified, which allowed nearly half of the genome to be assembled into a protein interaction map. Although no protein complexes have been defined biochemically in this species, homologous proteins in other bacteria have been studied in this manner. Interestingly, only about half of the interactions that are known to occur in *E. coli* were identified when homologous *H. pylori* proteins were used as baits in the two-hybrid system.

Reliability of two-hybrid interaction screening

As a genomic tool, the Y2H system suffers from several problems, including the high frequency of false positive and false negative results. False positives occur where the reporter gene is expressed in the absence of any specific interaction between the bait and prey. This may reflect 'autoactivation', where the bait or prey can activate the reporter gene unassisted, or a phenomenon known as 'sticky prey' where a particular prey protein can interact non-specifically with a series of baits. In large-scale two-hybrid screens, these types of false positives are quite easy to detect and can be eliminated. Other false positives occur through spontaneous mutations and can be more difficult to identify. Typically, researchers using matrix format screens use *reproducibility* as a measure of confidence in their results. Uetz *et al.* (2000) carried out two independent screens of their matrix and only accepted interactions occurring in both screens. Ito *et al.* (2001) regarded as plausible interactors only those proteins identified by three or more independent hits. In the analysis of vaccinia virus (McCraith *et al.* 2000), each assay was carried out four times and only those interactions occurring in three or four of the assays were accepted. Confidence in random library screens is increased by independent hits from overlapping clones. Fromont-Racine *et al.* (1997) devised a system in which confidence in a prey was assessed according to the number of overlapping fragments interacting with a specific bait, the size of the fragments, and how many times the specific interaction was recorded. False positives can also be caused by genuine but irrelevant interactions. For example, it is possible that two proteins normally found in separate cell compartments could, by chance, interact when they are both expressed in the yeast nucleus.

False negatives are revealed when known protein interactions are not detected and when similar studies reveal different sets of interacting proteins with little overlap. Uetz *et al.* (2000) carried out two different types of screen, one involving discrete baits and one involving pools of baits. Although a large number of potential interactors was found in each screen, only 12 were common to both screens. The similar proteome-wide study carried out by Ito *et al.* (2000, 2001) identified nearly 3300 interacting proteins but, of the 841 proteins found to be involved in three or more interactions, only 141 were in common with the set of 692 interacting proteins catalogued by Uetz and colleagues (Hazbun &

Fields 2001). Similarly, Fromont-Racine *et al.* (1997) screened their random genomic library with three of the same splicing proteins that Uetz *et al.* used to screen the arrayed yeast prey ORFs. Interestingly, about 10 high-confidence prey were identified by each bait in each screen but, for two of the baits, only two prey were found to be common to both screens. For the remaining bait, there were six overlaps (Uetz & Hughes 2000).

The tendency for similar screens to identify different sets of interacting proteins probably has several causes. First, the selection strategy used in each library may influence the interactions that take place. Secondly, the interactions may well be extremely complex, i.e. each screen only reveals a subset of the interactions taking place. Thirdly, the matrix and random library methods have been shown to differ in their sensitivity. Direct comparison of the two strategies reveals that random library screening produces more candidate interactors than the matrix method, i.e. the matrix method suffers from a higher incidence of false negatives. This may be because of incorrect folding of the fusion protein, caused perhaps by the presence of the fusion partner, or may reflect undetected PCR errors that occur during clone construction. The problem of false negatives is not so severe in the case of random libraries because each prey clone is represented by an overlapping series of fragments, giving much more scope for interacting functional domains to form. This certainly seems to have been an advantage in the case of Rain *et al.* (2001) who used a prey library of protein fragments and succeeded in linking many of the *H. pylori* proteins into a functional network. An extreme example of the difference between matrix and random library screening is shown in the study of Flajolet *et al.* (2000) who investigated interactions among the 10 mature polypeptides produced by hepatitis C virus. Constructs expressing these 10 polypeptides in a matrix format revealed no interactions at all, not even the well-characterized interactions among the capsid proteins. However, a library screening using random genomic fragments revealed all the expected interactions as well as some novel ones. It is likely that the prey constructs generated in the matrix strategy failed to fold properly and therefore could not behave as the normal proteins would *in vivo*. The use of fragment libraries rather than intact ORFs has a further advantage. Where

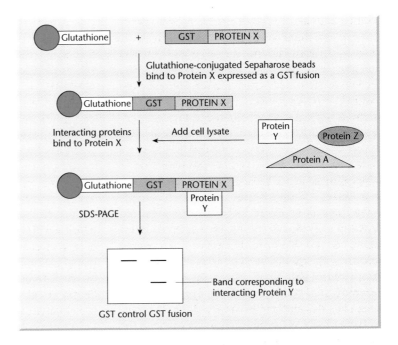

Fig. 11.6 Principle of GST-pulldown.

interactions occur between a bait and a series of overlapping prey fragments, the common sequence shared by a number of interacting prey fragments can, in principle, identify the particular domain of the protein that interacts with the bait (see e.g. Siomi *et al.* 1998; Flores *et al.* 1999; Fig. 11.5).

Alternative methods for global interaction analysis

GST-pulldown

While it is clear that the Y2H system currently provides the most amenable high-throughput format for studying protein interactions, the high frequency of false positive and false negative results is a disadvantage. Therefore, other methods that were initially devised for use with individual proteins have been developed for genomic-scale applications.

Glutathione-*S*-transferase (GST) is a bacterial enzyme with high affinity for its substrate, glutathione. The enzyme is widely used as a fusion tag in the expression of recombinant proteins, allowing affinity purification through binding to glutathione-coated Sepharose beads. If a specific protein bait is expressed as a fusion with GST, then other proteins that interact with the bait can be copurified (pulled

down) from a cell lysate containing a background of non-interacting proteins, by passing the lysate through a glutathione Sepharose column (Kaelin *et al.* 1991; Fig. 11.6). GST-pulldown has been applied on a genomic scale by Martzen *et al.* (1999) who generated 6144 yeast strains each carrying a different ORF fused to GST. Pools of GST fusions were assayed for biochemical activities and the source strains containing the corresponding ORFs were identified. Three previously uncharacterized genes were functionally annotated using this approach including a cyclic phosphodiesterase and a cytochrome *c* methyltransferase.

Mass spectrometry to characterize protein complexes

Mass spectrometry (MS) was discussed in detail in Chapter 9 as a method for the high-throughput annotation of proteins isolated from two-dimensional protein gels or eluted from protein chips. However, the technique can also be applied to the analysis of protein interactions if protein complexes can be isolated intact from whole cell preparations. This can be achieved by the affinity purification of one component. The enriched sample then has a low enough complexity to be fractionated on a standard

Protein complex	Reference
Yeast anaphase promoting complex	Zachariae *et al.* (1996)
Yeast spindle pole body	Wigge *et al.* (1998)
Human anaphase promoting complex	Grossberger *et al.* (1999)
Yeast ribosome	Link *et al.* (1999)
Human interchromatin granule cluster	Mintz *et al.* (1999)
Yeast nuclear pore complex	Rout *et al.* (2000)

Table 11.1 A selection of protein complexes which have been investigated by coaffinity purification followed by mass spectrometry (MS). In some cases the complexes were fractionated by electrophoresis prior to MS, but in other cases this was unnecessary.

polyacrylamide gel. Individual bands can be excised from the gel and analysed by MS to identify interacting proteins that have been copurified (Lamond & Mann 1997; Neubauer *et al.* 1997; Blackstock & Weir 1999).

Affinity purification can be carried out using antibodies raised against one of the proteins in the complex, but this makes the procedure dependent on the physiological levels of that target protein. It is sometimes more convenient to overexpress the protein as a fusion with an epitope tag, such as GST, because this allows interacting proteins to be trapped even if the 'bait' is not normally expressed at high levels under the growth conditions used. The resolved protein bands are stained and excised from the gel, digested with a protease such as trypsin and then screened by matrix-assisted laser desorption ionization time of flight (MALDI-TOF) MS (Box 9.5, p. 168) to determine accurate molecular masses. These results are compared to theoretical masses derived from protein sequence databases to identify the individual polypeptides. Where such analysis is uninformative, other MS methods such as electrospray ionization coupled with collision-induced dissociation (Box 9.5), can be used to derive peptide sequences. These can be screened against EST databases (Choudhary *et al.* 2001).

Examples of this type of analysis are currently limited to abundant protein complexes, such as the yeast spliceosome (Neubauer *et al.* 1997; Gottschalk *et al.* 1998; Rigaut *et al.* 1999) and the human spliceosome (Neubauer *et al.* 1998). Interestingly, none of the components of the spliceosome identified by MS were identified in the Y2H screens discussed above. The novel interacting proteins revealed in the MS studies were confirmed by various biochemical assays, including splicing assays, and their colocal-ization was shown by expressing each component as a fusion with green fluorescent protein. A selection of protein complexes that have been analysed by mass spectroscopy is listed in Table 11.1. An interesting variation of this technique was used by Pandey *et al.* (2000) to identify proteins in the epidermal growth factor (EGF) signalling pathway. HeLa cells were treated with EGF and then cell lysates were precipitated with antiphosphotyrosine antibodies. As well as recovering the signalling complex that forms around the EGF receptor, any phosphorylated proteins acting further down the pathway – but not necessarily physically associated with the membrane complex – would also be precipitated. Nine proteins were identified in this experiment: seven were known members of the pathway, one was a known protein but not thus far known to be involved in EGF signalling, and one was uncharacterized.

Informatics tools and resources for protein interaction data

Protein interaction data from a number of international collaborations are being assimilated in databases that can be accessed over the Internet (Box 11.2). These databases are mostly derived from large-scale two-hybrid screens but much more data exists from traditional protein-by-protein studies than the two-hybrid screens currently provide. A significant challenge is how to extract such data from the scientific literature and assimilate them with the data emerging from library-based screens. Several informatics tools have been developed to extract information from literature databases such as MEDLINE, but these currently have a rather

Box 11.2 A selection of protein interaction databases and resources

Some useful protein interaction websites are listed below. These contain searchable lists of interacting proteins, interaction maps and other resources. Some of the databases are available both as academic and commercial versions. A detailed discussion of protein interaction databases on the Internet can be found in recent reviews (Legrain *et al.* 2001; Tucker *et al.* 2001; Xenarios & Eisenberg 2001).

http://www.bind.ca (Biomolecular Interaction Network Database, BIND). A database designed to list protein interactions with a variety of molecules including other proteins and nucleic acids. Contains resources for molecular complexes and pathways as well as individual interactions. About 6000 interactions are currently listed (Bader & Hogue 2000; Bader *et al.* 2001).

http://dip.doe-mbi.ucla.edu (Database of Interacting Proteins, DIP). A comprehensive database of about 1000 protein interactions (Xenarios *et al.* 2000, 2001).

http://www.hybrigenics.com (Hybrigenics). A protein interaction database for *Helicobacter pylori*, based

on the study of Rain *et al.* (2001). This site includes software for graphically visualizing subsets of protein interactions.

http://mips.gsf.de/proj/yeast/CYGD/db/ index.html (Munich Information Center for Protein Sequences, MIPS). A protein sequence database, which also contains some interaction data for yeast (Mewes *et al.* 2000).

http://www.incyte.com/sequence/proteome/index. shtml (Incyte Genomics). Comprehensive yeast and worm protein databases including access to protein interaction resources (Costanzo *et al.* 2001).

http://portal.curagen.com/extpc/com.curagen. portal.servlet.PortalYeastList (Curagen). A yeast protein interaction database based on the study of Uetz *et al.* (2000).

http://genome.c.kanazawa-u.ac.jp/Y2H/ Comprehensive data sets from proteome-wide yeast two-hybrid screens, based on the studies of Ito *et al.* (2000, 2001).

low fidelity and require extensive human curation (reviewed by Xenarois & Eisenberg 2001).

Another challenge is to find a simple way to present protein interaction data in a readily accessible and understandable way. The yeast proteome is likely to consist of over 6000 basic proteins, not including variations generated by post-translational modifications, which could increase this number substantially. Each protein is thought to interact, on average, with three others. The simplest way to represent interacting components in a system is a chart with interacting proteins joined by lines. Depicting the entire yeast proteome in such a way is likely to yield a map of incredible complexity and intricacy, and it is easy to imagine the information becoming lost in the mass of detail. Schwikowski *et al.* (2000) have assimilated interaction data for about 2500 yeast proteins and generated an interaction map that included just over 1500 of them. The map is reproduced in Fig. 11.7 and initially appears very complex. However, if proteins with

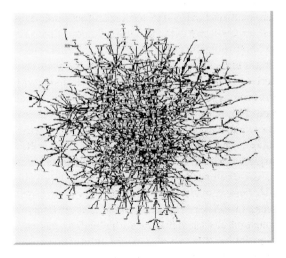

Fig. 11.7 An interaction map covering 1500 proteins (one-quarter of the yeast proteome). Interacting proteins are joined by lines, and each protein undergoes on average three interactions. Initially the map appears incredibly complex. (Reprinted from Tucker *et al.* 2001 by permission of Elsevier Science.)

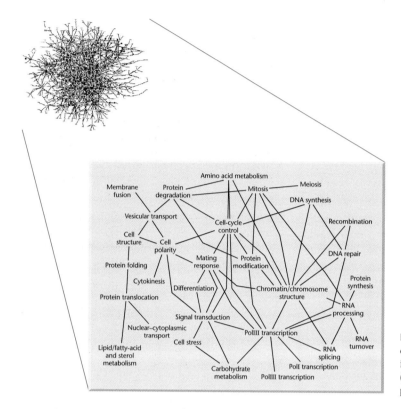

Fig. 11.8 The map from Fig. 11.7 can be simplified if proteins are grouped into functional interaction families. (Reprinted from Tucker *et al.* 2001 by permission of Elsevier Science.)

particular functions in the cell are highlighted they tend to cluster into regional interaction centres. This can be further simplified to give a functional interaction map in which basic cellular processes are linked together by virtue of protein interactions (Fig. 11.8). Thus, proteins involved in cell cycle control interact not only with each other, but also with proteins involved in related processes such as cell polarity, cytokinesis, DNA replication and mitosis. Proteins involved in DNA recombination interact among themselves and also with proteins involved in DNA repair and chromosome maintenance.

The existence of such a map is not only a valuable basic resource but, as it grows, it will provide a basis to *define* novel interactions. The interaction map provides a benchmark with which to judge the plausibility of newly discovered interactions, and helps to eliminate false positives. Statistical analysis of the map shows that nearly three-quarters of all protein interactions occur within the same functional protein group while most others occur with related functional groups. An unexpected interaction between proteins involved in, for example, Pol I transcription and vesicular transport should be regarded with suspicion and tested by rigorous biochemical and physical assays. Many will be disproved although some implausible interactions are inevitable.

Protein interaction 'knockouts'

We mentioned at the beginning of the chapter that gene function is a modular concept, each module representing an interaction between the protein product of that gene and another molecule inside or outside the cell. Techniques such as insertional mutagenesis and gene targeting (Chapter 10) may provide useful information on gene function at a cellular or organism level. However, in reality they are rather coarse methods because they abolish all the modules at once. A range of different mutant alleles is generally more informative than a single knockout because the former may preserve some interactions while abolishing others. However, this is a rather hit and miss approach and it would be much more satisfactory if particular interactions could be *selectively* disrupted.

Selective disruption is entirely possible using a variation of the two-hybrid screen called the *reverse two-hybrid system* (Vidal *et al.* 1996). In this technique, a bait that normally interacts with multiple prey is subject to *in vitro* mutagenesis to produce a range of artificial mutant alleles. These are then systematically tested against the available prey and interaction defective alleles (IDAs) are selected that fail to interact with one type of prey but successfully interact with all others (Shih *et al.* 1996; Endoh *et al.* 2000). In yeast, panels of IDAs for each bait protein can then be functionally tested by gene targeting and it should be possible to generate a range of specific phenotypes resulting from each 'interaction knockout'. The effects of the interaction knockout should represent a subset of the phenotypes caused by fully knocking out the corresponding gene.

Suggested reading

Boulton S.J., Vincent S. & Vidal M. (2001) Use of protein interaction maps to formulate biological questions. *Current Opinion in Chemical Biology* **5**, 57–62.

Pelletier J. & Sidhu S. (2001) Mapping protein–protein interactions with combinatorial biology methods. *Current Opinion in Biotechnology* **12**, 340–347.

Uetz P. (2001) Two-hybrid arrays. *Current Opinion in Chemical Biology* **6**, 57–62.

Vorm O., King A., Bennett K.L., Leber T. & Mann M. (2000) Protein interaction mapping for functional proteomics. In: *Proteomics: A Trends Guide 2000* (Supplement to *Trends in Biochemical Sciences*), 43–47.
Four excellent reviews discussing how high-throughput interaction technology is being used to construct protein interaction maps.

Legrain P. & Selig L. (2000) Genome-wide protein interaction maps using two-hybrid systems. *FEBS Letters* **480**, 32–36. *A good minireview covering two-hybrid methodology and its application to functional genomics.*

Phizicky E.M. & Fields S. (1995) Protein–protein interactions: methods for detection and analysis. *Microbiological Reviews* **59**, 94–123. *An excellent and comprehensive review covering all forms of protein interaction methodology.*

Xenarios I. & Eisenberg D. (2001) Protein interaction databases. *Current Opinion in Biotechnology* **12**, 334–339. *A recent review covering the state-of-the-art in protein interaction bioinformatics.*

Useful websites

http://www.biacore.com
BIACore, a company that manufactures chips for the detection of protein interactions.

http://www.bind.ca
http://dip.doe-mbi.ucla.edu
BIND and DIP, two databases listing protein interactions with proteins and other molecules.

CHAPTER 12

Applications of genome analysis and genomics

Introduction

The basic techniques for the manipulation of genes *in vitro* were first developed in the 1970s and quickly led to a major shift in emphasis in basic biological and medical research. Commercial exploitation of gene manipulation quickly followed and led to the rapid development of a global modern biotechnology industry. Genome analysis and genomics looks set to follow a similar pattern. Already genomics techniques are being used to answer fundamental questions in biology and are stimulating the formation of the next generation of biotechnology companies. The reason for this is simple: genomics has applications in every field of biology. Describing all these applications would take a complete book in itself. Consequently, in this chapter we have opted to focus on six themes which reflect both the successes achieved to date and the likely successes in the next 5–10 years. Three of these themes are devoted to medical applications (complex genetic disorders, drug responses and bacterial disease), two are on agricultural topics (quantitative traits, plant breeding) and the sixth is a review of the impact on developmental biology.

Theme 1: Understanding genetic diseases of humans

Introduction to theme 1

Genetic diseases in humans can be divided into two types: simple Mendelian traits and complex (polygenic) disorders. The former describes diseases caused by a single major gene defect and the inheritance pattern (dominant, recessive, sex-linked) is usually easily discerned. A complicating factor is incomplete penetrance where not all genetically predisposed individuals manifest the disease. Similarly, there can be considerable variation in severity

of disease between individuals with the same allelic variation because of the presence of modifier genes. Differences in the severity of disease also can be caused by different alleles of the same gene or by different genes. Examples of the latter are the *BRCA1* and *BRCA2* genes which play a part in early onset familial breast cancer.

Mendelian traits are relatively easy to study but account for only a small proportion of human disease. Most human diseases are polygenic in nature and these include cardiovascular disease, asthma, cancer, diabetes, rheumatoid arthritis, obesity, alcoholism and schizophrenia. Such complex diseases involve multiple genes, environmental effects and their interactions. Rather than being caused by specific and relatively rare mutations, complex diseases and traits may result principally from genetic variation that is relatively common in the population. The fact that a large number of genes, many with small effects, are involved in many complex diseases greatly complicates efforts to identify genetic regions involved in the disease process and makes replication of results difficult. The distinction in terminology between Mendelian and complex traits is not meant to imply that complex diseases do not follow the rules of Mendelian inheritance; rather, it is an indication that the inheritance pattern of complex traits is difficult to discern.

There are three main approaches to mapping the genetic variants involved in a disease: functional cloning, the candidate gene strategy and positional cloning (see Chapter 6). In functional cloning, a knowledge of the underlying protein defect leads to localization of the responsible gene. In the candidate gene approach, genes with known or proposed function with the potential to influence the disease phenotype are investigated for a direct role in disease. Positional gene cloning is used when the biochemical nature of the disease is unknown (the norm!). The responsible gene is mapped to the correct location on the chromosome and successive narrowing of the candidate interval eventually results in the identification of the correct gene.

The gene-finding methods described above are used in conjunction with two other analytical methods: linkage analysis and linkage disequilibrium (LD) mapping. In linkage analysis one looks at the inheritance of the disease gene and selected markers in several generations of the same family. In contrast, in LD mapping one looks at co-inheritance in populations of unrelated individuals. There is another difference. Linkage analysis can be used only for coarse mapping (e.g. only 10% recombination will be observed in a region of 10 Mb), whereas LD can be used for fine mapping as resolution is limited only by the spacing of the markers used.

Linkage mapping

Any kind of genetic marker can be used in linkage mapping and in classical genetics these markers are other phenotypic traits. In practice, it is difficult to detect linkages for loci more than 25 centimorgans (cM) apart. Thus, to be useful, markers need to be distributed throughout the genome at a frequency of at least 1 marker every 10 cM. In humans there are not enough phenotypic traits that have been mapped to give anything like the desired marker density. For this reason, physical markers are very attractive. The first such markers to be described were restriction fragment length polymorphisms (RFLPs, see p. 2) but the ones favoured today are the single nucleotide polymorphisms (SNPs, see p. 51) because they occur once every 1000 bp. When large multigeneration pedigrees are available (e.g. the Centre d'Etude du Polymorphisme Humain (CEPH) families, p. 70) linkage analysis is a powerful technique for locating disease genes and has been applied to a number of simple Mendelian traits. The probability of linkage is calculated and expressed as a logarithm$_{10}$ of odds (LOD) score with a value above 3 being significant (Box 12.1). If linkage to a marker is observed then the chromosomal location of that marker is also the location of the disease gene.

Conventional linkage analysis seldom works for complex diseases. The involvement of many genes and the strong influence of environmental factors means that large multigeneration pedigrees are seen only rarely. Consequently, analysis is undertaken of families in which both parents and at least two chil-

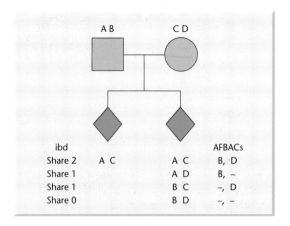

Fig. 12.1 Affected sib pair families. A nuclear family pedigree is shown with the father (■) and mother (●) in the first row and the two affected children of either sex (♦) in the second row. Assume for simplicity that we can distinguish all four parental alleles, denoted A, B, C and D in the genetic region under study, with the parental alleles ordered such that A and C are transmitted from the father and mother, respectively, to the first affected child. Four possible configurations among the two offspring with respect to the alleles inherited from the parents are possible: they can share both parental alleles (AC); they can share an allele from the father (A) but differ in the alleles received from the mother (C and D); they can share an allele from the mother (C) but differ in the alleles received from the father (A and B); or they can share no parental alleles in common. These four configurations are equally likely if there is no influence of the genetic region under consideration on the disease. The parental alleles that are never transmitted to the affected sib pair in each family type are used as a control population in association studies using nuclear family data, the so-called affected family-based control (AFBAC) sample. (Redrawn with permission from Thomson 2001.)

dren (sib pairs) have the disease in question. These are known as *nuclear families*. The way this analysis is undertaken is shown in Fig. 12.1. Suppose that we believe that a certain region of the genome is implicated in a disease state and that we can distinguish the four parental chromosomes (A, B, C, D). If the region under test does not carry a gene predisposing to disease then the chances of two affected sibs having two, one or no parental chromosome regions in common are 25, 50 and 25%, respectively. On the other hand, deviation from this Mendelian random expectation indicates that the affected sibs have chromosome regions that are *identical by descent* (ibd) suggesting the presence of genes predisposing to the disease in question. Physical markers, particularly microsatellites, are ideal for distinguishing the

Box 12.1 **Logarithm of odds scores** (Adapted from Connor & Ferguson-Smith 1997)

Figure B12.1 shows pedigrees for two families affected by an autosomal dominant disorder. In family A the affected man in the second generation has received the disease allele together with RFLP allele 1 from his father. Similarly, he has received the normal allele and RFLP allele 2 from his mother. If these two loci are on the same chromosome then it follows that he must have one chromosome that carries the disease allele together with allele 1 and the other carries the normal allele and RFLP allele 2. Consequently, the arrangement of the disease and marker alleles, also known as the *phase*, can be deduced with certainty in this individual. If the loci are linked it will be apparent in the next generation as a tendency for the disease allele to segregate with RFLP allele 1 and the normal allele to segregate with RFLP allele 2. This is indeed the case in family A, where four affected offspring carry RFLP allele 1 and the five unaffected children only carry RFLP allele 2.

If the loci described above are not linked, the probability of such a striking departure from independent assortment occurring by chance in nine offspring is the probability of correctly calling heads or tails for nine consecutive tosses of a coin. That is:

$(0.5)^9 = 0.002.$

However, if these two loci are linked such that there is only a 10% chance of crossing over (i.e. a recombination fraction, or θ, of 0.1), the probability of the disease segregating with RFLP allele 1 or the normal allele with RFLP allele 2 is:

$(0.9)^9 = 0.4.$

It follows that linkage at 10% recombination is 200 times (0.4/0.002) more likely than no linkage. Similarly, if the disease allele and the RFLP allele are identical, then no recombination could occur and the recombination fraction would be zero. For this family, this is 500 times (1/0.002) more likely than no linkage.

The usual way of representing these probability ratios is as logarithms, referred to as logarithm of odds (LOD) or Z scores. For family A at a recombination fraction of 10% the LOD score is $\log_{10} 200 = 2.3$ and at a recombination fraction of zero it is $\log_{10} 500 = 2.5$.

The figure also shows a two-generation pedigree for family B. In this case, all four affected siblings carry RFLP allele 2 and another four healthy siblings do not. As in family A, this signifies a marked disturbance of independent assortment and suggests linkage between the disease and RFLP allele 1. If this is the case, the youngest child must represent a recombinant because he has inherited RFLP allele 1 from his father but not the disease. However, it could be that the youngest child is non-recombinant and all the other children represent crossovers between the two loci. Although this is much less likely, it cannot be excluded in the absence of phase information from the grandparents. Calculation of the LOD scores for such a family is more complicated because the two possible phases must be taken into account. Whichever phase is considered, at least one

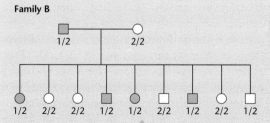

Fig. B12.1 Two families with an autosomal dominant trait showing results of DNA analysis for a marker restriction fragment length polymorphism (RFLP) with alleles 1 and 2.

continued

Box 12.1 *continued*

Table B12.1 Logarithm of odds (LOD) scores at values of the recombination fraction from 0 to 40% for the two families shown in Fig. 1.

	Recombination fraction (%)				
	0	**10**	**20**	**30**	**40**
Family A	2.7	2.3	1.8	1.3	0.7
Family B	$-\infty$	1.0	0.9	0.6	0.3
Total	$-\infty$	3.3	2.7	1.9	1.0

recombination event must have taken place and hence the recombination fraction cannot be zero (see Table B12.1).

Analysis of the combined data from the two families shows that the maximum LOD score is 3.3 and this occurs at 10% recombination. The ease with which data from phase-known and phase-unknown families can be combined in this way is the reason why the use of LOD scores has become universal for the analysis of linkage data. The maximum value of the LOD score gives a measure of the statistical significance of the result. A value greater than 3 is usually accepted as demonstrating that linkage is present and in most situations it corresponds to the 5% level of significance used in conventional statistical tests. Conversely, if LOD scores below − 2 are obtained, this indicates that linkage has been excluded at the corresponding values of the recombination fraction.

The relationship between the recombination fraction and the actual physical distance between the loci depends on several factors. A recombination fraction of 0.1 (10% recombination) corresponds to a map distance of 10 cM. However, with increasing distance between the loci the recombination fraction falls as a result of the occurrence of double crossovers. In humans, 1 cM is equivalent to 1 Mb of DNA on average.

chromosome regions derived from each parent. Not only are microsatellites highly polymorphic, a sufficient number of them have been placed throughout the genome.

Type I diabetes was the first complex disease for which genome-wide linkage scans in affected sib pairs was carried out (Field *et al.* 1994; Hashimoto *et al.* 1994). Since then, a number of other complex diseases have been mapped including bipolar mood disorder (McInnes *et al.* 1996) and Crohn's disease where a number of different loci have been identified (Hugot *et al.* 1996; Rioux *et al.* 2000). A similar methodology has been used to identify quantitative trait loci (QTLs, see p. 218) controlling adult height (Hirschhorn *et al.* 2001).

Association (linkage disequilibrium) mapping

Association studies compare marker frequencies in unrelated cases and controls and test for the co-occurrence of a marker and disease *at the population level*. A significant association of a marker with disease may implicate a candidate gene in the aetiology of a disease. Alternatively, an association can be caused by LD of marker allele(s) with the gene predisposing to disease. LD implies close physical linkage of the marker and the disease gene. As might be expected, LD is not stable over long time periods because of the effects of meiotic recombination. Thus, the extent of LD decreases in proportion to the number of generations since the LD-generating event. In general, the closer the linkage of two SNPs then the longer the LD will persist in the population but other factors do have an influence, e.g. extent of inbreeding, presence of recombination hotspots, etc. Reich *et al.* (2001) have shown that, in a US population of northern European descent, LD typically extends for about 60 kb. By contrast, LD in a Nigerian population extends for a much shorter distance (5 kb) reflecting the fact that northern Europeans are of more recent evolutionary origin.

So, it should be apparent that the ideal population for LD studies will be one that is isolated, has a narrow population base and can be sampled not too many generations from the event causing the disease mutation. The Finnish and Costa Rican populations are considered ideal because they are relatively homogenous and show LD over a much wider distance than US populations. This is particularly important because it influences the number of markers that need to be used. In a typical *linkage* analysis one uses markers every 10 Mb (10 cM) but for LD studies one needs many more markers. For a US population of northern European descent the markers would need to be every 20–50 kb on average but this could be extended to every 200–500 kb for Finnish or Costa Rican populations.

Genome-wide LD scans have been undertaken to locate simple Mendelian traits. Lee *et al.* (2001) were able to use such a methodology to localize the critical region for a rare genetic disease (SLSJ cytochrome oxidase deficiency) in a close-knit isolated community. More typically, LD analysis is used to fine-map traits following initial localization to chromosomal regions by linkage analysis as described in the previous section. SNPs are ideal for this purpose because over 1.4 million of them have been mapped at an average density of one every 1.9 kb (Sachidanandam *et al.* 2001). For studies on complex diseases, evidence for LD is sought in nuclear families or trios because this avoids possible ethnic mismatching between patients and randomly ascertained controls. In such cases, the parental alleles that are never transmitted to the affected offspring are used as the controls, the so-called affected family-based control (AFBAC) sample.

Two studies on Crohn's disease illustrate how LD can be used in fine mapping (Fig. 12.2). In the first study (Rioux *et al.* 2001), linkage analysis had shown that susceptibility to Crohn's disease mapped to an 18-Mb region of chromosome 5 with a maximal LOD score at marker D5S1984. Using 56 microsatellites, LD was detected between Crohn's disease and two other markers, IRF1p1 and D5S1984, which are 250 kb apart. All the known genes in this region were examined for allelic variants that could confer increased susceptibility to Crohn's disease but no candidate genes were identified. This was a little surprising because this genomic region encodes the cytokine gene cluster that includes many plausible candidate genes for inflammatory disease. Because no obvious candidates had emerged, a detailed SNP map was prepared with the markers spaced every 500 bp. Many of these SNPs showed LD with susceptibility to Crohn's disease, confirming the presence of a gene predisposing to Crohn's disease in the area under study.

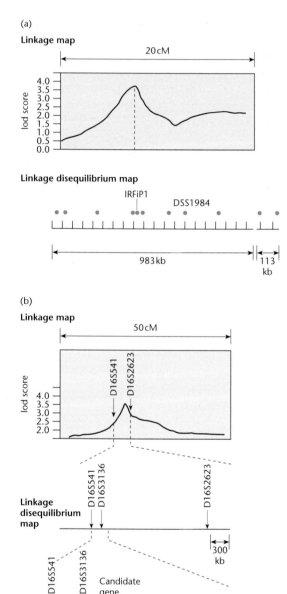

Fig. 12.2 Details of the mapping of two loci associated with Crohn's disease. (a) Mapping of a locus on chromosome 5 by Rioux *et al.* (2001). (b) Mapping of a locus on chromosome 16 by Hugot *et al.* (2001). The numbers along the bottom line correspond to the SNPs used in fine mapping. All the SNPs except 10 and 11 showed tight linkage (see text for further details).

In the second study (Hugot *et al.* 2001), linkage analysis had mapped a susceptibility locus for Crohn's disease to chromosome 16. With the aid of 26 microsatellites the locus was mapped to a 5-Mb region between markers D16S541 and D16S2623. LD analysis showed a weak association of Crohn's disease with marker D16S3136, which lies between the other two markers. A 260-kb region around marker D16S3136 was sequenced but only one characterized gene was identified and this did not appear to be a likely Crohn's disease candidate. Sequencing also identified 11 SNPs and three of these showed strong LD with Crohn's disease in 235

affected families indicating that the susceptibility locus was nearby. By using the GRAIL program and an expressed sequence tag (EST) homology search (p. 51) a number of putatively transcribed regions were identified and one of these (*NOD2*) was identified as the susceptibility locus. Further analysis showed that some of the SNPs used in the LD study were the causative mutations.

Significance of haplotypes

The pattern of SNPs in a stretch of DNA is known as the haplotype. Figure 12.3 shows the evolution of a

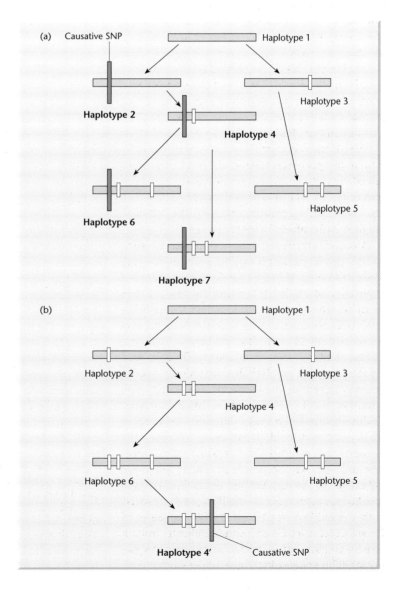

Fig. 12.3 The two ways in which a causative single nucleotide polymorphism (SNP) can become associated with a particular haplotype. In (a) the causative SNP arises early, whereas in (b) it arises late. (Redrawn with permission from Judson *et al.* 2000.)

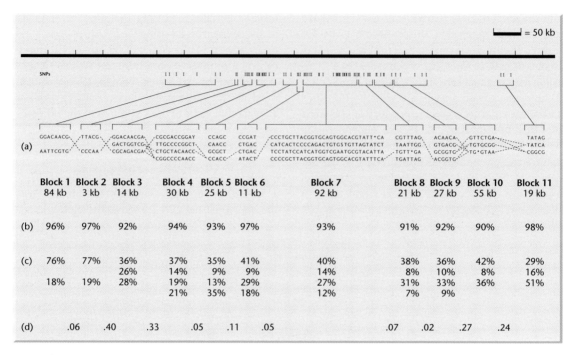

Fig. 12.4 Block-like haplotype diversity at 5q31. (a) Common haplotype patterns in each block of low diversity. Dashed lines indicate locations where more than 2% of all chromosomes are observed to transition from one common haplotype to a different one. (b) Percentage of observed chromosomes that match one of the common patterns exactly. (c) Percentage of each of the common patterns among untransmitted chromosomes. (d) Estimated rate of haplotype exchange between the block. (Reprinted from Daly *et al.* 2001 by permission of Nature Publishing Group, New York.)

number of theoretical haplotypes, some of which include an SNP causing disease. From this figure, it is clear that not all SNPs would be predictive of the disease. Also, not all haplotypes are informative and their detection in an LD association study would complicate interpretation of data. This is exactly the situation encountered in the two studies on Crohn's disease described above. To understand haplotype structure better, Daly *et al.* (2001) undertook a detailed analysis of 103 SNPs within the 500-kb region on chromosome 5q31 associated with Crohn's disease. Their results showed a picture of discrete haplotype blocks of tens to hundreds of kilobases, each with limited diversity punctuated by apparent sites of recombination (Fig. 12.4). In a corresponding study, Johnson *et al.* (2001) genotyped 122 SNPs in nine genes from 384 individuals and found a limited number of haplotypes (Fig. 12.5).

The existence of haplotype blocks should greatly simplify LD analysis. Rather than using all the SNPs in a region, we can identify exactly which SNPs will be redundant and which will be informative in asso-

ciation studies. The latter are referred to as haplotype tag SNPs (htSNPs) and they are markers that capture the haplotype of a genomic region. Thus, once the haplotype blocks in any given region are identified they can be treated as alleles and tested for LD. This not only simplifies the analysis but it reduces the number of SNPs that need to be genotyped. However, it should be noted that the current version (late 2001) of the EST database (dbSNP) does not contain sufficient SNPs to be of value in constructing a haplotype map of the entire genome. Such a map is essential if significant progress is to be made in understanding the genetic and biochemical basis of polygenic disorders.

Nature of complex diseases

For a long time, geneticists have debated whether complex diseases are caused by mutations of small effect in a large number of genes or by mutations of large effect in a small number of genes. The answer is that we still do not know. The work on Crohn's

CFLAR

C/T	A/T	A/G*	T/–	G/T	G/A	Freq.
C	A	A	T	G	G	46.00%
.	.	G	.	T	A	44.25%
T	.	G	.	T	A	8.75%
.	T	G	.	.	A	0.50%

CASP10

C/T	C/T	A/G	A/G	C/T	G/A	A/G	T/A	G/A	G/C	G/A	Freq.
C	C	A	C	G	A	T	G	G	G		44.00%
T	.	G	.	.	A	.	.	C	A		39.00%
T	.	G	.	.	A	.	A	.	C	A	7.00%
T	.	G	.	.	G	A	.	.	C	A	6.25%
.	T	G	G	T	.	.	.	A	.	A	1.75%
.	T	G	G	.	.	.	A	.	C	A	0.50%

GAD2

A/G	C/T	C/A	A/G	A/G	C/A	G/A	A/G	C/A/T	T/G	C/T	G/C*	T/A*	Freq.
A	C	C	A	G	C	G	A	C	T	C	G	T	45.00%
.	.	.	.	.	.	.	.	T	.	.	.	.	28.00%
.	T	A	G	A	.	.	.	T	.	.	.	A	12.75%
G	.	.	.	.	.	G	A	G	T	C	.	.	8.25%
.	.	.	.	.	.	.	.	T	.	.	.	A	1.75%
.	.	.	.	.	A	.	T	.	.	.	.	.	1.00%
.	T	A	G	A	.	.	.	T	.	T	C	.	0.75%
.	.	.	.	.	A	.	T	.	.	.	.	.	0.50%

H19

G/C	G/C	G/T	T/C	C/T	C/T	A/T	A/G	C/G	G/C	G/T	G/A	G/A	Freq.
G	G	G	T	C	C	A	A	C	G	G	G	G	34.75%
C	.	T	C	T	.	T	G	G	C	T	A	A	19.00%
.	C	.	C	T	.	T	G	G	C	T	A	A	15.00%
.	.	.	.	.	.	.	.	.	.	.	A	A	10.25%
.	.	T	C	T	.	T	G	G	C	T	A	A	5.75%
.	.	.	.	T	.	.	.	.	.	.	.	.	4.50%
.	C	.	C	T	.	T	G	.	C	T	A	A	1.00%
.	.	.	C	T	.	T	G	G	C	T	A	A	1.00%
.	.	.	.	.	.	G	C	T	A	A			1.00%
C	.	T	C	T	.	T	G	.	C	T	A	.	0.50%
.	C	.	C	T	.	T	G	.	C	T	A	.	0.50%
.	C	.	C	T	.	T	G	G	C	T	.	A	0.50%
.	.	.	C	T	.	T	G	G	C	T	A	.	0.50%
.	.	.	.	.	.	.	.	.	.	C	.	.	0.50%
.	.	.	.	.	.	G	C	.	.	A	A		0.50%

INS

A/C	C/T	A/T	C/G	C/T	C/T	C/A	C/T	G/T	G/A	G/A	C/A	C/T	G/A	Freq.
A	C	A	C	C	C	C	C	G	G	G	C	C	G	45.00%
.	.	.	.	.	.	.	.	.	.	.	A	.	.	20.00%
C	T	T	G	.	T	A	T	T	A	.	.	.	.	13.25%
.	.	.	.	.	.	.	.	.	.	.	.	.	A	11.25%
C	.	T	.	T	.	A	.	.	.	A	A	.	.	3.75%
.	.	.	.	.	.	.	.	.	.	.	.	T	.	3.50%
C	.	.	.	.	.	.	.	.	.	.	.	.	.	1.50%
C	.	T	.	T	.	A	.	.	.	.	.	.	.	0.50%
.	T	T	G	.	T	A	T	T	A	.	.	.	.	0.50%

TCF8

G/A	A/G	C/T	T/C	T/C	T/G	C/G	T/C	A/G	T/C	A/G	G/A	T/C	A/G	Freq.
G	A	C	T	T	T	C	T	A	T	A	G	T	A	33.50%
.	.	T	.	C	.	.	.	.	.	G	.	.	.	13.75%
.	.	T	C	C	.	.	.	.	.	.	.	.	.	13.25%
.	.	T	C	C	.	.	C	.	.	.	.	C	.	8.25%
.	.	T	.	C	.	.	.	.	.	.	.	.	.	8.00%
.	.	T	C	C	.	.	.	.	.	.	.	C	.	5.25%
.	.	.	.	C	.	G	.	.	.	A	.	.	.	4.50%
A	.	.	.	.	.	.	.	.	.	.	.	.	.	3.75%
.	G	T	.	C	.	.	.	.	.	G	.	.	.	2.25%
.	.	.	.	.	.	.	.	G	.	.	.	.	.	1.75%
.	.	T	C	C	G	.	.	.	.	.	.	C	.	1.75%
.	G	T	T	C	.	.	.	.	C	G	.	.	.	1.25%
.	.	.	.	.	.	.	.	.	.	.	.	G	.	0.75%
.	.	.	.	C	.	.	.	.	.	.	.	.	.	0.75%
.	.	T	C	.	.	.	.	.	.	.	.	.	.	0.75%

CASP8

T/G	T/C	G/A	G/T	C/G*	G/A	G/A	C/G	C/T	G/C	A/G	A/G	C/A	Freq.
T	T	G	G	C	G	G	C	C	G	A	A	C	39.00%
.	.	A	.	.	A	A	G	.	.	.	.	A	18.25%
.	.	.	.	G	.	.	.	.	.	.	G	.	12.75%
.	C	.	.	.	.	.	.	.	.	.	.	.	9.25%
.	.	A	.	.	A	.	G	.	.	.	.	A	6.50%
.	.	A	.	.	A	.	G	.	.	G	.	A	3.75%
.	.	.	.	.	.	.	.	.	.	.	.	A	2.75%
G	.	A	T	.	.	.	.	.	.	.	.	.	1.75%
.	.	.	.	.	.	.	.	G	T	C	.	A	1.75%
.	.	.	.	.	.	.	.	G	T	.	.	A	1.25%
.	.	.	.	.	A	.	G	.	.	.	.	A	1.00%
G	.	A	T	.	.	.	.	.	G	.	.	A	0.75%

SDF1

G/A	A/G	C/T	G/C	G/A	G/A	C/T	C/T	A/G	G/C	A/T	T/C	+/–	T/C	G/A	+/–*	T/C*	G/A	C/T*	T/C*	C/T	G/G*	Freq.
G	A	C	G	G	G	C	C	A	G	A	T	+	T	G	+	T	G	C	T	C	C	30.75%
.	.	.	.	.	.	.	.	.	.	.	.	.	.	A	.	.	T	.	.	.	.	17.50%
.	.	.	.	.	.	T	.	.	C	.	.	.	A	.	.	.	T	.	.	.	G	15.00%
A	G	.	.	A	.	.	T	G	.	T	C	.	C	.	.	–	C	.	T	.	G	14.25%
.	G	.	C	.	A	.	T	.	.	.	.	.	.	.	.	–	.	.	T	.	.	10.00%
.	.	T	.	.	.	.	.	.	.	.	.	.	.	.	.	–	.	.	T	.	.	5.50%
.	.	.	.	.	.	.	.	.	.	.	.	.	.	.	.	–	.	.	T	.	.	1.50%
.	.	.	.	.	.	.	.	.	.	.	.	.	.	–	.	.	.	C	T	G	.	1.50%
A	G	.	.	A	.	.	T	G	.	T	C	.	.	.	.	–	C	.	T	.	G	0.75%
A	G	.	.	A	.	.	T	G	.	T	C	.	C	.	.	–	C	A	T	.	.	0.50%

Fig. 12.5 Common European haplotypes and their haplotype tag single nucleotide polymorphisms (htSNPs) observed at nine genes. Boxed SNPs represent the htSNPs that can capture the common haplotypes that are segregating in European populations. Dots represent the allele that is found on the most common haplotype. Asterisks indicate SNPs described in dbSNP. (Reprinted from Johnson *et al.* 2001 by permission of Nature Publishing Group, New York.)

disease cited above showed that a defective *NOD2* gene makes its bearer susceptible to Crohn's disease. However, it confers risk only, and disease occurrence is dependent on many other factors. This is not *the* gene for Crohn's disease and other chromosomal loci have been implicated. The number of genes involved and the way they interact to cause disease is unknown as yet. In this context, a study (Altshuler *et al.* 2000a) of the Pro12Ala polymorphism in the *PPARgamma* gene in 3000 individuals showed that

Box 12.2 Glossary of terms used in human genetics

Ascertainment bias

This is the difference in the likelihood that affected relatives of the cases will be reported to the geneticist as compared with the affected relatives of controls.

Concordance

If two related individuals in a family have the same disease they are said to be concordant for the disorder (cf. discordance).

Discordance

If only one member of a pair of relatives is affected with a disorder then the two relatives are said to be discordant for the disease (cf. concordance).

Familial aggregation

Because relatives share a greater proportion of their genes with one another than with unrelated individuals in the population, a primary characteristic of diseases with complex inheritance is that affected individuals tend to cluster in families. However, familial aggregation of a disease does not necessarily mean that a disease has a genetic basis as other factors could be at work.

Founder effect

If one of the founders of a new population happens to carry a relatively rare allele, that allele will have a far higher frequency than it had in the larger group from which the new population was derived. The founder effect is well illustrated by the Amish in Pennsylvania, the Afrikaners in South Africa and the French-Canadians in Quebec. An early Afrikaner brought the gene for variegate porphyria and the incidence of this gene in South Africa is 1 in 300 compared with 1 in 100 000 elsewhere.

Genome scan

This is a method whereby DNA of affected individuals is systematically analysed using hundreds of polymorphic markers in a search for regions that are shared by the two sibs more frequently than on a purely random basis. When elevated levels of allele-sharing are found at a polymorphic marker it suggests that a locus involved in the disease is located close to the marker. However, the more polymorphic the loci studied the more likely it is that elevated allele-sharing occurs by chance alone and hence one looks for high LOD scores.

Index case

See 'proband'.

Multiplex family

A family with two or more affected members.

Non-parametric (model-free) analysis

This method makes no assumption concerning the number of loci or the role of environment and chance in causing lack of penetrance (q.v.). Instead, it depends solely on the assumption that two affected relatives will have disease-predisposing alleles in common.

Parametric (model-based) linkage analysis

This method of analysis assumes that there is a particular mode of inheritance (autosomal dominant, X-linked, etc.) that explains the inheritance pattern. Therefore one looks for evidence of a genetic locus that recombines with a frequency that is less than the 50% expected with unlinked loci.

Penetrance

In clinical experience, some disorders are not expressed at all even though the individuals in

continued

Box 12.2 *continued*

question carry the mutant alleles. Penetrance is the probability that such mutant alleles are phenotypically expressed.

Proband

The member through whom a family with a genetic disorder is first brought to attention (ascertained) is the proband or index case if he or she is affected.

Relative risk

The familial aggregation (cf.) of a disease can be measured by comparing the frequency of the disease in the relatives of an affected individual with its frequency in the general population. The relative risk ratio is designated by the symbol λ. In practice, one measures λ for a particular class of relative, e.g. sibs, parents.

Sibs

Brothers and sisters are sibs.

Simplex family

A family in which just one member has been diagnosed with a particular disease.

Transmission disequilibrium test

This tests whether any particular alleles at a marker are transmitted more often than they are not transmitted from heterozygous parents to affected offspring. The benefit of this test is that it only requires trios (cf.).

Trio

An affected child plus both parents.

the more common proline allele (85% frequency) led to a modest (1.25-fold) but significant increase in diabetes risk. Because the risk allele occurs at such high frequency, its modest effect translates into a large population attributable risk.

Glossary

Any reader who is not familiar with the methods used in the study of human genetics will have great difficulty in understanding the primary literature. The reason for this is the widespread use of specialist terminology. To assist such readers a glossary of commonly used terms is given in Box 12.2.

Theme 2: Understanding responses to drugs (pharmacogenomics)

Introduction to theme 2

The perfect therapeutic drug is one that effectively treats a disease and is free of unwanted side-effects.

Over the past 25 years many important new classes of drugs have been launched. However, even the most successful and effective of these provide optimal therapy only to a subset of those treated. Some individuals with a particular disease may receive little or no benefit from a drug while others may experience drug-related adverse effects. Such individual variations in response to a drug are responsible for the high failure rates of new drug candidates at the clinical trials stage.

Pharmacogenomics is the study of the association between genomic, genetic and proteomic data on the one hand and drug response patterns on the other. The objective is to explain interpatient variability in drug response and to predict the likely response in individuals receiving a particular medicine. As such, pharmacogenomics has the potential to influence the way approved medicines are used as well as have an impact on how clinical trials are designed and interpreted during the drug development process. Relevant information may be derived during the clinical trial recruitment phase, following the treatment phase, or both.

Causes of variation in drug response

There are two fundamental causes of individual responses to drugs. The first of these is variation in the structure of the target molecule. If a drug acts by blocking a particular receptor then it may be that the receptor is not identical in all individuals. A good example is the variation observed in the response of patients with acute promyelocytic leukaemia (APL) to all-*trans*-retinoic acid (ATRA). Some patients who contract APL do so because of a balanced translocation between chromosomes 15 and 17 that results in the formation of a chimaeric PML-RARα receptor gene. Other patients have a translocation between chromosomes 11 and 17 that results in the formation of a chimaeric PLZF-RARα receptor. These chimaeras are believed to cause APL by interference with RAR function. Clearly, the RAR receptor is different in the two types of APL and this is reflected in the fact that only the first type responds to ATRA (He *et al.* 1998).

Another good example of receptor variation is the polymorphism exhibited by the β2-adrenergic receptor. Nine naturally occurring polymorphisms have been identified in the coding region of the receptor and one of these (Arg16Gly) has been well studied. Asthmatic patients who are homozygous for the Arg16 form of the receptor are 5.3 times more likely to respond to albuterol (salbutamol) than those who are homozygous for the Gly16 form. Heterozygotes give an intermediate response, being 2.3 times more likely to respond than Gly16 homozygotes (Martinez *et al.* 1997). Also of interest is the ethnic variation in frequency of this polymorphism: in white people, Asians and those of Afro-Caribbean origin it is 0.61, 0.40 and 0.50, respectively.

The second cause of variation in drug response is differences in pharmacokinetics: differences in the way that a particular drug is adsorbed, distributed, metabolized and excreted (ADME) by the body. A drug that is not absorbed or that is metabolized too quickly will not be effective. On the other hand, a drug that is poorly metabolized could accumulate and cause adverse effects (Fig. 12.6). Obviously, variations in ADME can have multiple causes and the best studied are polymorphisms in drug transport and drug metabolism.

An important polymorphism associated with drug transport is a variant in the multidrug-resistance

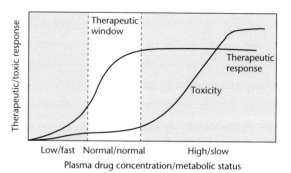

Fig. 12.6 Plasma drug concentrations in different patients after receiving the same doses of a drug metabolized by a polymorphic enzyme.

gene *MDR-1*. The product of this gene is an ATP-dependent membrane efflux pump whose function is the export of substances from the cell, presumably to prevent accumulation of toxic substances or metabolites. A mutation in exon 26 of the *MDR-1* gene (*C3435T*) correlated with alterations in plasma levels of certain drugs. For example, individuals homozygous for this variant exhibited fourfold higher plasma levels of digoxin after a single oral dose and an increased maximum concentration on chronic dosage (Hoffmeyer *et al.* 2000). Other substrates for this transporter are a number of important drugs that have a narrow therapeutic window; i.e. there is little difference between the drug concentration that gives the desired effect and that which is toxic. Therefore, this polymorphism could have a major impact on the requirement for individual dose adjustments for carriers of this mutation. In this context it should be noted that the mutation is probably very common, given that 40% of the German population are homozygous for it.

Most drugs are altered before excretion, either by modification of functional groups or by conjugation with molecules that enhance water solubility (e.g. sugars). Most functional group modifications are mediated by the cytochrome P450 group of enzymes and the genes that encode them are highly polymorphic (for review see Ingelman-Sundberg *et al.* 1999). Over 70 allelic variants of the *CYP2D6* locus have been described and, of these, at least 15 encode non-functional gene products. Phenotypically, four different types can be recognized (Table 12.1). The significance of these phenotypes is illustrated by the response of patients to the drug nortriptyline. Most

Table 12.1 The four drug metabolism phenotypes.

Phenotype	Frequency (%)	Cause
Extensive metabolizers	75–85	
Intermediate metabolizers	10–15	Both gene copies encode decreased enzyme with lower activity than normal
Poor metabolizers	5–10	Homozygous for two low-activity or non-functional alleles
Ultrarapid metabolizers	2–7	Amplification of the *CYP2D6* locus

patients require 75–150 mg per day in order to reach a steady-state plasma concentration of 50–150 µg per litre. However, poor metabolizers need only 10–20 mg per day, whereas ultrarapid metabolizers need 300–500 mg per day. If the genotype or phenotype of the patient is not known then there is a significant risk of overdosing and toxicity (poor metabolizers) or underdosing (ultrarapid metabolizers). The polymorphism of the *CYP2D6* locus is particularly important because it is implicated in the metabolism of over 100 drugs in addition to nortriptyline.

For many genetic polymorphisms affecting drug efficacy there is no evident phenotype in the absence of a drug challenge. This brings an unwanted element of chance into the selection of appropriate therapies for patients and the selection of patients for clinical trials of new drugs. In both cases there are very significant cost implications. However, if the phenotype or, better still, the genotype of an individual was known in advance then better clinical decisions could be made. Because many of the polymorphisms described above are the result of single nucleotide changes, i.e. they are SNPs (see p. 51), detecting them should be relatively easy. Using the hybridization arrays described earlier (p. 151), it should be possible to create an SNP profile for all the polymorphisms known to affect drug safety and efficacy (Fig. 12.7). This profiling could be extended to multiple base insertions and deletions.

Personalized medicine

From the material presented in theme 1 it should be clear that we are beginning to get a handle on the genetic and biochemical causes of common but complex diseases. Initially this will lead to better classification of the different sub-types of a disease and hence to better diagnoses. This in turn will facilitate selection of the most appropriate therapies. Later, when we know the exact cause of each disease, we should be able to develop drugs that will treat the cause rather than the symptoms as we do at present. Ultimately, genetic analysis of affected individuals will suggest what drugs *could* be used to treat the disease and a second, pharmacogenomic analysis will determine which drugs *should* be used.

A different way of using pharmacogenomic data has been suggested by a study on the chemosensitivity of different cancer cell lines. Staunton *et al.* (2001) used oligonucleotide chips to study the expression levels of 6817 genes in a panel of 60 human cancer cell lines for which the chemosensitivity profiles had been determined. Their objective was to determine if the gene expression signatures of untreated cells were sufficient for the prediction of chemosensitivity. Gene expression-based classifiers of sensitivity or resistance for 232 compounds were generated and in independent tests were found to be predictive for 88 of the compounds, irrespective of the tissue of origin of the cells. These results could open the door to the development of more effective chemotherapy regimes for cancer patients.

Theme 3: Understanding and combatting bacterial pathogenicity

Introduction to theme 3

Bacteria use a variety of biochemical mechanisms to gain access to their niche on or within a host, to colonize that niche, to escape the host defences and to exclude competing organisms. Many of these

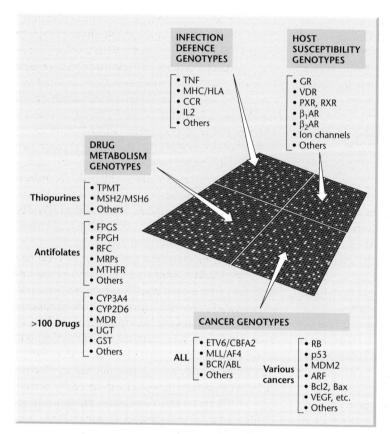

Fig. 12.7 A possible design of microarray for the diagnosis of pharmacogenomic traits. (Reprinted from Evans & Relling 1999 by permission of the American Association for the Advancement of Science.)

virulence factors and their regulatory elements can be divided into a smaller number of groups based on the conservation of similar mechanisms. These common themes are found throughout bacterial virulence factors (for review see Finlay & Falkow 1997). With the exception of those few bacteria that synthesize toxins in the absence of eukaryotic cells, e.g. *Clostridium botulinum*, contact between a bacterium and its site of infection is an essential prerequisite for pathogenicity. Bacteria possess specialized factors required for attachment to the host and these are either type IV pili (fimbriae) or non-pilus adhesins. Only a few basic designs of these molecules have been observed. There is considerable sequence similarity between the non-pilus adhesins of *Xylella* (a plant pathogen) and *Haemophilus* and *Moraxella* (human pathogens). Similarly, the fimbrial sequences of *Pseudomonas*, *Bordetella* and *Haemophilus* are related. Adhesins also contribute to cell–cell interactions between bacteria which are often a prerequisite for invasion.

Some bacteria possess the means to enter host cells and replicate there and this requires special determinants called invasion factors. These factors frequently subvert the signalling systems of the host, thereby promoting internalization of the bacterium by cells not normally proficient for phagocytosis, e.g. epithelial cells. Two general mechanisms of internalization have been discovered and both involve structural mimicry (Stebbins & Galan 2001). One of these mechanisms involves the production by the pathogen of homologues of host proteins; e.g. *Yersinia* sp. and *Salmonella* sp. produce tyrosine phosphatases with considerable sequence and structural homology to eukaryotic phosphatases. Because tyrosine phosphorylation does not commonly occur in bacteria it is probable that these molecules have evolved specifically to modulate host cellular functions. The second mechanism involves effectors that have no obvious amino acid sequence similarity to host cell proteins but which have the same three-dimensional structural similarity. This kind of mimicry is very common in the enteric bacteria (*Escherichia coli, Salmonella, Shigella, Yersinia*) as well as *Listeria, Helicobacter* and *Pseudomonas* sp. A good example is provided by *Yersinia pseudotuberculosis*. This bacterium uses the envelope protein invasin to bind host cell β1-integrin surface receptors, thereby manipulating signal transduction pathways in the host and stimulating internalization. The potency of invasin is such that it will out-compete natural host substrates such as fibronectin for β_1-integrin binding.

Part of the pathogenic process involves avoiding destruction by the host defences. In some cases this is achieved by adopting an intracellular life style, whereas in others the bacterium synthesizes extracellular polysaccharides that are poor immunogens and have antiphagocytic properties. If bacteria cannot avoid attack by immunoglobulin molecules then they may synthesize specific enzymes to destroy them. Should the pathogen be engulfed by phagocytes it may be able to synthesize protective factors to overcome the conditions of oxidative stress, low pH and attack by antimicrobial defence peptides.

Toxins are the best known of the bacterial virulence factors. In some cases these are produced to enable the pathogen to degrade the membrane of the phagocytic vacuole allowing it to escape into the host cytoplasm. In other cases, a non-invasive pathogen may use its toxins to effect the release of nutrients from the host. Again, there is a commonality of mechanisms. In Gram-negative plant and animal pathogens, these toxins are delivered into host cells via a complex protein secretion system termed type III (Lahaye & Bonas 2001; Plano *et al.* 2001) or an ancestral conjugation system termed type IV (Christie 2001). Similarly, there are only a few types of toxins, despite a large number of different host targets. Repeats in toxins (RTX) are produced by a wide variety of Gram-negative bacteria and fall into two categories: the haemolysins which affect a variety of cell types and the leukotoxins which are more cell- and species-specific. The genes encoding RTX toxins have a common four gene operon structure and the toxins themselves are characterized by 6–40 repeats of a nonapeptide (Lally *et al.* 1999).

Other virulence factors that are easily detected during whole genome sequencing projects are proteins involved in iron sequestration and antibiotic synthesis. Iron is essential for virulence but is never free in the cytoplasm or the extracellular matrix. In the case of antibiotics, their function is believed to be the suppression of other bacteria that could compete for nutrients released as a result of host cell destruction. Polyketide synthesis genes are found in *Xylella fastidiosa* and *E. coli* O157 and β-lactam synthesis is a property of other pathogens such as *Erwinia* sp.

The commonality of virulence mechanisms extends to bacteria–insect interactions. Gram-negative bacteria that are pathogenic for insects encode three gene families for insecticidal toxins and homologues of these genes are found in *Yersinia pestis*, the plague bacterium which is spread by fleas (Parkhill et al. 2001b).

Comparative genetics can be used to recognize virulence determinants in a newly sequenced organism. Genes with significant homology to well-characterized virulence genes in other organisms are likely to be involved in virulence as well. However, caution must be exercised if the homology is to a 'putative' virulence gene because many of these subsequently turn out to have no role in infection (Wassenaar & Gaastra 2001). Another indicator is if a gene is present in virulent strains but absent from non-virulent or less virulent isolates of the same species. Evidence for a role in virulence also is suggested if a gene product has numerous amino acid variations in different isolates.

Most pathogens do not have a single virulence determinant; rather, pathogenicity is dependent on a combination of many different factors as shown for the plant pathogen *X. fastidiosa* in Fig. 12.8. Nor does every pathogen carry out every type of interaction with the host. In reality, virulence genes can be subdivided into three categories: true virulence genes, virulence-associated genes and virulence life style genes (Fig. 12.9).

Horizontal transfer of virulence determinants

The fact that there is considerable protein sequence homology for each type of pathogenicity determinant suggests that there has been extensive horizontal transfer of virulence genes. Support for this comes from the identification of pathogenicity islands (PAIs) in organisms whose genomes have been completely sequenced. PAIs are large chromosomal regions (35–200 kb) encoding several virulence gene clusters

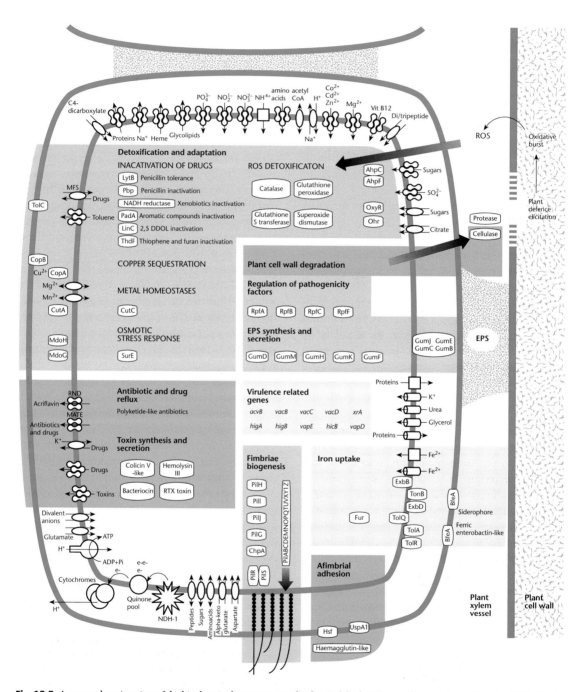

Fig. 12.8 A comprehensive view of the biochemical processes involved in *Xylella fastidiosa* pathogenicity and survival in the host xylem. The principal functional categories are shown in bold, and the bacterial genes and gene products related to that function are arranged within the coloured section containing the bold heading. Transporters are indicated as follows: cylinders, channels; ovals, secondary carriers, including the MFS family; paired dumb-bells, secondary carriers for drug extension; triple dumb-bells, ABC transporters; bulb-like icon, F-type ATP synthase; squares, other transporters. Icons with two arrows represent symporters and antiporters (H^+ or Na^+ porters, unless noted otherwise). 2,5DDOL, 2,5-dichloro-2,5-cyclohexadiene-1,4-dol; EPS, exopolysaccharides; MATE, multiantimicrobial extrusion family of transporters multidrug efflux gene (*XF2686*); MFS, major facilitator superfamily of transporters; pbp, β-lactamase-like penicillin-binding protein (*XF1621*); RND, resistance-nodulation-cell division superfamily of transporters; ROS, reactive oxygen species. (Redrawn with permission from Simpson *et al.* 2000.)

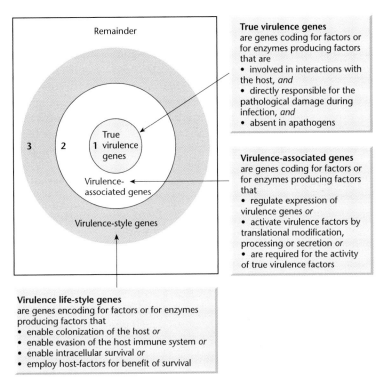

Fig. 12.9 Depending on the definition of virulence, more or fewer genes are called 'virulence genes'. The number of virulence genes and virulence-associated genes included in a given definition is represented by concentric circles. In collection 1, only those virulence factors are included that are directly involved in causing disease ('true virulence genes'). The addition of 'virulence-associated genes' increases the number of identified virulence genes and thus the size of circle 2. The gene pool identified by inactivation and phenotypic characterization includes all genes that lead to an attenuated phenotype as 'virulence life-cycle genes' (circle 3). The remaining genes are housekeeping genes, structural genes and essential genes. The border between collection 3 and the remainder cannot be exactly defined. (Reprinted from Wassenaar & Gaastra 2001 by permission of Elsevier Science.)

that are present in all pathogenic isolates and generally absent from non-pathogenic isolates. They can be recognized in genome sequences by the fact that their percentage G + C content differs from that of the rest of the genome (Karlin 2001). Of particular interest is that they encode an integrase, are flanked by direct repeats and insert into the chromosome adjacent to tRNA genes (Hacker *et al.* 1997). In this respect, PAIs resemble temperate bacteriophages and could have been acquired by new hosts as a result of transduction (Boyd *et al.* 2001). Alternatively, spread could have been achieved by conjugative transposons.

Genomics and the development of new antibiotics

Bacterial resistance to antibiotics is a major problem in human healthcare. There are two types of antibiotic resistance: innate and acquired. Acquired resistance arises by mutation or by acquisition of a plasmid carrying genes encoding enzymes that modify antibiotics and render them ineffective. Innate resistance is encoded in the chromosome and

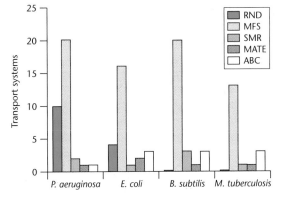

Fig. 12.10 Comparison of the number of predicted drug efflux systems in different bacteria. ABC, ATP-binding cassette family; MATE, multidrug and toxic compound extrusion family; MFS, major facilitator superfamily; RND, resistance–nodulation–cell division family; SMR, small multidrug resistance family. (Reproduced with permission from Stover *et al.* 2000.)

the most common form is the presence of multiple genes for antibiotic efflux systems (Fig. 12.10). Orthologues of these systems are widespread in bacteria and within a genus there can be many paralogues, as in *Pseudomonas aeruginosa* and *X. fastidiosa* (Simpson *et al.* 2000; Stover *et al.* 2000). Because of the prevalence of these efflux pumps some companies have embarked on programmes to find inhibitors of them (Renau *et al.* 1999).

Using bioinformatics it should be possible to separate the genes of bacteria into four types:
1 genes that are conserved (orthologous) in most or all living organisms;
2 genes that are conserved within a particular phylogenetic kingdom;
3 genes conserved within an order or genus; and
4 organism-specific genes.

Genes of the first type will not make good targets for antibiotics, whereas the other three types offer a degree of selectivity. Analysis of genome sequence data indicates that roughly 25% of bacterial genes are required for growth on a nutrient-rich medium. Of these, 40% have no mammalian or other eukaryotic orthologues and are potential targets for new antimicrobials. Some of these genes will be orthologous in all bacteria and their gene products could be the target for broad-spectrum antibiotics. Other genes may only have orthologues in a more restricted range of bacteria and could be used for the selection of narrow-spectrum antibiotics. Some bacteria, such as *Chlamydia* and *Borrelia*, use novel pathways for synthesis of aminoacyl-tRNAs (Raczniak *et al.* 2001). For the ultimate in specificity one would select as a target genes that are non-orthologous gene replacements (see p. 113). The advantage of this latter approach is that if a suitable antibiotic was identified, its use would be restricted to a small number of clinical indications and this would minimize misuse and the associated development of resistance. Alternatively, one could target an essential pathway or protein that only occurs in a very restricted group of bacteria. Examples of these are the enzymes involved in lipogenesis and lipolysis and the novel glycine-rich proteins in *Mycobacterium tuberculosis* (Cole *et al.* 1998).

An example of the use of genomics to facilitate antibiotic development is provided by the work of Jomaa *et al.* (1999). Using sequence data they identified two genes in the malarial parasite *Plasmodium falciparum* that encode key enzymes in the DOXP pathway of isoprenoid biosynthesis. This pathway functions in some bacteria, algae and higher plants but not mammals and is used by the malarial parasite when growing in red blood cells. Inhibitors of one of the two gene products, DOXP reductoisomerase, cured mice infected with a related species of *Plasmodium* and exhibited low toxicity.

Bioinformatics could be used in a different way to facilitate antibiotic development. Currently, many antibiotics are ineffective because they are unable to penetrate the cell wall and membrane. One solution is to administer the antibiotic as a prodrug that is activated by cytoplasmic enzymes with a known distribution within bacteria. Another is to modify the antibiotic by coupling it to a molecule that is the substrate for particular transporters.

One disadvantage of many antibiotics is that they kill beneficial bacteria as well as harmful ones. A good example is the loss of gut microflora following administration of broad-spectrum antibiotics, an event that can be accompanied by unwanted fungal colonization. Furthermore, because antibiotics are bactericidal there is very strong selective pressure for antibiotic resistance. Both these unwanted effects arise because the antibiotics are targeted at proteins whose function is essential for growth of the bacteria. An alternative approach would be to target proteins that are essential for virulence but not for growth. Previously, it was difficult to identify these proteins and to develop high-throughput screens but with genomics the task is relatively simple. Potential targets are first identified from genome analysis and their suitability assessed by determining the virulence of isolates carrying mutations in that gene. For example, *Yersinia* strains with a defect in type III secretion are severely attenuated (Parkhill *et al.* 2001b). This secretion system thus is a candidate target for a new antibiotic.

Genomics and the development of new vaccines

Whereas antibiotics are used to treat bacterial infections, vaccines are used to prevent infections taking hold in the first place. However, there are many common but serious bacterial diseases for which efficacious vaccines do not exist. For these diseases, the traditional approaches to vaccine discovery have so far failed and new strategies are required. One approach that was thought promising was recombinant DNA technology. This has been used to develop three very efficacious vaccines for protection

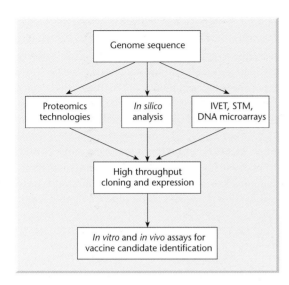

Fig. 12.11 The different genomics approaches to the development of vaccines for bacterial pathogens. (Reprinted from Grandi 2001 by permission of Elsevier Science.)

against rabies, hepatitis B virus and *Bordetella pertussis* (whooping cough). The specific antigens used in the vaccine were selected on the basis of immunological data from patients and the appropriate genes were cloned and expressed in other hosts. However, this approach has several limitations. The immunogenic proteins will not necessarily be protective and, even if they are protective, might not be suitable for use in vaccines because of sequence variability. In addition, using this approach, only a few antigens can be analysed simultaneously. In the event of negative results from animal models or clinical trials the whole process must begin again.

Genomics offers an alternative route to vaccine development through the use of three different strategies. As shown in Fig. 12.11, each strategy involves the same four basic steps and differs only in the method of selecting candidate genes from the genome. If the *in silico* strategy is adopted then several bioinformatics tools can be used to search whole genome sequences for vaccine candidates. A program such as BLAST (p. 102) can be used to compare protein-coding sequences with known virulence proteins in other organisms. New virulence factors also can be identified by searching for unknown genes that are co-regulated with known virulence genes. Other programs can be used to

identify secreted or surface-associated proteins by searching for motifs associated with transmembrane sequences. Once candidate genes are selected, they are cloned, expressed and purified and used for immunological studies. By using robotics and high-throughput screening techniques, in 6 months it is possible to screen over 500 proteins for their suitability for use in vaccines (Grandi 2001).

A good illustration of this approach has been presented by Pizza *et al.* (2000). Attempts to develop a vaccine for group B *Neisseria meningitidis* using the conventional approach had failed and there were two reasons for this. First, the capsular polysaccharide of group B meningococci is a poor immunogen and has the potential to cause autoimmunity. Secondly, the surface-exposed proteins that had been tested were able to induce protective immunity but only offered protection against a limited number of isolates because of high sequence variability. Pizza *et al.* (2000) identified 570 putative cell surface or secreted proteins and cloned the corresponding genes. Of these, 61% were expressed successfully and the purified proteins were used to immunize mice. Immune sera were then screened for bactericidal activity and for the ability to bind to the surface of meningococci. Seven representative proteins were selected for further study and were evaluated for their degree of sequence variability among multiple isolates of the bacteria. The eventual outcome was the identification of two highly conserved proteins with good potential as vaccine candidates.

The second approach to identifying candidate proteins for vaccines is to identify those proteins that are specifically induced during infection. Two techniques that have been developed to do this are *in vivo* expression technology (IVET) and signature-tagged mutagenesis (STM). Both are described in our sister book (Primrose *et al.* 2001) and are reviewed in Chiang *et al.* (1999). Both these technologies were developed before the genomics era and neither is dependent on sequence information. By contrast, DNA microarray technology is dependent on genome sequence data. Microarrays are used to identify virulence genes by growing the pathogens in appropriate *in vivo* models and then determining which genes are newly expressed compared with cells grown *in vitro*. Using DNA microarrays carrying the 2156 annotated genes of a group B meningococcus, 348 and 324 genes were identified that were induced by contact with epithelial and endothelial cells, respectively (Grandi 2001).

The third strategy for identifying vaccine candidates is to use proteomics. The proteins that are most likely to produce a protective immune response are localized on the cell surface. Although there are algorithms for predicting protein localization from genome sequence data, they give no indication of the protein composition of the cell surface in qualitative and quantitative terms. However, this information can be obtained by using a variety of protein analysis techniques to elucidate the protein composition of the cell surface after growth *in vitro*. Of particular value in this regard is MALDI-TOF mass spectrometry. It is used to generate a peptide-mass fingerprint which is then compared with theoretical fingerprints of all the proteins predicted from analysis of the genome sequence. By using automated procedures, which include excision of protein spots from two-dimensional gels, enzymatic digestion, mass spectrometry and database searching (p. 166), it is possible to identify hundreds of proteins in only a few days (Pandey & Mann 2000).

Theme 4: Getting to grips with quantitative traits

Quantitative trait loci

As noted in theme 1, the term complex trait refers to any phenotype that does not exhibit classic Mendelian recessive or dominant inheritance attributable to a single gene locus. Most, but not all, complex traits can he explained by polygenic inheritance; i.e. these traits require the simultaneous presence of mutations in multiple genes. Polygenic traits may be classified (Lander & Schork 1994) as discrete traits, measured by a specific outcome (e.g. development of diabetes or cleft palate), or quantitative traits measured by a continuous variable (e.g. grain yield, body weight).

Many important biological characteristics are inherited quantitatively but, because these effects have not generally been resolvable individually, quantitative geneticists have used biometrical procedures to characterize them (Falconer & McKay 1996). However, many issues in quantitative genetics and evolution are difficult to address without

additional information about the genes that underlie continuous variation. The identification of such *quantitative trait loci* (QTLs) first became possible with the advent of RFLPs as genetic markers and the increasing availability of complete RFLP maps in many organisms. The development of new types of markers has allowed dense marker frameworks to be assembled for many species. These genetic maps, together with other genomic resources such as physical clone sets, ESTs, annotated genome sequence data and functional genomics tools, have now made the task of identifying QTLs considerably easier.

Mapping quantitative trait loci

The basis of all QTL detection, regardless of the species to which it is applied, is the identification of association between genetically determined phenotypes and specific genetic markers. However, there is a special problem in mapping QTLs and other complex trait genes and that is penetrance; i.e. the degree to which the transmission of a gene results in the expression of a trait. For a single gene trait, biological or environmental limitation accounts for penetrance, but in a multigenic trait the genetic context is important. Hence, the consequences of inheriting one gene rely heavily on the co-inheritance of others. For this reason careful thought needs to be given to the analysis methodology.

In inbreeding populations, i.e. crops and laboratory and domestic animals, the kinds of cross used to dissect QTLs are shown in Fig. 12.12. Populations are typically generated by crosses between inbred strains, usually the first generation backcross (N_2) or intercross (F_2). Higher generation crosses (N_3), panels of recombinant inbred lines (RIL), recombinant congenic strains (RC) or inbred strains themselves may also be used. The chromosomal content in these panels is the heart of the study. They define which alleles are inherited in individuals, so that chromosomal associations can be made, and provide genetic recombination information so that location within a chromosome can be deduced. F_1 hybrids are genetically identical to each other but individuals in subsequent generations are not. Backcross progeny reveal recombination events on only one homologue, the one inherited from the F_1 parent, but intercross progeny reveal themselves on

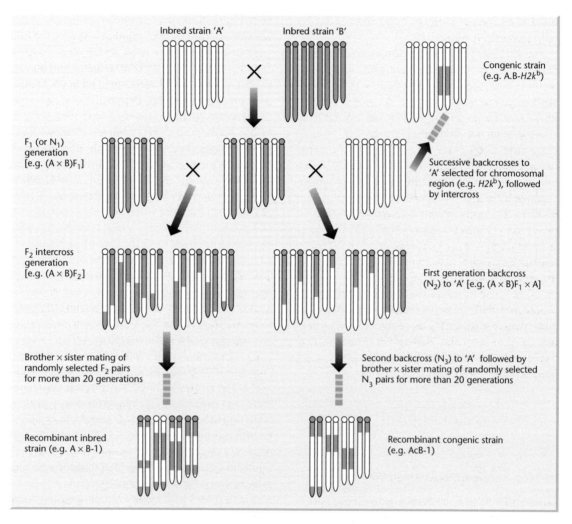

Fig. 12.12 Crosses used in the analysis of complex traits. (Adapted from Frankel 1995 by permission of Elsevier Science.)

both homologues. RILs and RC strains also harbour recombinations but, unlike backcross and intercross progeny, these are homozygous at all loci as a result of inbreeding. A congenic strain has only one chromosomal region that distinguishes it from a parental strain. Because they have an unchanging genotype, RILs and congenic strains offer an elegant way of discriminating between the role of the environment and of genetic factors in the expression of a phenotype.

There are a number of approaches that have been used to map QTLs. All involve arranging a cross between two inbred strains differing substantially in a quantitative trait. Segregating progeny, of the type

shown in Fig. 12.12 and described above, are scored both for the trait and for a number of genetic markers. In the case of plants, RFLPs or amplified fragment length polymorphisms (AFLPs) have been traditionally used, and in the case of animals the markers would be microsatellites and SNPs.

In the method of Edwards *et al.* (1987), linear regression was used to examine the relationship between the performance for the quantitative trait and the genotypes at the marker locus. If there is a statistically significant association between the trait performance and the marker locus genotypes (Fig. 12.13) it is inferred that a QTL is located near the marker locus. Lander and Botstein (1989) have

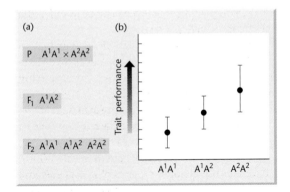

Fig. 12.13 Relationship between the performance for a quantitative trait and genotypes at a marker locus for an F_2 interbreeding population. (a) Genetic composition of the F_2 population sampled. (b) Quantitative performance of different types of F_2 genotypes.

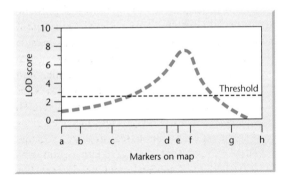

Fig. 12.14 Interval mapping to determine whether a quantitative trait locus (QTL) is located near a particular genetic marker. The distances between the genetic markers represent genetic distance. Another example of interval mapping is shown in Fig. 12.2.

pointed out that there are a number of disadvantages with this method. In particular, it cannot distinguish between tight linkage to a QTL with small effect, and loose linkage to a QTL with large effect. This is not a problem with the second method, *interval mapping*, which is an adaptation of the LOD score analysis used in human genetics. As noted in Box 12.1, a LOD score of 3 or higher is usually accepted as demonstrating that linkage is present, although lower LOD scores can still indicate linkage. For interval mapping one evaluates the LOD score for different marker loci along the chromosome, e.g. from marker loci a to e in Fig. 12.14. On the basis of the size of the genome and the number of marker loci analysed, a threshold value or significance level is

determined. In the example shown in Fig. 12.14 this threshold value has been set at 2.5. Where the LOD score exceeds the threshold, a QTL is likely to be found. Haley *et al.* (1994) described a method for adapting the interval mapping approach to map QTLs in outbreeding populations.

These methods have been used to map QTLs in both crops and livestock animals. Paterson *et al.* (1988) mapped 15 QTLs in tomato affecting such traits as fruit mass, the concentration of soluble solids and fruit pH. Andersson *et al.* (1994) generated an F_2 intercross between wild boars and domestic pigs. Two hundred offspring were genotyped for 105 polymorphic markers. Their growth rates were recorded from birth until they weighed 70 kg and, after slaughter, fat levels and intestinal lengths were determined. Fatness, whether measured as the percentage of fat in the abdominal cavity or as backfat thickness, mapped to the proximal region of chromosome 4. The QTLs for intestinal length and growth rate were located distal to the fatness QTL.

The development of methods for mapping and characterizing QTLs can also be used to study evolution at the molecular level. A good example is provided by maize and its relatives, the teosintes. Although they differ in both ear (corn cob) and plant growth morphology (Fig. 12.15), it is believed that maize is a domestic form of teosinte. Indeed, a small number of genes selected by the prehistoric peoples of Mexico may have transformed teosinte into maize within the past 10 000 years. Morphologically there are five points of difference between the two plants and analysis by classical genetics has suggested that five major and independently inherited gene differences distinguish the two species. Doebley (1992) has confirmed this by molecular methods and has shown that the major differences map to five restricted regions of the genome, each on a different chromosome. A putative QTL also has been identified by high-resolution backcrossing (Doebley *et al.* 1997).

Impact of genomics on quantitative trait locus analysis

Genome projects are in progress for a number of commercially important crop species and domestic animals (Box 12.3). One immediate benefit arising from such projects is the availability of dense marker

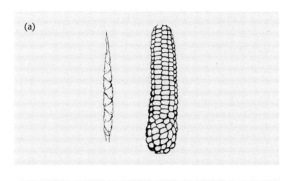

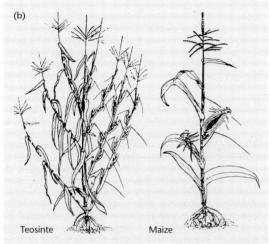

Teosinte   Maize

Fig. 12.15 Comparison of (a) ear structure, and (b) plant morphology for maize and its ancestor teosinte. (Adapted and redrawn from Doebley 1992 by permission of Elsevier Science.)

maps, which allow high-resolution mapping of QTLs. In plants, this has been achieved through the use of specialized genetic populations, such as chromosome segment substitution lines (CSSLs) and near isogenic lines (NILs), which are effectively equivalent to congenic mouse strains. Recently, such lines have been used in backcrosses to map a number of QTLs in tomato (Alpert & Tanksley 1996) and cereals (Vladutu *et al.* 1999; Yamamoto *et al.* 2000). In one of the most successful reports, a QTL affecting the amount of soluble solids in tomatoes was refined to a candidate region less than 1 kb in length (Fridman *et al.* 2000). In animals, dense marker maps are being used to study the inheritance of quantitative traits in a number of commercially important species, e.g. pigs and sheep. More information about the current state of these projects and

access to maps and genome data can be found in the websites listed in Box 12.3.

The ability to map QTLs as Mendelian factors at high resolution defines a candidate region of the genome, potentially allowing the corresponding genes to be positionally cloned (Box 6.1, p. 96). In the case of the tomato QTL affecting fruit quality discussed above, the defined candidate region was so small that it actually fell within a gene, *Brix9-2-2*, which encodes invertase (Fridman *et al.* 2000). Progress in the genome projects of *Arabidopsis* and various cereals provides a number of invaluable resources to help identify QTLs (Yamamoto & Sasaki 1997; Lukowitz *et al.* 2000; Sasaki & Burr 2000). As well as dense genetic maps, there are growing EST collections, panels of P1-derived artificial chromosome (PAC) and bacterial artificial chromosome (BAC) contigs and increasing amounts of genomic sequence data which can be mined for relevant information. By fully exploiting such resources, it is possible to produce a list of candidate genes within the defined region of the genome, and develop frameworks of region-specific markers for high-resolution mapping. For example, a rice QTL that affects sensitivity to day-length has been identified using such a map-based strategy. Linkage mapping showed that the candidate region was about 100 kb, and physical mapping showed that this region was contained within a single PAC clone. This clone was systematically sequenced and a high-resolution marker framework established using cleaved amplified polymorphic sequences (CAPS, p. 58). Further mapping using this regional map thus narrowed the candidate interval to 12 kb, which contained two genes, one of which was a rice homologue of the photoperiod-regulated *Arabidopsis* gene *CONSTANS* (Yano *et al.* 2000).

Functional genomics as a resource for cloning quantitative trait loci

Functional genomics resources, such as mutant collections and microarrays, are also useful for identifying QTLs (Coelho *et al.* 2000; Richmond & Summerville 2000; Walbot 2000). This is particularly the case where linkage mapping is unable to identify a candidate region tightly enough to allow direct map-based cloning. In such cases, information on protein structure, expression profiles or

Box 12.3 Internet resources for domestic animal and crop plant genome projects

The following websites provide current information on the progress of mapping and sequencing the genomes of a variety of domesticated species. Many of the sites also provide links to further resources, including gene mutation and sequence databases and functional analysis tools.

Domestic animals

Dog

http://mendel.berkeley.edu/dog.html Dog Genome Project (University of California, Berkeley, University of Oregon, Fred Hutchinson Cancer Research Center)

Cow

http://www.marc.usda.gov/genome/cattle/cattle.html Meat Animal Research Centre

Pig

http://www.marc.usda.gov/genome/swine/swine.html Meat Animal Research Centre
http://www.projects.roslin.ac.uk/pigmap/pigmap.html Pig genome resources at the Roslin Institute, including links to the PigMAP linkage map and PIGBASE pig genome database

Sheep

http://www.marc.usda.gov/genome/sheep/sheep.html Meat Animal Research Centre
http://www.projects.roslin.ac.uk/sheepmap/front.html UK Sheep Genome Mapping Project, Roslin Institute

Horse

http://www.vgl.ucdavis.edu/~lvmillon/ Veterinary Genetics Laboratory, UC Davis

Chicken

http://www.ri.bbsrc.ac.uk/chickmap/ChickMapHomePage.html ChickMaP project, Roslin Institute
http://poultry.mph.msu.edu/ US Poultry Genome Project, Michigan State University

Crop plants

Cotton

http://algodon.tamu.edu/htdocs-cotton/cottondb.html USDA Agricultural research centre

Sorghum

http://algodon.tamu.edu/sorghumdb.html USDA Agricultural research centre

Rice*

http://btn.genomics.org.cn/rice/ Rice Genome Database
http://portal.tmri.org/rice/ Torrey Mesa Research Institute Rice Genome Project

Barley

http://ukcrop.net/barley.html Scottish Crop Research Institute

Maize

http://www.agron.missouri.edu/ Comprehensive maize genomics and functional genomics database

Wheat and oats

http://ars-genome.cornell.edu/cgi-bin/WebAce/webace?db=graingenes

* As this book went to press, the rice genome sequence was published [see Yu, J. *et al*. (2002) A draft sequence of the rice genome (*Oryza sativa* L. ssp. *indica*). *Science* **296**, 79–92 and Goff, S.A. *et al*. (2002) A draft sequence of the rice genome (*Oryza sativa* L. ssp. *japonica*). *Science* **296**, 92–100]. This is the second plant genome to be completed and the first genome of a domestic species. A special issue of the journal *Science* was devoted to the rice genome project and contains many useful articles and links to appropriate WWW resources.

mutant phenotypes can provide valuable supplementary information that allow QTLs to be isolated. As an example of such a candidate gene approach, we consider the identification of plant disease-resistance genes. Sequence analysis has shown that many such genes contain conserved features such as leucine-rich repeats, and it is clear that expression is induced when plants are challenged with a particular pathogen. One applicable strategy is therefore to map such genes first, on the basis of their appropriate structural and functional properties, and then compare their map positions with those predicted for QTLs. Examples where this approach has been used include the isolation of stress- and disease-response genes as reported by Pelleschi *et al.* (1999) and Pflieger *et al.* (1999).

Theme 5: The impact of genomics on agriculture

Introduction to theme 5

The first completed plant genome sequence, that of *Arabidopsis thaliana*, was published in the year 2000 (*Arabidopsis* Genome Initiative 2000). Genome mapping and sequencing projects are ongoing in a number of other plant species, including many commercially important crops (Box 12.3). These projects are complemented by the availability of increasing numbers of plant ESTs, currently over 250 000, representing a wide variety of species (Bouchez & Hofte 1998; Somerville & Somerville 1999; Terryn *et al.* 1999). For the first time, therefore, it is possible to begin to look at all aspects of plant biology on a global scale. This includes aspects that are of particular interest in agriculture, such as growth and development, tolerance or resistance to stress and disease, and metabolic activity.

Traditionally, there have been two approaches to improving the quality of cultivated plants. The first is by conventional breeding, which involves exploiting varietal or specific differences and using crosses and tissue-culture-based techniques to concentrate favourable traits in particular plant lines. One problem with this approach is that many of the most valued traits in agriculture – disease resistance, stress resistance, increased yield etc. – are controlled by QTLs, so complex and extensive breeding pro-

grammes are required (see Theme 4). The alternative is genetic engineering, because single genes from other species can often confer desirable traits on plants. Although this approach to agricultural improvement is comparatively rapid, the behaviour of the transgenes is unpredictable and there is much public concern about the long-term effects of such plants on health and the environment. The technology and application of genetic engineering in plants is discussed in detail in our sister text, *Principles of Gene Manipulation* (Primrose *et al.* 2001). Some of the issues of public concern over the use of transgenic plants and food products derived therefrom have been discussed (Firn & Jones 1999; Malik & Saroha 1999; Smith *et al.* 2001).

Genomic resources now provide an additional route to the improvement of plants, through the rapid identification of genes and pathways controlling important plant traits. Some of these resources have been discussed in detail elsewhere in this book and will not be repeated here (see Chapter 10 for discussion of genome-scale mutagenesis programmes in plants, using T-DNA and transposon tagging). Below, we discuss how the rapidly accumulating genome, EST and proteomic data for a number of important plant species now makes it possible to carry out genome-scale expression profiling. This brings tremendous advantages to agriculture because it provides a holistic view of plant development and how plants respond to abiotic and biotic stresses in the environment, and thus allows the functional annotation and exploitation of desirable genes (Richmond & Somerville 2000; Schaffer *et al.* 2000).

Global expression profiling in agriculture

Initial studies of global gene expression in plants were small in scale. The random sampling of EST sequences from rice suspension cells and *Arabidopsis* plants exposed to osmotic stress helped to reveal a number of stress-induced genes (Umeda *et al.* 1994; Pih *et al.* 1997). The first array-based study involved a glass microarray containing 48 *Arabidopsis* ESTs and compared gene expression in two tissues – roots and leaves (Schena *et al.* 1995). A similar study with a larger number of ESTs was performed by Ruan *et al.* (1998) and, in the same year, *Arabidopsis* microarrays containing about 800 ESTs were used to study light-induced gene expression (Desprez *et al.* 1998). Interestingly, the first plant microarray experiment with direct agricultural implications involved not *Arabidopsis* but strawberry, where 200 genes were shown to undergo changes in expression during ripening (Lemieux *et al.* 1999).

Over the last 2 years, an increasing number of expression profiling experiments have been carried out to investigate gene expression changes associated with traits of agricultural value. EST sampling projects in two naturally salt-tolerant species (*Mesembryanthemum crystallinum* and *Dunaliella salina*) and two species resistant to desiccation (*Tortula ruralis* and *Craterostigmata plantagineum*) have been used to identify drought stress-related genes. Sampling differences between well-watered and salt-stressed or desiccated plants were used to identify ESTs either induced or repressed by these extreme conditions (Bockel *et al.* 1998; Cushman *et al.* 1999; Cushman & Bohnert 2000; Wood & Oliver 1999). Large-scale microarray-based screening projects have been undertaken to identify drought stress-related genes in *Arabidopsis*, rice and *M. crystallinum* (Cushman & Bohnert 2000; Kawasaki *et al.* 2001; Seki *et al.* 2001) and oxidative stress-related genes in *Arabidopsis* (Desikan *et al.* 2001).

Microarrays have also been used to study the response of the plant to biotic stresses. Schenk *et al.* (2000) have recently investigated disease defence responses in *Arabidopsis thaliana* using microarrays containing about 2400 cDNAs. They found that over 700 transcripts showed significantly increased or reduced expression levels, some of which were specific to particular treatments (e.g. exposure to chemicals such as methyl jasmonate and salicylic acid which are known defence-related signalling molecules) and some of which were general (i.e. affected by multiple treatments). These data suggested that there was considerable cross-talk between different defence signalling pathways. The phenomenon of cross-talk may represent a problem when devising strategies to combat disease, because artificially modifying the response to one stimulus could alter the response to several others in an undesirable way. However, as well as identifying common themes in transcriptional profiles, microarray analysis can also separate individual responses. Reymond *et al.* (2000) studied the effect of mechanical wounding and insect feeding on gene expression

profiles in *Arabidopsis*. They found that mechanical wounding induced a number of genes also known to be involved in the response to water stress, while such genes were only minimally affected by insect feeding. Thus, strategies to counteract the effects of insect feeding could be targeted to a specific subset of genes so as not to interfere with water stress tolerance. As well as local defence responses, the transcriptional changes accompanying systemic acquired resistance have also been monitored at the global level in *Arabidopsis* (Maleck *et al.* 2000; Dong 2001).

Proteomics and plant breeding

Two-dimensional gel electrophoresis (2DE) was discussed in Chapter 9 as a technique for the high-resolution separation of proteins, allowing systematic protein characterization and the identification of proteins that are differentially expressed in alternative samples. The technique has also been used to characterize proteins that are involved in stress response pathways, e.g. the response to drought (Moons *et al.* 1997; Rey *et al.* 1998), cold (Danyluk *et al.* 1991; Cabane *et al.* 1993) and heat shock (Lund *et al.* 1998).

2DE has also been used as a diagnostic tool in plant breeding because different lines or cultivars often show polymorphisms in terms of the spots generated on two-dimensional gels. Such differences have allowed unambiguous genotyping in many plant species, including rice (Abe *et al.* 1996), wheat (Picard *et al.* 1997) and barley (Gorg *et al.* 1992). Similarly, 2DE has been used to distinguish between intraspecific variants and to investigate the taxonomy of closely related species, e.g. in the genus *Triticum* (Thiellement *et al.* 1999).

Polymorphism occurs at three levels. Firstly position shifts (PS), which often correspond to mutations that change the mass and/or charge of the protein. Secondly presence/absence polymorphisms (P/A), where a protein is present in one sample but absent in another. Finally quantitative variants, corresponding to so-called *protein quantity loci* (PQLs). The first two types of polymorphism are generally Mendelian characters and represent useful genetic markers (de Vienne *et al.* 1996). These have been used in wheat, maize and pine, for example, to generate comprehensive genetic maps also containing

DNA markers such as RFLPs, RAPDs and AFLPs (e.g. Costa *et al.* 2000). Quantitative polymorphisms are useful for identifying QTLs that cannot be pinned down by traditional map-based cloning or functional candidate approaches (Theme 4). The basis of this method is that PQLs that map coincidentally with QTLs can be used to validate candidate genes. An example is provided by the study of de Vienne *et al.* (1999) who identified a candidate gene on chromosome 10 of maize for a QTL affecting drought response. The candidate gene was *ASR1*, known to be induced by water stress and ripening. Verification of the association was made possible because a PQL found by the comparison of 2DE gels from control and drought stressed maize plants mapped to the same region; the PQL controlled levels of the ASR1 protein under different drought stress conditions.

Theme 6: Developmental genomics

Introduction to theme 6

Developmental biology is the area of research which seeks to determine the molecular and cellular mechanisms involved in the coordination of growth, cell division, cell differentiation, pattern formation and morphogenesis, i.e. the processes required to convert a single cell into a complex multicellular organism (Twyman 2001). An understanding of development is becoming increasingly important in fields such as medicine, biotechnology and agriculture. Many human diseases can be traced back to faults in development, and gene products expressed in development could represent valuable therapeutic agents or potential drug targets. In agriculture, it is clear that many valuable traits are essentially developmental phenotypes. For example, grain yield in cereal plants is a consequence of seed endosperm development, and meat quality in cows, pigs, sheep and chickens is related to muscle development.

The impact of genomics on the study of development is seen in a number of areas. First, an important theme in developmental biology is the study of gene expression patterns. In particular, the comparison of expression patterns in normal development and when the process is perturbed, either experimentally or by mutation, can be very informative as this can lead to the elucidation of regulatory

hierarchies and signalling pathways. As in other fields, developmental gene expression has traditionally been investigated on a gene-by-gene basis but the advent of high-throughput expression profiling now allows the analysis of large numbers of genes simultaneously. Secondly, developmental biology has traditionally relied heavily on the study of mutants, and large-scale mutagenesis screens have had a critical role in the elucidation of developmental processes (Jurgens *et al.* 1984; Nusslein-Volhard *et al.* 1984; Wieschaus *et al.* 1984). Comprehensive mutagenesis projects are now in progress for a number of species (Chapter 10), which should facilitate the production of a full catalogue of developmental mutants. Thirdly, one of the most remarkable discoveries to come from the study of development is the close conservation in developmental gene function across diverse species. This, together with the availability of complete genome sequences for a number of lower eukaryotes, means that model organisms can be used as a valuable resource for functionally annotating the human genome and testing the effects of drugs. Finally, as noted in a recent review by Ko (2001), a large number of genes are expressed solely during development, so studying development on a genomic scale is essential in order to assemble comprehensive EST resources, cDNA clone sets and functional data.

Benefits of global expression profiling in development

Transcriptional profiling using microarrays and direct sequence sampling are being used increasingly to study changes in gene expression during development. Ko *et al.* (2000) have sampled EST databases to define patterns of gene expression during pre-implantation mouse development. Marguiles *et al.* (2001) have used the serial analysis of gene expression (SAGE) technique (p. 160) to compare genes expressed in the developing mouse forelimb and hindlimb. Others have used spotted cDNA arrays or Affymetrix GeneChips to study differences in gene expression profiles between embryonic tissues and developmental time stages in animals (Harrison *et al.* 1995; Miki *et al.* 2001) and plants (Girke *et al.* 2000; Zhu *et al.* 2001). The advantage of such multiplex analysis is that catalogues of differentially expressed genes can be identified in one series of experiments. This can lead to the rapid elucidation of regulatory pathways, hierarchies and networks, and can also help to functionally annotate anonymous genes on the basis of co-expression or co-regulation. One problem with the use of array technology in the study of development, however, is obtaining sufficient starting material to prepare the required amount of poly(A)$^+$ mRNA for probe synthesis. This is particularly the case in the study of mammalian development where it is difficult to obtain large numbers of embryos. A number of techniques have been published recently that address this issue by allowing the amplification of cDNA from small amounts of material, even single cells (Kacharmina *et al.* 1999; Brady 2000).

Despite these difficulties, expression profiling remains an extremely valuable tool in developmental biology. One application is the elucidation of regulatory networks by identifying genes lying downstream of a particular master regulator. This can be achieved by comparing the transcriptional profiles of normal embryos with those in which the master regulator gene is either knocked out (loss of function mutation) or overexpressed (gain of function mutation). Livesey *et al.* (2000) used this methodology to identify genes regulated by the homeodomain protein Crx, while Callow *et al.* (2000) identified a number of genes with altered expression profiles in HDL-knockout mice.

Another application of global expression profiling is the characterization of groups of functionally related genes. A frustrating phenomenon often seen in vertebrate development is genetic redundancy, where one gene compensates fully or partially for the function of another, so that gene knockout strategies do not reveal an informative phenotype. As an example, we consider the muscle-specific regulator genes *myoD* and *myf-5*. The *myoD* gene, when expressed in transfected fibroblasts, causes differentiation into muscle cells. Therefore this is an excellent candidate key regulator of myogenesis. However, *myoD* knockout mice showed no phenotype, indicating that *myoD* is not necessary for the development of a viable animal (Rudnicki *et al.* 1992). The basis of this result is that loss of *myoD* results in upregulation of a related gene, *myf-5*, which is sufficient for normal muscle development. Only when both genes are knocked out simultaneously does muscle development fail, leading to

perinatal lethality (Rudnicki *et al.* 1993). Global expression profiling has the potential to identify such regulatory networks in one series of experiments because it provides an overall view of the developmental pathway rather than a view biased by the analysis of a single gene.

Although sequence sampling and microarray hybridization provide valuable expression data, the complex and dynamic gene expression patterns seen in development can only be fully appreciated when viewed at the level of the whole embryo. Techniques such as *in situ* hybridization and immunohistochemistry (Chapter 9) provide the basis for this analysis but the full benefits cannot be realized without bioinformatics resources that allow gene expression patterns to be observed, compared and manipulated *in silico*.

A number of databases exist which provide accessible expression data for developing organisms. The Interactive Fly provides comprehensive and extensively cross-referenced descriptions of *Drosophila* genes involved in development and includes, where appropriate, information on how mutating one gene can alter the expression of others (http://flybase.bio.indiana.edu/allied-data/lk/interactive-fly/aimain/1aahome.htm). The Kidney Development Database fulfils a similar function for the developing mammalian kidney, and allows searches based on gene name, developmental stage or gene expression pattern (http://golgi.ana.ed.ac.uk/kidhome.html). However, in each case expression patterns are provided as text descriptions, which makes cross-referencing and networking difficult, particularly where the gene concerned has a dynamic expression pattern. Some databases include pictures, e.g. the expression pattern search resource at wormbase (http://www.wormbase.org/db/searches/expr_search), but these are not presented in a standardized manner and cannot be super-imposed or easily cross-referenced.

A solution to this problem is the creation of a database where developmental gene expression patterns are modelled in four dimensions, allowing both spatial and temporal expression patterns to be viewed (Davidson & Baldock 2001). Ideally, multiple genes could be shown simultaneously using different colours and mRNA and protein expression profiles compared. The beginnings of such a project have been described recently for the mouse. Ringwald *et al.* (2000) describe the Gene Expression

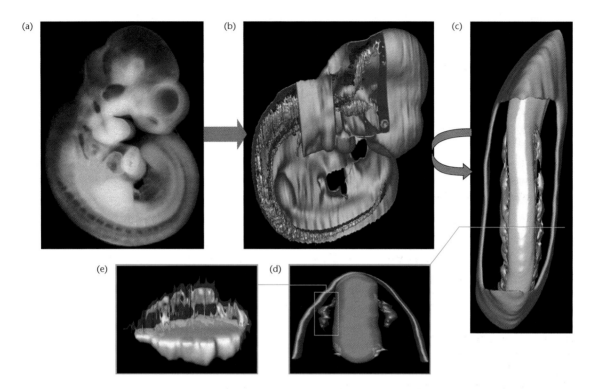

Fig. 12.16 An example from the GeneEMAC expression database, showing how data from *in situ* hybridization experiments (a) can be used to construct three-dimensional models of embryos that can be manipulated *in silico* to show details of the expression pattern (b–e). In this case, the expression of *Scgn10* (a neuron-specific gene expressed in the superior cervical ganglia) is demonstrated. (Images reproduced with permission from the GeneEMAC database.)

Information Resource, which combines a three-dimensional atlas of mouse development with a gene expression database. The eventual aim of this project is to map expression data onto the expression atlas to provide an interactive digital map of developmental gene expression. Details of the progress of this project can be found at the following website: http://www.informatics.jax.org/mgihome/GXD/gxdgen.shtml. Streicher *et al.* (2000) have developed a novel image manipulation technique called GeneEMAC to reconstruct whole embryo expression patterns from serial sections. Other methods include three-dimensional confocal reconstruction of sections (Hecksher-Sorensen & Sharpe 2001) and non-invasive techniques such as magnetic resonance imaging (Louie *et al.* 2000). Currently, the GeneEMAC database contains the expression profiles of a small number of mouse genes (Fig. 12.16). It has the potential to become another valuable

integrated resource, helping to assimilate the large amount of expression data already in the literature and emerging from current high-throughput *in situ* hybridization projects (Komiya *et al.* 1997; Neidhardt *et al.* 2000). Details of the project are available at the following website: http://www.univie.ac.at/GeneEMAC.

Developmental genomics and drug discovery

Drug discovery is another area in which developmental biology and genomics are beginning to have a significant impact, because many gene products expressed uniquely during early development are growth factors and other signalling proteins that are either drug targets or potential therapeutic agents. Development therefore has an impact on drug discovery in two ways: first, by providing test systems for novel drugs and, secondly, by providing em-

Embryonic gene product	Potential therapeutic use
Basic fibroblast growth factor (bFGF)	Peripheral vascular disease
Nerve growth factor (NGF)	Peripheral nerve disease
Sonic hedgehog (Shh)	Alzheimer's disease
Indian hedgehog (Ihh)	Osteoporosis, fractures
Bone morphogenetic protein 2 (BMP2)	Bone repair
Bone morphogenetic protein 7 (BMP7)	Bone repair
Erythropoeitin	Renal failure
Granulocyte colony stimulating factor (G-CSF)	Neutropenia

Table 12.2 Some embryonic proteins that are potential therapeutic agents in humans. Erythropoietin and G-CSF are marketed under the names Epogen and Neupogen, respectively, by the US biotechnology company Amgen.

bryonic proteins that may be therapeutic in adults. A selection of embryonic gene products currently under development as therapeutics is listed in Table 12.2.

An important consideration is that disease genes in humans are often very good potential drug targets, and many of these genes have homologues in model organisms such as *Drosophila*, *Caenorhabditis elegans* and the mouse. This conservation allows animal models and, in some cases, even unicellular model organisms such as yeast to be used as experimental systems to determine gene function and drug activity. The validity of this approach can be tested by using the human gene to 'rescue' the mutant phenotype of another organism that lacks the putative homologue (Hengartner & Horvitz 1994). If this approach is successful it suggests that the human gene product and its counterpart are functionally equivalent thereby allowing novel drugs to be tested on the human protein in its surrogate environment. Indeed, the functional similarity between developmental gene products in humans and other animals is so conserved that many biotechnology companies are using a factory-style screening approach. For example, thousands of mutant strains of *C. elegans* can be tested with a panel of potential drugs to identify interactions that rescue the mutant phenotype. The mutant gene can then be identified, used to find human homologues, and this may provide a useful target for drug development. Functional genomics strategies such as array hybridization and interaction screening can also be exploited to identify gene products that act downstream of the disease gene, revealing further potential drug targets (Rhodes & Smith 1998).

Suggested reading

Theme 1: Understanding genetic diseases of humans

Nussbaum R.L., McInnes R.R. & Willard H.F. (2001) *Genetics in Medicine*, W.B. Saunders, Philadelphia. *An excellent text which has very comprehensible sections on complex traits and population genetics.*

The 26 April 2002 issue of the journal *Science* was devoted to the study of complex diseases. The informative articles can be accessed at the following URL: http://www.sciencemag.org/content/vol296/issue5568/index.shtml#specialintro.

Theme 2: Understanding responses to drugs

Judson R., Stephens J.C. & Windemuth A. (2000) The predictive power of haplotypes in clinical response. *Pharmacogenomics* **1**, 15–26. *This review stands out as a model of clarity in a large field of papers that are almost impossible to understand.*

Theme 3: Understanding and combatting bacterial pathogenicity

Black T. & Hare R. (2000) Will genomics revolutionize antimicrobial drug discovery? *Current Opinion in Microbiology* **3**, 522–527. *This paper presents information that is complementary to that given in this chapter, with a particular emphasis on the drug discovery process.*

Shankar N., Baghdayan A.S. & Gilmore M.S. (2002) Modulation of virulence within a pathogenicity island in vancomycin-resistant *Enterococcus faecalis*. *Nature* **417**, 746–50. *A fascinating paper describing how a commensal organism can become a pathogen through the acquisition of new traits.*

Theme 4: Getting to grips with quantitative traits

Trends in Genetics **11**, issue 12 (December 1995). *This special issue was devoted to the use of molecular methods for dissecting complex traits. Although a little out of date now, this is a very useful collection of reviews covering the study of quantitative inheritance in* Drosophila *(genetic variation in bristle content), mice, various crop species, domestic animals and humans. Specialist methodologies for QTL mapping are discussed in a very accessible format.*

Jannick J-L, Bink B.C.A.M. & Jansen R.C. (2001) Using complex plant pedigrees to map valuable genes. *Trends in Plant Science* **6**, 337–342.

Yano M. (2001) Genetic and molecular dissection of naturally occurring variation. *Current Opinion in Plant Biology* **4**, 130–135.

These two recent reviews provide excellent coverage of current methods for QTL mapping in plants, taking advantage of genomic resources.

Theme 5: The impact of genomics on agriculture

Bouchez D. & Höfte H. (1998) Functional genomics in plants. *Plant Physiology* **118**, 725–732.

Cushman J.C. & Bohnert H.J. (2000) Genomic approaches to plant stress tolerance. *Current Opinion in Plant Biology* **3**, 117–124. *This paper succinctly demonstrates how such resources can be used to benefit agriculture, using the investigation of stress tolerance as an example.*

Richmond T. & Somerville S. (2000) Chasing the dream: plant EST microarrays. *Current Opinion in Plant Biology* **3**, 108–116.

Somerville C. & Somerville S. (1999) Plant functional genomics. *Science* **285**, 380–383.

Schaffer R., Landgraf J., Perez-Amador P. & Wisman E. (2000) Monitoring genome-wide expression in plants. *Current Opinion in Biotechnology* **11**, 162–167.

Terryn N., Rouze P. & Van Montagu M. (2000) Plant genomics. *FEBS Letters* **452**, 3–6.

A selection of useful reviews which cover genomics and functional genomics resources in plants.

Theme 6: Developmental genomics

Ko M.S.H. (2001) Embryogenomics: developmental biology meets genomics. *Trends in Biotechnology* **19**, 511–518.

Rhodes S.J. & Smith R.C. (1998) Using the power of developmental biology for drug discovery. *Drug Discovery Today* **3**, 361–369.

Schlessinger D. & Ko M.S.H. (1998) Developmental genomics and its relation to ageing. *Genomics* **52**, 113–118.

These three reviews provide a good summary of the synergy between developmental biology and genomics, summarizing the benefits for both fields.

References

Abe T., Gusti R.S., Ono M. & Sasahara T. (1996) Variations in glutelin and high molecular weight endosperm proteins among subspecies of rice (*Oryza sativa* L.) detected by two-dimensional gel electrophoresis. *Genes and Genetic Systems* **71**, 63–68.

Abola E., Kuhn P., Earnest T. & Stevens R.C. (2000) Automation of X-ray crystallography. *Nature Structural Biology* **7**, 973–977.

Adams K.L. *et al.* (2000a) Repeated, recent and diverse transfers of a mitochondrial gene to the nucleus in flowering plants. *Nature* **408**, 354–357.

Adams K.L., Rosenblueth M., Qui Y.L. & Palmer J.D. (2001) Multiple losses and transfers to the nucleus of two mitochondrial succinate dehydrogenase genes during angiosperm evolution. *Genetics* **158**, 1289–1300.

Adams M.D. *et al.* (1991) Complementary DNA sequencing: expressed sequence tags and human genome project. *Science* **252**, 1651–1656.

Adams M.D. (1996) Serial analysis of gene expression: ESTs get smaller. *Bioassays* **18**, 261–262.

Adams M.D. *et al.* (2000b) The genome sequence of *Drosophila melanogaster*. *Science* **287**, 2185–2195.

Aitman T.J. *et al.* (1999) Identification of *CD36* (*Fat*) as an insulin-resistant gene causing defective fatty acid and glucose metabolism in hypertensive rats. *Nature Genetics* **21**, 76–83.

Alford R.L. & Caskey C.T. (1994) DNA analysis in forensics, disease and animal/plant identification. *Current Opinion in Biotechnology* **5**, 29–33.

Alizadeh A.A. *et al.* (2000) Distinct types of diffuse large B-cell lymphoma identified by gene expression profiling. *Nature* **403**, 503–511.

Alm R.A. *et al.* (1999) Genomic-sequence comparison of two unrelated isolates of the human gastric pathogen *Helicobacter pylori*. *Nature* **397**, 176–180.

Alpert K.B. & Tanksley S.D. (1996) High-resolution mapping and isolation of a yeast artificial chromosome contig containing *fw2.2*: a major fruit weight quantitative trait locus in tomato. *Proceedings of the National Academy of Sciences, USA* **93**, 15503–15507.

Altschul S.F., Gish W., Miller W., Myers E.W. & Lipmann D.J. (1990) Basic local alignment search tool. *Journal of Molecular Biology* **215**, 403–410.

Altschul S.F. *et al.* (1997) Gapped BLAST and PSI-BLAST: a new generation of protein database search programs. *Nucleic Acids Research* **25**, 3389–3402.

Altshuler D. *et al.* (2000a) The common PPARgamma Pro12Ala polymorphism is associated with decreased risk of type 2 diabetes. *Nature Genetics* **26**, 76–80.

Altshuler D. *et al.* (2000b) An SNP map of the human genome generated by reduced representation shotgun sequencing. *Nature* **407**, 513–516.

Amsterdam A. *et al.* (1999) A large-scale insertional mutagenesis screen in zebrafish. *Genes and Development* **13**, 2713–2724.

Anand R., Riley J.H., Butler R., Smith J.C. & Markham A.F. (1990) A 3.5 genome equivalent multi-access YAC library: construction, characterisation, screening and storage. *Nucleic Acids Research* **18**, 1951–1956.

Andersen J.S. & Mann M. (2000) Functional genomics by mass spectrometry. *FEBS Letters* **480**, 25–31.

Anderson N.G. & Anderson L. (1982) The human protein index. *Clinical Chemistry* **28**, 739–748.

Andersson J.O., Doolittle W.F. & Nesbo C.L. (2001) Are there bugs in our genome? *Science* **292**, 1848–1850.

Andersson L. *et al.* (1994) Genetic mapping of quantitative trait loci for growth and fatness in pigs. *Science* **263**, 1771–1774.

Andersson S.G.E. *et al.* (1998) The genome sequence of *Rickettsia prowazekii* and the origin of mitochondria. *Nature* **396**, 133–140.

Antson D-O. *et al.* (2000) PCR-generated padlock probes detect single nucleotide variation in genomic DNA. *Nucleic Acids Research* **28**, E58.

Appel R.D. *et al.* (1996) Federated two-dimensional electrophoresis database: a simple means of publishing two-dimensional electrophoresis data. *Electrophoresis* **17**, 540–546.

Apweiler R. *et al.* (2001a) The InterPro database, an integrated documentation resource for protein families, domains and functional sites. *Nucleic Acids Research* **29**, 37–40.

Apweiler R. *et al.* (2001b) Proteome analysis database: online application of InterPro and CluSTr for the functional classification of proteins in whole genomes. *Nucleic Acids Research* **29**, 44–48.

Arabidopsis Genome Initiative (2000) Analysis of the genome sequence of the flowering plant *Arabidopsis thaliana*. *Nature* **408**, 796–813.

Aravind L. *et al.* (1998) Evidence for massive gene exchange between archaeal and bacterial hyperthermophiles. *Trends in Genetics* **14**, 442–444.

Aronson H.E., Royer W.E. Jr & Hendrickson W.A. (1994) Quantification of tertiary structural conservation despite primary sequence drift in the globin fold. *Protein Science* **3**, 1706–1711.

Ashburner M. *et al.* (2000) Gene ontology: tool for the unification of biology. *Nature Genetics* **25**, 25–29.

Aston C., Mishra B. & Schwartz D.C. (1999) Optical mapping and its potential for large-scale sequencing projects. *Trends in Biotechnology* **17**, 297–302.

Attwood T.K. & Parry-Smith D.J. (1999) *Introduction to Bioinformatics*. Prentice Hall, London.

Auch D. & Reth M. (1990) Exon trap cloning: using PCR to rapidly detect and clone exons from genomic DNA fragments. *Nucleic Acids Research* **18**, 6743–6744.

Audic S. & Claverie J. (1997) The significance of digital gene expression profiles. *Genome Research* **7**, 986–995.

Azpiroz-Leehan R. & Feldmann K.A. (1997) T-DNA insertion mutagenesis in *A. thaliana*: going back and forth. *Trends in Genetics* **13**, 146–152.

Babiychuk E., Fuanghthong M., VanMontagu M., Inze D. & Kushnir S. (1997) Efficient gene tagging in *Arabidopsis thaliana* using a gene trap approach. *Proceedings of the National Academy of Sciences, USA* **94**, 12722–12727.

Bader G.D. & Hogue C.W. (2000) BIND: a data specification for storing and describing biomolecular interactions, molecular complexes and pathways. *Bioinformatics* **16**, 465–477.

Bader G.D., Donaldson I., Wolting C., Ouellette B.F., Pawson T. & Hogue CW (2001) BIND: the biomolecular interaction network database. *Nucleic Acids Research* **29**, 242–245.

Baer R. *et al.* (1984) DNA sequence and expression of the B95.8 Epstein–Barr virus genome. *Nature* **310**, 207–211.

Bains W. & Smith G.C. (1988) A novel method for nucleic acid sequence determination. *Journal of Theoretical Biology* **135**, 303–307.

Baker H.M., Norris G.E., Morgan W.T., Smith A. & Baker E.N. (1993) Crystallization of the C-terminal domain of rabbit serum hemopexin. *Journal of Molecular Biology* **229**, 251–252.

Baner J., Nilsson M., Isaksson A., Mendel-Hartvig M., Antson, D.O. & Landegren, U. (2001) More keys to padlock probes: mechanisms for high-throughput nucleic acid analysis. *Current Opinion in Biotechnology* **12**, 11–15.

Barbazuk W.B. *et al.* (2000) The syntenic relationship of the zebrafish and human genomes. *Genome Research* **10**, 1351–1358.

Bartel P.L., Roecklein J.A., SenGupta D. & Fields S. (1996) A protein linkage map of *Escherichia coli* bacteriophage T7. *Nature Genetics* **12**, 72–77.

Baudin A., Ozier-Kalogeropoulos O., Denouel A., Lacroute F. & Cullin C. (1993) A simple and efficient method for direct gene deletion in *Saccharomyces cerevisiae*. *Nucleic Acids Research* **21**, 3329–3330.

Baulcombe D.C. (1999) Fast forward genetics based on virus-induced gene silencing. *Current Opinion in Plant Biology* **2**, 109–113.

Baxevanis A.D. (2002) The molecular biology database collection: 2002 update. *Nucleic Acids Research* **30**, 1–12.

Bechtold N. & Pelletier G. (1998) *In planta Agrobacterium*-mediated transformation of adult *Arabidosis thaliana* plants by vacuum infiltration. *Methods in Molecular Biology* **82**, 259–266.

Bechtold N., Ellis J. & Pelletier G. (1993) In planta *Agrobacterium*-mediated gene transfer by infiltration of adult *Arabidopsis thaliana* plants. *Comtes Rendus de l'Academy des Sciences Paris, Life Sciences* **316**, 1194–1199.

Beckers J. & Hrabe de Angelis M. (2001) Large-scale mutational analysis for the annotation of the mouse genome. *Current Opinion in Chemical Biology* **6**, 17–23.

Behr M.A. *et al.* (1999) Comparative genomics of BCG vaccines by whole-genome DNA microarray. *Science* **284**, 1520–1523.

Bellanné-Chantelot C. *et al.* (1992) Mapping the whole human genome by fingerprinting yeast artificial chromosomes. *Cell* **70**, 1059–1068.

Bender W., Spierer P. & Hogness D.S. (1983) Chromosome walking and jumping to isolate DNA from the *Ace* and *rosy* loci and the bithorax complex in *Drosophila melanogaster*. *Journal of Molecular Biology* **168**, 17–33.

Benos P.V. *et al.* (2001) From first base: the sequence of the tip of the X chromosome of *Drosophila melanogaster*, a comparison of two sequencing strategies. *Genome Research* **11**, 710–730.

Bensen R.J. *et al.* (1995) Cloning and characterization of the maize *An1* gene. *Plant Cell* **7**, 75–84.

Bertucci F. *et al.* (1999) Sensitivity issues in DNA array based expression measurements: advantages of nylon membranes. *Human Molecular Genetics* **8**, 1715–1722.

Bianchi M.M. *et al.* (1999) How to bring orphan genes into functional families. *Yeast* **15**, 513–526.

Bihoreau M.T. *et al.* (2001) A high-resolution consensus linkage map of the rat, integrating radiation hybrid and genetic maps. *Genomics* **75**, 57–69.

Bird A. (1995) Gene number, noise reduction and biological complexity. *Trends in Genetics* **11**, 94–99.

Bittner M., Meltzer P. & Trent J. (1999) Data analysis and integration: of steps and arrows. *Nature Genetics* **22**, 213–215.

Bittner M. *et al.* (2000) Molecular classification of cutaneous malignant melanoma by gene expression profiling. *Nature* **406**, 536–540.

Blackstock W.P. & Weir M.P. (1999) Proteomics: quantitative and physical mapping of cellular proteins. *Trends in Biotechnology* **17**, 121–127.

Blackwood E.M. & Eisenman R.N. (1991) Max: a helix–loop–helix zipper protein that forms a sequence-specific DNA-binding complex with c-Fos. *Science* **256**, 1014–1018.

Blanchard A.P., Kaiser R.J. & Hood L.E. (1996) High-density oligonucleotide arrays. *Biosensors Bioelectronics* **6/7**, 687–690.

Bockel C., Salamini F. & Bartels D. (1998) Isolation and characterization of genes expressed during early events of the dehydration process in the resurrection plant *Craterostigama plantagineum*. *Journal of Plant Physiology* **152**, 158–166.

Bode V.C., McDonald J.D., Guenet J.L. & Simon D. (1988) *hph–1*: A mouse mutant with hereditary hyperphenylalaninemia induced by ethylnitrosourea mutagenesis. *Genetics* **118**, 299–305.

Bork P. & Bairoch A. (1996) Go hunting in sequence databases but watch out for the traps. *Trends in Genetics* **12**, 425–427.

Bork P. & Copley R. (2001) Filling in the gaps. *Nature* **409**, 818–820.

Bork P. & Koonin E.V. (1998) Predicting functions from protein sequences: where are the bottlenecks? *Nature Genetics* **18**, 313–318.

Borodovsky M. & McIninch J. (1993) Recognition of genes in DNA sequence with ambiguities. *Biosystems* **30**, 161–171.

Bosch, I. *et al.* (2000) Identification of differentially expressed genes from limited amounts of RNA. *Nucleic Acids Research* **28**, e27.

Botstein D., White R.L., Skolnick M. & Davis R.W. (1980) Construction of a genetic linkage map in man using restriction fragment length polymorphisms. *American Journal of Human Genetics* **32**, 314–331.

Bouchez D. & Höfte H. (1998) Functional genomics in plants. *Plant Physiology* **118**, 725–732.

Bouchez D., Camilleri C. & Caboche M. (1993) A binary vector based on Basta resistance for in planta transformation of *Arabidopsis thaliana*. *Comtes Rendus de l'Academy des Sciences Paris III: Vie* **316**, 1188–1193.

Bowater R.P. & Wells R.D. (2000) The intrinsically unstable life of DNA triplet repeats associated with human hereditary disorders. *Progress in Nucleic Acid Research and Molecular Biology* **66**, 159–202.

Bowes C. *et al.* (1993) Localization of a retroviral element within the *rd* gene coding for the β-subunit of cGMP phosphodiesterase. *Proceedings of the National Academy of Sciences, USA* **90**, 2955–2959.

Bowtell D.D.L. (1999) Options available – from start to finish – for obtaining expression data by microarray. *Nature Genet Supplement* **21**, 25–32.

Boyd E.F., Davis B.M. & Hochhut B. (2001) Bacteriophage–bacteriophage interactions in the evolution of pathogenic bacteria. *Trends in Microbiology* **9**, 137–144.

Brady, G. (2000) Expression profiling of single mammalian cells: small is beautiful. *Yeast* **17**, 211–217.

Brazma A. & Vilo J. (2000) Gene expression data analysis. *FEBS Letters* **480**, 17–24.

Breatnach R., Mandel J.L. & Chambon P. (1977) Ovalbumin gene is split in chicken DNA. *Nature* **270**, 314–319.

Brenner S., Chothia C. & Hubbard T. (1997) Population statistics of protein structures. *Current Opinion in Structural Biology* **7**, 369–376.

Brenner S. *et al.* (2000) Gene expression analysis by massively parallel signature sequencing (MPSS) on microbead arrays. *Nature Biotechnology* **18**, 630–634.

Brenner S.E. (1999) Errors in genome annotation. *Trends in Genetics* **15**, 132–133.

Brenner S.E. (2000) Target selection for structural genomics. *Nature Structural Biology* **7**, 967–969.

Brenner S.E. (2001) A tour of structural genomics. *Nature Reviews Genetics* **2**, 801–809.

Brenner S.E., Chothia C. & Hubbard T.J. (1998) Assessing sequence comparison methods with reliable structurally identified distant evolutionary relationships. *Proceedings of the National Academy of Sciences, USA* **95**, 6073–6078.

Briem H. & Kuntz I.D. (1996) Molecular similarity based on DOCK-generated fingerprints. *Journal of Medicinal Chemistry* **39**, 3401–3408.

Britten R.J. & Kohne D.E. (1968) Repeated sequences in DNA. *Science* **161**, 529–540.

Broome S. & Gilbert W. (1978) Immunological screening method to detect specific translation products. *Proceedings of the National Academy of Sciences, USA* **75**, 2746–2749.

Brown M.P. *et al.* (2000) Knowledge-based analysis of microarray gene expression data by using support vector machines. *Proceedings of the National Academy of Sciences, USA* **97**, 262–267.

Bryant S.H. & Lawrence C.E. (1993) An empirical energy function for threading protein-sequence through the folding motif. *Proteins* **16**, 92–112.

Buckler A.J. *et al.* (1991) Exon amplification: a strategy to isolate mammalian genes based on RNA splicing. *Proceedings of the National Academy of Sciences, USA* **88**, 4005–4009.

Buetow K.H., Edmonson M.N. & Cassidy A.B. (1999) Reliable identification of large numbers of candidate SNPs from public EST data. *Nature Genetics* **21**, 323–325.

Bult C.J. *et al.* (1996) Complete genome sequence of the methanogenic archaeon, *Methanococcus jannaschii*. *Science* **273**, 1058–1073.

Bulyk M.L. *et al.* (2001) Exploring the DNA binding specificities of zinc fingers with DNA microarrays. *Proceedings of the National Academy of Sciences, USA* **98**, 7158–7163.

Burge C. & Karlin S. (1997) Prediction of complete gene structures in human genomic DNA. *Journal of Molecular Biology* **268**, 78–94.

Burge C.B. & Karlin S. (1998) Finding the genes in genomic DNA. *Current Opinion in Structural Biology* **8**, 346–354.

Burke D.T., Carle G.F. & Olson M.V. (1987) Cloning of large segments of exogenous DNA into yeast by means of artificial chromosome vectors. *Science* **236**, 806–813.

Burley S.K. (2000) An overview of structural genomics. *Nature Structural Biology* **7**, 932–934.

Burley S.K. *et al.* (1999) Structural genomics: beyond the human genome project. *Nature Genetics* **23**, 151–157.

Burset M. & Guigó R. (1996) Evaluation of gene structure prediction programs. *Genomics* **34**, 353–367.

Bussow K. *et al.* (1998) A method for global protein expression and antibody screening on high-density filters of an arrayed cDNA library. *Nucleic Acid Research* **26**, 5007–5008.

Cabane M., Calvet P., Vincens P. & Boudet A.M. (1993) Characterisation of chilling-acclimation-related proteins in soybean and identification of one as a member of the heat shock protein (HSP 70) family. *Planta* **190**, 346–353.

Caenorhabditis elegans Sequencing Consortium (1998) Sequence and analysis of the genome of *C. elegans*. *Science* **282**, 2012–2018.

Cahill D.J. (2000) Protein arrays: a high-throughput solution for proteomics research? *Proteomics: A Trends Guide 2000* (Supplement to *Trends in Biochemical Science*), 47–51.

Cai D. *et al.* (1997) Positional cloning of a gene for nematode resistance in sugar beet. *Science* **275**, 832–838.

Callow M.J., Dudoit S., Gong E.L., Speed T.P. & Rubin E.M. (2000) Microarray expression profiling identifies genes with altered expression in HDL-deficient mice. *Genome Research* **10**, 2022–2029.

Campisi L. *et al.* (1999) Generation of enhancer trap lines in *Arabidopsis* and characterization of expression patterns in the inflorescence. *Plant Journal* **17**, 699–707.

Cantor C.R., Smith C.L. & Mathew M.K. (1988) Pulsed-field gel electrophoresis of very large DNA molecules. *Annual Review of Biophysical Chemistry* **17**, 287–304.

Capecchi M.R. (2000) Choose your target. *Nature Genetics* **26**, 159–161.

Cargill M. *et al.* (1999) Characterization of single-nucleotide polymorphisms in coding regions of human genes. *Nature Genetics* **22**, 231–238.

Casjens S. *et al.* (2000) A bacterial genome in flux: the twelve linear and nine circular extrachromosomal DNAs in an infectious isolate of the Lyme disease spirochete *Borrelia burgdorferi*. *Molecular Microbiology* **35**, 490–516.

Celis J.E. *et al.* (1995) The human keratinocyte two-dimensional gel protein database (update 1995): mapping components of signal transduction pathways. *Electrophoresis* **16**, 2177–2140.

Chait B.T., Wang R., Beavis R.C. & Kent S.B.H. (1993) Protein ladder sequencing. *Science* **262**, 89–92.

Chatterjee P.K. & Coren J.S. (1997) Isolating large nested deletions in bacterial and P1 artificial chromosomes by *in vivo* P1 packaging of products of Cre-catalysed recombination between the endogenous and a transposed *lox*P site. *Nucleic Acids Research* **25**, 2205–2212.

Chatterjee P.K. *et al.* (1999) Direct sequencing of bacterial and P1 artificial chromosome-nested deletions for identifying position-specific single-nucleotide polymorphisms. *Proceedings of the National Academy of Sciences, USA* **96**, 13276–13281.

Chee M. *et al.* (1996) Accessing genetic information with high-density arrays. *Science* **274**, 610–614.

Chen J.J. *et al.* (1998) Profiling expression patterns and isolating differentially expressed genes by cDNA microarray system with colorimetry detection. *Genomics* **51**, 313–324.

Chen Y.Z. *et al.* (2001) A bac-based sts-content map spanning a 35-mb region of human chromosome 1p35–p36. *Genomics* **74**, 55–70.

Cheung V.G., Morley M., Aguilar F., Massimi A., Kucherlapati R. & Childs G. (1999) Making and reading microarrays. *Nature Genetics* **21** (Suppl.), 15–19.

Cheung V.G. *et al.* (2001) Integration of cytogenetic landmarks into the draft sequence of the human genome. *Nature* **409**, 953–958.

Chiang S.L. *et al.* (1999) *In vivo* genetic analysis of bacterial virulence. *Annual Review of Microbiology* **53**, 129–154.

Chissoe S.L., Marra M.A., Hillier L., Brinkman R., Wilson R.K. & Waterson R.H. (1997) Representation of cloned genomic sequences in two sequencing vectors: correlation of DNA sequence and subclone distribution. *Nucleic Acids Research* **25**, 2960–2966.

Cho R.J. *et al.* (1998) A genome-wide transcriptional analysis of the mitotic cell cycle. *Molecular Cell* **2**, 65–73. [http://genomics.stanford.edu]

Choudhary J.S., Blackstock W.P., Creasy D.M. & Cottrell J.S. (2001) Matching peptide mass spectra to EST and genomic DNA databases. *Trends Biotechnology* (Suppl. A: Trends Guide to Proteomics II) **19**, S17–S22.

Christendat D. *et al.* (2000) Structural proteomics of an archaeon. *Nature Structural Biology* **7**, 903–909.

Christie P.J. (2001) Type IV secretion: intercellular transfer of macromolecules by systems ancestrally related to conjugation machines. *Molecular Microbiology* **40**, 294–305.

Chu G., Vollrath D. & Davis R. (1986) Separation of large DNA molecules by contour clamped homogenous electric fields. *Science* **234**, 1582–1585.

Chu S. *et al.* (1998) The transcriptional program of sporulation in budding yeast. *Science* **282**, 699–705. [http://cmgm.stanford.edu/pbrown/sporulation]

Church D.M. & Buckler A.J. (1999) Gene identification by exon amplification. *Methods in Enzymology* **303**, 83–99.

Clarke L. & Carbon J. (1976) A colony bank containing synthetic Col E1 hybrid plasmids representative of the entire *E. coli* genome. *Cell* **9**, 91–99.

Claudio J.O. *et al.* (1998) Identification of sequence tagged transcripts differentially expressed within the human hematopoeitic hierarchy. *Genomics* **50**, 44–52.

Claverie J.M. (1997) Computational methods for the identification of genes in vertebrate genomic sequences. *Human Molecular Genetics* **6**, 1735–1744.

Cliften P.F. *et al.* (2001) Surveying *Saccharomyces* genomes to identify functional elements by comparative DNA sequence analysis. *Genome Research* **11**, 1175–1186.

Clough S.J. & Bent A. (1998) Floral dip: a simplified method for *Agrobacterium*-mediated transformation of *Arabidopsis thaliana*. *Plant Journal* **16**, 735–743.

Coelho P.S.R., Kumar A. & Snyder M. (2000) Genome-wide mutant collections: toolboxes for functional genomics. *Current Opinion in Microbiology* **3**, 309–315.

Cole S.T. & Saint Girons I. (1994) Bacterial genomes. *FEMS Microbiology Reviews* **14**, 139–160.

Cole S.T. *et al.* (1998) Deciphering the biology of *Mycobacterium tuberculosis* from the complete genome sequence. *Nature* **393**, 537–544.

Cole S.T. *et al.* (2001) Massive gene decay in the leprosy bacillus. *Nature* **409**, 1007–1011.

Collins F.S. (1992) Positional cloning: let's not call it reverse any more. *Nature Genetics* **1**, 3–6.

Connor J.M. & Ferguson-Smith M.A. (1997) *Essential Medical Genetics*, 5th edn. Blackwell Science, Oxford.

Cooley L., Berg C. & Spralding A. (1988) Controlling P element insertional mutagenesis. *Trends in Genetics* **4**, 254–258.

Copeland N.G. & Jenkins N.A. (1991) Development of applications of a molecular genetic linkage map of the mouse genome. *Trends in Genetics* **7**, 113–118.

Coren J.S. & Sternberg N. (2001) Construction of a PAC vector system for the propagation of genomic DNA in bacterial and mammalian cells and subsequent generation of nested deletions in individual library members. *Gene* **264**, 11–18.

Cort J.R., Yee A., Edwards A.M., Arrowsmith C.H. & Kennedy M.A. (2000) Structure-based functional classification of hypothetical protein MTH538 *from Methanobacterium thermoautotrophicum*. *Journal of Molecular Biology* **302**, 189–203.

Costa P. *et al.* (2000) A genetic map of Maritime pine based on AFLP, RAPD and protein markers. *Theoretical and Applied Genetics* **100**, 39–48.

Costanzo M.C. *et al.* (2001) YPD™, PombePD™ and WormPD™: model organism volumes of the BioKnowledge™ library, an integrated resource for protein information. *Nucleic Acids Research* **29**, 75–79.

Coulson A., Sulston J., Brenner S. & Karn L. (1986) Toward a physical map of the genome of the nematode *Caenorhabditis elegans*. *Proceedings of the National Academy of Sciences, USA* **83**, 7821–7825.

Cox D.R., Burmeister M., Price E.R., Kim S. & Myers R.M. (1990) Radiation hybrid mapping: a somatic cell genetic method for constructing high-resolution maps of mammalian chromosomes. *Science* **250**, 245–250.

Cox D.R., Green E.D., Lander E.S., Cohen D. & Myers R.M. (1994) Assessing mapping progress in the human genome project. *Science* **265**, 2031–2032.

Craig J.M. & Bickmore W.A. (1994) The distribution of CpG islands in mammalian chromosomes. *Nature Genetics* **7**, 376–381.

Craig J.M., Earnshaw W.C. & Vagnarelli P. (1999) Mammalian centromeres: DNA sequence, protein composition, and role in cell cycle progression. *Experimental Cell Research* **246**, 249–262.

Cregan P.B. *et al.* (1999) Target isolation of simple sequence repeat markers through the use of bacterial artificial chromosomes. *Theoretical and Applied Genetics* **98**, 919–928.

Cross S.H. & Bird A.P. (1995) CpG islands and genes. *Current Opinion in Genetics and Development* **5**, 309–314.

Cushman J.C. & Bohnert H.J. (2000) Genomic approaches to plant stress tolerance. *Current Opinion in Plant Biology* **3**, 117–124.

Cushman M.A. *et al.* (1999) An expressed sequence tag (EST) database for the common ice plant *Mesembryanthemum crystallinum*. *Plant Physiology* **120S**, 145.

D'Eustachio P. & Ruddle F.H. (1983) Somatic cell genetics and gene families. *Science* **220**, 919–928.

Daly M.J. *et al.* (2001) High-resolution haplotype structure in the human genome. *Nature Genetics* **29**, 229–232.

Dandekar T., Snel B., Huynen M. & Bork P. (1998) Conservation of gene order: a fingerprint of proteins that physically interact. *Trends in Biochemical Science* **23**, 324–328.

Dandekar T. *et al.* (2000) Re-annotating the *Mycoplasma pneumoniae* genome sequence: adding value, function and reading frames. *Nucleic Acids Research* **28**, 3278–3288.

Danyluk J., Rassart E. & Sarhan F. (1991) Gene expression during cold and heat shock in wheat. *Biochemistry and Cell Biology* **69**, 383–391.

Das L. & Martienssen R. (1995) Site-selected transposon mutagenesis at the *hcf106* locus in maize. *Plant Cell* **7**, 287–294.

Datson N.A., van de Vosse E., Dauwerse H.G., Bout M., van Ommen G-J.B. & den Dunnen J.T. (1996) Scanning for genes in large genomic regions: cosmid-based exon trapping of multiple exons in a single product. *Nucleic Acids Research* **24**, 1105–1111.

Datson N.A., van der Perk-de Jong J., van den Berg M.P., de Kloet E.R. & Vreugdenhil E. (1999) MicroSAGE: a modified procedure for serial analysis of gene expression in limited amounts of tissue. *Nucleic Acids Research* **27**, 1300–1307.

Dausset J., Cann H., Cohen D., Lathrop M., Lalouel J.M. & White R.L. (1990) Collaborative genetic mapping of the human genome. *Genomics* **6**, 575–577.

Davidson D. & Baldock R. (2001) Bioinformatics beyond sequence: mapping gene function in the embryo. *Nature Reviews Genetics* **2**, 409–417.

Davidson E.H. & Britten R.J. (1973) Organization, transcription and regulation in the animal genome. *Quarterly Review of Biology* **48**, 565–613.

Davies K.E., Young B.D., Elles R.G., Hill M.E. & Williamson R. (1981) Cloning a representative genomic library of the human X chromosome after sorting by flow cytometry. *Nature* **293**, 374–376.

Day R.N. (1998) Visualization of Pit-1 transcription factor interactions in the living cell nucleus by fluorescence resonance energy transfer microscopy. *Molecular Endrocrinology* **12**, 1410–1419.

Dayhoff M.O. ed. (1978) *Atlas of Protein Sequence and Structure*, vol. 5. National Biomedical Research Foundation, Washington DC.

De Risi J. *et al.* (1996) Use of a cDNA microarray to analyse gene expression patterns in human cancer. *Nature Genetics* **14**, 457–460.

De Risi J.L., Iyer V.R. & Brown P.O. (1997) Exploring the metabolic and genetic control of gene expression on a genomic scale. *Science* **278**, 680–686.

de Vienne D. *et al.* (1996) Two-dimensional electrophoresis of proteins as a source of monogenic and codominant markers for population mapping the expressed genome. *Heredity* **76**, 166–177.

de Vienne D., Leonardi A., Damerval C. & Zivy M. (1999) Genetics of proteome variation as a tool for QTL characterisation: application to drought stress responses in maize. *Journal of Experimental Botany* **50**, 303–309.

de Waard V., van den Berg B.M.M., Veken J., Schultz-Heienbrok R., Pannekoek H. & van Zonneveld A.J. (1999) Serial analysis of gene expression to assess the endothelial cell response to an atherogenic stimulus. *Gene* **226**, 1–8.

de Wildt R.M.T., Mundy C.R., Gorick B.D. & Tomlinson I.M. (2000) Antibody arrays for high-throughput screening of antibody–antigen interactions. *Nature Biotechnology* **18**, 989–994.

Dear P.H. & Cook P.R. (1993) Happy mapping: linkage mapping using a physical analogue of meiosis. *Nucleic Acids Research* **11**, 13–20.

DeBerardinis R., Goodier J., Ostertag E. & Kazazian H. (1998) Rapid amplification of a retrotransposon subfamily is evolving the mouse genome. *Nature Genetics* **20**, 288–290.

DeBry R.W. & Seldin M.F. (1996) Human–mouse homology relationships. *Genomics* **33**, 337–351. Updated regularly on the On-Line Mendelian Inheritance in Man web site, http://www.ncbi.nlm.nih.gov/Omim/Homology

Dehal P. *et al.* (2001) Human chromosome 19 and related regions in mouse: conservative and lineage-specific evolution. *Science* **293**, 104–111.

den Dunnen J.T. (1999) Cosmid-based exon trapping. *Methods in Enzymology* **303**, 100–110.

Denny P. & Justice M.J. (2000) Mouse as the measure of man? *Trends in Genetics* **16**, 283–287.

Desikan R., Mackerness S.A.H., Hancock J.T. & Neill S.J. (2001) Regulation of the *Arabidopsis* transcriptome by oxidative stress. *Plant Physiology* **127**, 159–172.

Desprez T., Amselem J., Caboche M. & Hofte H. (1998) Differential gene expression in Arabidopsis monitored using cDNA arrays. *Plant Journal* **14**, 643–652.

Dimmock N.J., Easton A.J. & Leppard K.N. (2001) *Introduction to Modern Virology*, 5th edn. Blackwell Science, Oxford.

Doebley J. (1992) Mapping the genes that made maize. *Trends in Genetics* **8**, 302–307.

Doebley J., Stec A. & Hubbard L. (1997) The evolution of apical dominance in maize. *Nature* **386**, 485–488.

Dogget N. (1992) *The Human Genome Project*. Los Alamos Science No. 20. University Science Books, Sausalito, CA.

Dong S. *et al.* (2001) Flexible use of high-density oligonucleotide arrays for single-nucleotide polymorphism discovery and validation. *Genome Research* **11**, 1418–1424.

Dong X.N. (2001) Genetic dissection of systemic acquired resistance. *Current Opinion in Plant Biology* **4**, 309–314.

Donis-Keller H. *et al.* (1987) A genetic linkage map of the human genome. *Cell* **51**, 319–337.

Doolittle W.F. (1995) The multiplicity of domains in proteins. *Annual Review of Biochemistry* **64**, 287–314.

Dovichi N.J. & Zhang J.-Z. (2001) DNA sequencing by capillary array electrophoresis. *Methods in Molecular Biology* **167**, 225–239.

Drmanac R., Labat I., Brukner I. & Crkvenjakov R. (1989) Sequencing of megabase plus DNA by hybridization: theory of the method. *Genomics* **4**, 114–128.

Drmanac S. *et al.* (1998) Accurate sequencing by hybridization for DNA diagnostics and individual genomics. *Nature Biotechnology* **16**, 54–58.

Duggan D.J., Bittner M., Chen Y., Meltzer P. & Trent J.M. (1999) Expression profiling using cDNA microarrays. *Nature Genetics* **21** (Suppl.), 10–14.

Dujon B. (1996) The yeast genome project: what did we learn? *Trends in Genetics* **12**, 263–270.

Dujon B. (1998) European functional analysis network (EUROFAN) and the functional analysis of *Saccharomyces cerevisiae*. *Electrophoresis* **19**, 617–624.

Dujon B. *et al.* (1989) Mobile introns: definition of terms and recommended nomenclature. *Gene* **82**, 115–118.

Dunn J.J. & Studier F.W. (1983) Complete nucleotide sequence of bacteriophage T7 DNA and the locations of T7 genetic elements. *Journal of Molecular Biology* **166**, 477–535.

Duyk G.M., Kim S.W., Myers R.M. & Cox D.R. (1990) Exon trapping: a genetic screen to identify candidate transcribed sequences in cloned mammalian genomics DNA. *Proceedings of the National Academy of Sciences, USA* **87**, 8995–8999.

Ebersole T.A. *et al.* (2000) Mammalian artificial chromosome formation from circular alphoid input DNA does not require telomere repeats. *Human Molecular Genetics* **9**, 1623–1631.

Eddy S. (1999) Noncoding RNA genes. *Current Opinion in Genetics and Development* **9**, 695–699.

Edgell D.R., Belfort M. & Shub D.A. (2000) Barriers to intron promiscuity in bacteria. *Journal of Bacteriology* **182**, 5281–5289.

Edwards A.M. *et al.* (2000) Protein production: feeding the crystallographers and NMR spectroscopists. *Nature Structural Biology* **7** (Suppl), 970–972.

Edwards M.D., Stuber C.W. & Wendel J.F. (1987) Molecular-marker-facilitated investigations of quantitative-trait loci in maize. I. Numbers, genomic distribution and types of gene action. *Genetics* **116**, 113–125.

Eichler E.E. (2001) Recent duplication, domain accretion and the dynamic mutation of the human genome. *Trends in Genetics* **17**, 661–669.

Eickbush T. (1999) Exon shuffling in retrospect. *Science* **283**, 1465–1467.

Eisen J.A., Heidelberg J.F., White O. & Salzberg S.L. (2000) Evidence for symmetric chromosomal inversions around the replication origin in bacteria. *Genome Biology* **1**, Research, 0011.1–0011.9.

Eisen M.B., Wiley D.C., Karplus M. & Hubbard R.E. (1994) HOOK: a program for finding novel molecular architectures that satisfy the chemical and steric requirements of a macromolecule binding site. *Proteins* **19**, 199–221.

Eisen M.B., Spellman P.T., Brown P.O. & Botstein D. (1998) Cluster analysis and display of genome-wide expression patterns. *Proceedings of the National Academy of Sciences, USA* **95**, 14863–14868.

Elkin C.J. *et al.* (2001) High-throughput plasmid purification for capillary sequencing. *Genome Research* **11**, 1269–1274.

Endoh H., Walhout A.J.M. & Vidal M. (2000) A green fluorescent protein-based reverse two-hybrid system: application to the characterization of large numbers of potential protein–protein interactions. *Methods in Enzymology* **328**, 74–88.

Enright A.J., Iliopoulos I., Kyrpides N.C. & Ouzounis C.A. (1999) Protein interaction maps for complete genomes based on gene fusion events. *Nature* **402**, 86–90.

Etkins R.P. (1998) Ligand assays: from electrophoresis to miniaturised microarrays. *Clinical Chemistry* **44**, 2015–2030.

Evans M.J., Carlton M.B.L. & Russ A.P. (1997) Gene trapping and functional genomics. *Trends in Genetics* **13**, 370–374.

Evans W.E. & Relling M.V. (1999) Pharmacogenomics: translating functional genomics into rational therapeutics. *Science* **286**, 487–491.

Ewing B. & Green P. (1998) Base-calling of automated sequencer traces using Phred II: error probabilities. *Genome Research* **8**, 186–194.

Ewing B., Hillier L., Wendl M.C. & Green P. (1998) Base-calling of automated sequencer traces using Phred I: accuracy assessment. *Genome Research* **8**, 175–185.

Falconer D.S. & MacKay T.F.C. (1996) *Introduction to Quantitative Genetics* (4th edn.), Longman, Harlow Essex, UK.

Fan J-B., Chikashige Y., Smith C.L., Niwa O., Yanagida M. & Cantor C.R. (1989) Construction of a *Not* I restriction map of the fission yeast *Schizosaccharomyces pombe* genome. *Nucleic Acids Research* **17**, 2801–2818.

Feldmann K.A. (1991) T-DNA insertion mutagenesis in *Arabidopsis*: mutational spectrum. *Plant Journal* **1**, 71–82.

Feldmann K.A. & Marks M.D. (1987) *Agrobacterium*-mediated transformation of germinating seeds of *Arabidopsis thaliana*: a non-tissue culture approach. *Molecular and General Genetics* **208**, 1–9.

Ferrin L.J. & Camerini-Otero R.D. (1991) Selective cleavage of human DNA: RecA-assisted restriction endonuclease (RARE) cleavage. *Science* **254**, 1494–1497.

Fickett J.W. (1996) Finding genes by computer: the state of the art. *Trends in Genetics* **12**, 316–320.

Fickett J.W. & Hatzigeorgiou A.G. (1997) Eukaryotic promoter recognition. *Genome Research* **7**, 861–878.

Fickett J.W. & Tung C.S. (1992) Assessment of protein coding measures. *Nucleic Acids Research* **20**, 6441–6450.

Field L.L., Tobias R. & Magnus T. (1994) A locus on chromosome 15q26 (IDDM3) produces susceptibility to insulin-dependent diabetes mellitus. *Nature Genetics* **8**, 189–194.

Fields C., Adams M.D., White O. & Venter J.C. (1994) How many genes in the human genome. *Nature Genetics* **7**, 345–346.

Fields S. & Song O. (1989) A novel genetic system to detect protein–protein interactions. *Nature* **340**, 245–246.

Finkelstein A.V. & Ptytsin O.B. (1987) Why do globular proteins fit the limited set of folding patterns. *Progress in Biophysics and Molecular Biology* **50**, 171–190.

Finlay B.B. & Falkow S. (1997) Common themes in microbial pathogenicity revisited. *Microbiology and Molecular Biology Reviews* **61**, 136–169.

Finley R. & Brent R. (1994) Interaction mating reveals binary and ternary interactions between Drosophila cell cycle regulators. *Proceedings of the National Academy of Sciences, USA* **91**, 12980–12984.

Fire A., Xu S., Montgomery M.K., Kostas S.A., Driver S.E. & Mehlo C.C. (1998) Potent and specific genetic interference by double-stranded RNA in *Caenorhabditis elegans. Nature* **391**, 806–811.

Firn R.D. & Jones C.G. (1999) Secondary metabolism and the risks of GMOs. *Nature* **400**, 13–14.

Fischer D. & Eisenberg D. (1997) Assigning folds to the proteins encoded by the genome of *Mycoplasma genitalium. Proceedings of the National Academy of Sciences, USA* **94**, 11929–11934.

Fischer D. & Eisenberg D. (1999a) Finding families for genomic ORFans. *Bioinformatics* **15**, 759–762.

Fischer D. & Eisenberg D. (1999b) Predicting structures for genome proteins. *Current Opinion in Structural Biology* **9**, 208–211.

Flajolet M. *et al.* (2000) A genomic approach of the hepatitis C virus generates a protein interaction map. *Gene* **241**, 369–379.

Fleischmann R.D. *et al.* (1995) Whole-genome random sequencing and assembly of *Haemophilus influenzae* Rd. *Science* **269**, 496–512.

Flores A. *et al.* (1999) A protein–protein interaction map of yeast RNA polymerase III. *Proceedings of the National Academy of Sciences, USA* **96**, 7815–7820.

Florijn R.J. *et al.* (1995) High resolution DNA fiberFISH genomic DNA mapping and colour barcoding of large genes. *Human Molecular Genetics* **4**, 831–836.

Fodor S.P.A. *et al.* (1991) Light-directed spatially addressable parallel chemical synthesis. *Science* **251**, 767–773.

Fodor S.P.A., Rava R.P., Huang X.C., Pease A.C., Holmes C.P. & Adams C.L. (1993) Multiplexed biochemical assays with bioiogical-chips. *Nature* **364**, 555–556.

Fonstein M. & Haselkorn R. (1995) Physical mapping of bacterial genomes. *Journal of Bacteriology* **177**, 3361–3369.

Fortna A. & Gardiner K. (2001) Genomic sequence analysis tools: a user's guide. *Trends in Genetics* **17**, 158–164.

Foury F., Roganti T., Lecrenier N. & Purnelle B. (1998) The complete sequence of the mitochondrial genome of *Saccharomyces cerevisiae. FEBS Letters* **440**, 325–331.

Frankel W.N. (1995) Taking stock of complex trait genetics in mice. *Trends in Genetics* **11**, 471–477.

Fraser A.G., Kamath R.S., Zipperlen P., Martinez-Campos M., Sohrmann M. & Ahringer J. (2000) Functional genomic analysis of *C. elegans* chromosome I by systematic RNA interference. *Nature* **408**, 325–330.

Fraser C.M. *et al.* (1995) The minimal gene complement of *Mycoplasma genitalium. Science* **270**, 397–403.

Frengen E. *et al.* (1999) A modular, positive selection bacterial artificial chromosome vector with multiple cloning sites. *Genomics* **58**, 250–253.

Frengen E. *et al.* (2000) Modular bacterial artificial chromosome vectors for transfer of large inserts into mammalian cells. *Genomics* **68**, 118–126.

Fridman E., Pleban T. & Zamir D. (2000) A recombination hotspot delimits a wild-species quantitative trait locus for tomato sugar content to 484 bp within an invertase gene. *Proceedings of the National Academy of Sciences, USA* **97**, 4718–4723.

Friedrich G. & Soriano P. (1991) Promoter traps in embryonic stem cells: a genetic screen to identify and mutate developmental genes in mice. *Genes and Development* **5**, 1513–1523.

Frijter A.C.J. *et al.* (1997) Construction of a bacterial artificial chromosome library containing large EcoRI and HindIII genomic fragments of lettuce. *Theoretical and Applied Genetics* **94**, 390–399.

Frohme M. *et al.* (2001) Directed gap closure in large-scale sequencing projects. *Genome Research* **11**, 901–903.

Fromont-Racine M., Rain J.C. & Legrain P. (1997) Toward a functional analysis of the yeast genome through exhaustive two-hybrid screens. *Nature Genetics* **16**, 277–282.

Fromont-Racine M. *et al.* (2000) Genome-wide protein interaction screens reveal functional networks involving Sm-like proteins. *Yeast* **17**, 95–110.

Fung E.T., Thulasiraman V., Weinberger S.R. & Dalmasso E.A. (2001) Protein biochips for differential profiling. *Current Opinion in Biotechnology* **12**, 65–69.

Gaasterland T. & Oprea M. (2001) Whole-genome analysis: annotations and updates. *Current Opinion in Structural Biology* **11**, 377–381.

Gaiano N., Amsterdam A., Kawakami K., Allende M., Becker T. & Hopkins N. (1996) Insertional mutagenesis and rapid cloning of essential genes in zebrafish. *Nature* **383**, 829–832.

Galitski T., Saldanha A.J., Styles C.A., Lander E.S. & Fink G.R. (1999) Ploidy regulation of gene expression. *Science* **285**, 251–254.

Galperin M.Y., Walker D.R. & Koonin E.V. (1998) Analagous enzymes: independent inventions in enzyme evolution. *Genome Research* **8**, 779–790.

Garcia-Vallve S., Romeu A. & Palau J. (2000) Horizontal gene transfer in bacterial and archaeal complete genomes. *Genome Research* **10**, 1719–1725.

Ge H. (2000) UPA, a universal protein array system for quantitative detection of protein–protein, protein–DNA, protein–RNA and protein–ligand interactions. *Nucleic Acids Research* **28**, e3, I–VII.

Geisler R. *et al.* (1999) A radiation hybrid map of the zebrafish genome. *Nature Genetics* **23**, 86–89.

Gelfand M.S. & Koonin E.V. (1997) Avoidance of palindromic words in bacterial and archaeal genomes: a close connection with restriction enzymes. *Nucleic Acids Research* **25**, 2430–2439.

Giacolone J. *et al.* (2000) Optical mapping of BAC clones from the human Y chromosome *DAZ* locus. *Genome Research* **10**, 1421–1429.

Giaever G. *et al.* (1999) Genomic profiling of drug sensitivities via induced haploinsufficiency. *Nature Genetics* **21**, 278–283.

Gibrat J.F., Madej T. & Bryant S.H. (1996) Surprising similarities in structure comparison. *Current Opinion in Structural Biology* **6**, 377–385.

Gierl A. & Saedler H. (1992) Plant transposable elements and gene tagging. *Plant Molecular Biology* **19**, 39–49.

Girke T., Todd J., Ruuska S., White J., Benning C. & Ohlrogge J. (2000) Microarray analysis of developing *Arabidopsis* seeds. *Plant Physiology* **124**, 1570–1581.

Glass J. *et al.* (2000) The complete sequence of the mucosal pathogen *Ureaplasma urealyticum*. *Nature* **407**, 757–762.

Gockel G. & Hachtel W. (2000) Complete gene map of the plastid genome of the nonphotosynthetic euglenoid flagellate *Astasia longa*. *Protist* **151**, 347–351.

Goffeau A. *et al.* (1996) Life with 6000 genes. *Science* **274**, 546–567.

Golub T.R. *et al.* (1999) Molecular classification of cancer: class discovery and class prediction by gene expression monitoring. *Science* **286**, 531–537.

Gonczy P. *et al.* (2000) Functional genomic analysis of cell division in *C. elegans* using RNAi of genes on chromosome III. *Nature* **408**, 331–336.

Gonnet G.H., Cohen M.A. & Benner S.A. (1992) Exhaustive matching of the entire protein-sequence database. *Science* **256**, 1443–1445.

Goodner B.W. *et al.* (1999) Combined genetic and physical map of the complex genome of *Agrobacterium tumefaciens*. *Journal of Bacteriology* **180**, 3816–3822.

Gorg A., Postel W., Baumer M. & Weiss W. (1992) Two-dimensional polyacrylamide gel electrophoresis, with immobilized pH gradients in the first dimension, of barley seed proteins: discrimination of cultivars with different mating grades. *Electrophoresis* **13**, 192–203.

Goss S.J. & Harris H. (1977) Gene transfer by means of cell fusion II. The mapping of 8 loci on human chromosome 1 by statistical analysis of gene assortment in somatic cell hybrids. *Journal of Cell Science* **25**, 39–57.

Gossler A., Joyner A.L., Rossant J. & Skarnes W.C. (1989) Mouse embryonic stem cells and reporter constructs to detect developmentally regulated genes. *Science* **244**, 463–465.

Gottschalk A. *et al.* (1998) A comprehensive biochemical and genetic analysis of the yeast U1 snRNP reveals five novel proteins. *RNA* **4**, 374–393.

Grandbastien M-A. (1992) Retroelements in higher plants. *Trends in Genetics* **8**, 103–108.

Grandi G. (2001) Antibacterial vaccine design using genomics and proteomics. *Trends in Biotechnology* **19**, 181–188.

Granjeaud S., Bertucci F. & Jordan B.R. (1999) Expression profiling: DNA arrays in many guises. *BioEssays* **21**, 781–790.

Graveley B.R. (2001) Alternative splicing: increasing diversity in the proteomic world. *Trends in Genetics* **17**, 100–107.

Gray M.W., Burger G. & Lang F. (1999) Mitochondrial evolution. *Science* **283**, 1476–1481.

Green P. (1997) Against a whole-genome shotgun. *Genome Research* **7**, 410–417.

Gress T.M. *et al.* (1992) Hybridization fingerprinting of high-density cDNA library arrays with cDNA pools derived from whole tissues. *Mammalian Genome* **3**, 609–610.

Grossberger R. *et al.* (1999) Characterization of the DOC1/APC10 subunit of the yeast and the human anaphase-promoting complex. *Journal of Biological Chemistry* **274**, 14500–14507.

Gygi S.P., Rochon Y., Franza B.R. & Aebersold R. (1999) Correlation between protein and mRNA abundance in yeast. *Molecular Cell Biology* **19**, 1720–1730.

Haab B.B., Dunham M.J. & Brown P.O. (2001) Protein microarrays for highly parallel detection and quantitation of specific proteins and antibodies in complex solutions. *Genome Biology* **2**, research0004.1–0004.13.

Hacia J.G. (1999) Resequencing and mutational analysis using oligonucleotide microarrays. *Nature Genetics* **21** (Suppl.), 42–47.

Hacia J.G. *et al.* (1996) Detection of heterozygous mutations in BRCA1 using high density oligonucleotide arrays and two-colour fluorescence analysis. *Nature Genetics* **14**, 441–447.

Hacia J.G. *et al.* (1998) Strategies for mutational analysis of the large multi-exon *ATM* gene using high-density oligonucleotide arrays. *Genome Research* **8**, 1245–1258.

Hacker J., Blum-Oehler G., Muhldorfer I. & Tschape H. (1997) Pathogenicity islands of virulent bacteria: structure, function and impact on microbial evolution. *Molecular Microbiology* **23**, 1089–1097.

Hadley, C. & Jones, D. (1999) A systematic comparison of protein structure classifications: SCOP, CATH and FSSP. *Structure with Folding and Design* **7**, 1099–1112.

Haley C.S., Knott S.A. & Elsen J-M. (1994) Mapping quantitative trait loci in crosses between outbred lines using least squares. *Genetics* **136**, 1195–1207.

Hallick R.B. *et al.* (1993) Complete sequence of *Euglena gracilis* chloroplast DNA. *Nucleic Acids Research* **21**, 3537–3544.

Hamaguchi M. *et al.* (1992) Establishment of a highly sensitive and specific exon-trapping system. *Proceedings of the National Academy of Sciences, USA* **89**, 9779–9783.

Hamer L., DeZwaan T.M., Montenegro-Chamorro M.V., Frank S.A. & Hamer J.E. (2001) Recent advances in large-scale transposon mutagenesis. *Current Opinion in Chemical Biology* **5**, 67–73.

Hammond S.M., Caudy A.A. & Hannon G.J. (2001) Post-transcriptional gene silencing by double-stranded RNA. *Nature Review Genetics* **2**, 110–119.

Hanahan D. & Meselson M. (1980) Plasmid screening at high colony density. *Gene* **10**, 63–67.

Hancock J.M. (1996) Simple sequences in a 'minimal' genome. *Nature Genetics* **14**, 14–15.

Harrington C.A., Rosenow C. & Retief J. (2000) Monitoring gene expression using DNA microarrays. *Current Opinion in Microbiology* **3**, 285–291.

Harrington J.J., Bokkelen G.V., Mays R.W., Gustashaw K. & Willard H.F. (1997) Formation of *de novo* centromeres and construction of first-generation human artificial microchromosomes. *Nature Genetics* **15**, 345–355.

Harrison S.M., Dunswoodie S.L., Arkell R.M., Lerach H. & Beddington R.S.P. (1995) Isolation of novel tissue-specific genes from cDNA libraries representing the individual tissue constituents of the gastrulating mouse embryo. *Development* **121**, 2479–2489.

Hartman P.E. & Roth J.R. (1973) Mechanisms of suppression. *Advances in Genetics* **17**, 1–105.

Hashimoto L. *et al.* (1994) Genetic mapping of a susceptibility locus for insulin-dependent diabetes mellitus on chromosome 11q. *Nature* **371**, 161–164.

Hauge B.M., Hanley S., Giraudat J. & Goodman H.M. (1991) Mapping the *Arabidopsis* genome. In: *Molecular Biology of Plant Development* (eds G. Jenkins & W. Schurch). Cambridge University Press, Cambridge.

Hayashi T. *et al.* (2001) Complete genome sequence of enterohemorrhagic *Escherichia coli* O157:H7 and genomic comparison with a laboratory strain K-12. *DNA Research* **8**, 11–22.

Hazbun T.R. & Fields S. (2001) Networking proteins in yeast. *Proceedings of the National Academy of Sciences, USA* **98**, 4277–4278.

He L-Z. *et al.* (1998) Distinct interactions of PML-RARα with co-repressors determine differential responses to RA in APL. *Nature Genetics* **18**, 126–135.

Heale S.M., Stateva L.L. & Oliver S.G. (1994) Introduction of YACs into intact yeast cells by a procedure which shows low levels of recombinagenicity and co-transformation. *Nucleic Acids Research* **22**, 5011–5015.

Hecksher-Sorensen J. & Sharpe J. (2001) 3D confocal reconstruction of gene expression in mouse. *Mechanisms of Development* **100**, 59–63.

Hegyi H. & Gerstein M. (1999) The relationship between protein structure and function: a comprehensive survey with application to the yeast genome. *Journal of Molcular Biology* **288**, 147–164.

Heinemann U. *et al.* (2000) An integrated approach to structural genomics. *Progress in Biophysics and Molecular Biology* **73**, 347–362.

Heinemann U., Illing G. & Oschkinat H. (2001) High-throughput three-dimensional protein structure determination. *Current Opinion in Biotechnology* **12**, 348–354.

Heller R. *et al.* (1997) Discovery and analysis of inflammatory disease-related genes using cDNA microarrays. *Proceedings of the National Academy of Sciences, USA* **94**, 2150–2155.

Helmuth L. (2001) Map of the human genome 3.0. *Science* **293**, 583–584.

Hendrickson W.A., Horton J.R. & LeMaster D.M. (1990) Selenomethionyl proteins produced for analysis by multiwavelength anomalous diffraction (MAD): a vehicle for direct determination of three-dimensional structure. *EMBO Journal* **9**, 1665–1672.

Hengartner M.O. & Horvitz H.R. (1994) *C. elegans* cell survival gene *ced-9* encodes a functional homolog of the mammalian proto-oncogene *Bcl-2*. *Cell* **76**, 665–676.

Henikoff S. & Henikoff J.G. (1993) Performance evaluation of amino acid substitution matrices. *Proteins* **17**, 49–61.

Henze K. & Martin W. (2001) How do mitochondrial genes get into the nucleus? *Trends in Genetics* **17**, 383–387.

Henzel W.J. *et al.* (1993) Identifying proteins from two-dimensional gels by molecular mass searching of peptide fragments in protein sequence databases. *Proceedings of the National Academy of Sciences, USA* **90**, 5011–5015.

Herbert B.R. *et al.* (2001) What place for polyacrylamide in proteomics? *Trends in Biotechnology* **19** (Suppl. *Trends Guide to Proteomics II*), S3–S9.

Herskowitz I. (1987) Functional analysis of genes by dominant negative mutations. *Nature* **329**, 219–222.

Hess K.R. *et al.* (2001) Microarrays: handling the deluge of data and extracting reliable information. *Trends in Biotechnology* **19**, 463–468.

Hicks G.G., Shi E.G., Li X.M., Li C.H., Pawlak M. & Ruley H.E. (1997) Functional genomics in mice by tagged sequence mutagenesis. *Nature Genetics* **16**, 338–344.

Himmelreich R., Hilbert H., Plagens H., Pirkl E., Li B-C. & Herrmann R. (1996) Complete sequence analysis of the genome of the bacterium *Mycoplasma pneumoniae*. *Nucleic Acids Research* **24**, 4420–4449.

Hinnebush J. & Tilley K. (1993) Linear plasmids and chromosomes in bacteria. *Molecular Microbiology* **10**, 917–922.

Hiratsuka J. *et al.* (1989) The complete sequence of the rice (*Oryza sativa*) chloroplast genome: inter-molecular recombination between distinct tRNA genes accounts for a major plastid DNA inversion during the evolution of the cereals. *Molecular and General Genetics* **217**, 185–194.

Hirschhorn J.N. *et al.* (2001) Genome-wide linkage analysis of stature in multiple populations reveals several regions with evidence of linkage to adult height. *American Journal of Human Genetics* **69**, 106–116.

Hoffmeyer S. *et al.* (2000) Functional polymorphisms of the human multidrug-resistance gene: multiple sequence variations and correlation of one allele with P-glycoprotein expression and activity *in vivo. Proceedings of the National Academy of Sciences, USA* **97**, 3473–3478.

Hofreiter M. *et al.* (2001) Ancient DNA. *Nature Reviews: Genetics* **2**, 353–359.

Hoheisel J.D. (1994) Application of hybridization techniques to genome mapping and sequencing. *Trends in Genetics* **10**, 79–83.

Hoheisel J.D. *et al.* (1993) High resolution cosmid and P1 maps spanning the 14 Mb genome of the fission yeast *S. pombe. Cell* **73**, 109–120.

Holm L. & Sander C. (1995) Dali: a network tool for protein structure comparison. *Trends in Biochemical Science* **20**, 478–480.

Holm, L. & Sander, C. (1996) Mapping the protein universe. *Science* **273**, 595–602.

Hrabe de Angelis M. & Balling R. (1998) Large-scale ENU screens in the mouse: genetics meets genomics. *Mutation Research* **400**, 25–32.

Hrabe de Angelis M. *et al.* (2000) Genome-wide, large-scale production of mutant mice by ENU mutagenesis. *Nature Genetics* **25**, 444–447.

Hu G.S., Yalpani N., Briggs S.P. & Johal G.S. (1998) A porphyrin pathway impairment is responsible for the phenotype of a dominant disease lesion mimic mutant of maize. *Plant Cell* **10**, 1095–1105.

Huffaker T.C., Hoyt M.A. & Botstein D. (1987) Genetic analysis of the yeast cytoskeleton. *Annual Review of Genetics* **21**, 259–284.

Hughes A.L. & Hughes M.K. (1995) Small genomes for better flyers. *Nature* **377**, 391.

Hughes T.R. *et al.* (2000) Functional discovery via a compendium of expression profiles. *Cell* **102**, 109–126.

Hugot J.P. *et al.* (1996) Mapping of a susceptibility locus for Crohn's disease on chromosome 16. *Nature* **379**, 772–773.

Hugot J-P. *et al.* (2001) Association of NOD2 leucine-rich repeat variants with susceptibility to Crohn's disease. *Nature* **411**, 599–603.

Huynen M., Dandekar T. & Bork P. (1998a) Differential genome analysis applied to the species-specific features of *Helicobacter pylori. FEBS Letters* **426**, 1–5.

Huynen M. *et al.* (1998b) Homology-based fold predictions for *Mycoplasma genitalium* proteins. *Journal of Molecular Biology* **280**, 323–326. [URL:http://dove.embl-heidelberg.de/3D/]

Huynen M.A., Dandekar T. & Bork P. (1999) Variation and evolution of the citric acid cycle: a genomic perspective. *Trends in Microbiology* **7**, 281–291.

Ingelman-Sundberg M., Oscarson M. & McLellan R.A. (1999) Polymorphic human cytochrome P450 enzymes: an opportunity for individualized drug treatment. *Trends in Pharmacological Science* **20**, 342–349.

International Human Genome Sequencing Consortium (2001) Initial sequencing and analysis of the human genome. *Nature* **409**, 860–933.

Ioannou P.A. *et al.* (1994) A new bacteriophage P1-derived vector for the propagation of large human DNA fragments. *Nature Genetics* **6**, 84–89.

Ioshikhes I.P. & Zhang M.Q. (2000) Large-scale human promoter mapping using CpG islands. *Nature Genetics* **26**, 61–63.

Irizarry K. *et al.* (2000) Genome-wide analysis of single-nucleotide polymorphisms in human expressed sequences. *Nature Genetics* **26**, 233–236.

Ito T., Seki M., Hayashida N., Shibata D. & Shinozaki K. (1999) Regional insertional mutagenesis of genes on *Arabidopsis thaliana* chromosome V using the *Ac/Ds* transposon in combination with a cDNA scanning method. *Plant Journal* **17**, 433–444.

Ito T. *et al.* (2000) Toward a protein–protein interaction map of the budding yeast: a comprehensive system to examine two-hybrid interactions in all possible combinations between the yeast proteins. *Proceedings of the National Academy of Sciences, USA* **97**, 1143–1147.

Ito T., Chiba T., Ozawa R., Yoshida M., Hattori M. & Sakaki Y. (2001) A comprehensive two-hybrid analysis to explore the yeast protein interactome. *Proceedings of the National Academy of Sciences, USA* **98**, 4569–4574.

Iyer V.R. *et al.* (1999) The transcriptional program in the response of human fibroblasts to serum. *Science* **283**, 83–87.

Iyer V.R. *et al.* (2001) Genomic binding sites of the yeast cell cycle transcription factors SBF and MBF. *Nature* **409**, 533–538.

Jacob F. & Monod J. (1961) Genetic regulatory mechanisms in the synthesis of proteins. *Journal of Molecular Biology* **3**, 318–356.

James P., Quadroni M., Carafoli E. & Gonnet G. (1993) Protein identification by mass profile fingerprinting. *Biochemical and Biophysical Research Communications* **195**, 58–64.

Jasienski M. & Bazzaz F.A. (1995) Genome size and high CO_2. *Nature* **376**, 559–560.

Jasin M. (1996) Genetic manipulation of genomes with rare-cutting endonucleases. *Trends in Genetics* **12**, 224–228.

Jeffreys A.J. & Flavell R.A. (1977) The rabbit beta-globin gene contains a large insert in the coding squence. *Cell* **12**, 1097–1108.

Jelinsky S. & Samson L. (1999) Global response of *Saccharomyces cerevisiae* to an alkylating agent. *Proceedings of the National Academy of Sciences, USA* **96**, 1486–1491.

Jeon J-S. *et al.* (2000) T-DNA insertional mutagenesis for functional genomics in rice. *Plant Journal* **22**, 561–570.

Ji H., Moore D.P., Blomberg M.A., Braiterman L.T., Voytas D.F., Natsoulis G. & Boeke J.D. (1993) Hotspots for unselected Ty1 transposition events on yeast chromosome III are near tRNA genes and LTR sequences. *Cell* **73**, 1007–1018.

Johnson G.C.L. *et al.* (2001) Haplotype tagging for the identification of common disease genes. *Nature Genetics* **29**, 233–237.

Jomaa H. *et al.* (1999) Inhibitors of the nonmevalonate pathway of isoprenoid biosynthesis as antimalarial drugs. *Science* **285**, 1573–1576.

Jones D.T. (1997) Successful *ab initio* prediction of the tertiary structure of NK-lysin using multiple sequences and recognized supersecondary structural motifs. *Proteins* **S1**, 185–191.

Jones D.T. (1999a) GenTHREADER: an efficient and reliable protein fold recognition method for genomic sequences. *Journal of Molecular Biology* **287**, 797–815.

Jones D.T. (1999b) Protein secondary structure prediction based on position-specific scoring matrices. *Journal of Molecular Biology* **292**, 195–202.

Jones D.T. (2000) Protein structure prediction in the post-genomic era. *Current Opinion in Structural Biology* **10**, 371–379.

Jones D.T., Taylor W.R. & Thornton J.M. (1992) A new approach to protein fold recognition. *Nature* **358**, 86–89.

Jones S. & Thornton J.M. (1995) Protein–protein interactions: a review of protein dimer structures. *Progress in Biophysics and Molecular Biology* **63**, 31–65.

Jones S. & Thornton J.M. (1997) Prediction of protein–protein interaction sites using patch analysis. *Journal of Molecular Biology* **272**, 133–143.

Jones S.J.M. *et al.* (2001) Changes in gene expression associated with developmental arrest and longevity in *Caenorhabditis elegans*. *Genome Research* **11**, 1346–1352.

Jones V.W., Kenseth J.R., Porter M.D., Mosher C.L. & Henderson E. (1998) Microminiaturized immunoassays using atomic force microscopy and compositionally patterned antigen arrays. *Analytical Chemistry* **70**, 1233–1241.

Joos T.O. *et al.* (2000) A microarray enzyme linked immunosorbent assay for autoimmune diagnostics. *Electrophoresis* **21**, 2641–2650.

Judson, R., Stephens, J.C. & Windemuth, A. (2000) The predictive power of haplotypes in clinical response. *Pharmacogenomics* **1**, 15–26.

Jurgens G., Wieschaus E., Nusslein-Volhard C. & Kluding H. (1984) Mutations affecting the pattern of the larval cuticle in *Drosophila melanogaster*. II. Zygotic loci on the third chromosome. *Wilheim Roux Archives in Developmental Biology* **193**, 283–295.

Justice M.J., Noveroske J.K., Weber J.S., Zheng B.H. & Bradley A. (1999) Mouse ENU mutagenesis. *Human Molecular Genetics* **8**, 1955–1963.

Kacharmina J.E., Crino P.B. & Eberwine J. (1999) Preparation of cDNA from single cells and subcellular regions. *Methods in Enzymology* **303**, 3–18.

Kaelin W.G. Jr, Krek W., Sellers W.R., DeCaprio J.A., Kayle F.J. & Livingston D.M. (1991) Identification of cellular proteins that can interact specifically with the T/E1A-binding region of the retinoblastoma gene product. *Cell* **64**, 521–532.

Kakimoto T. (1996) CKI1, a histidine kinase homolog implicated in cytokinin signal transduction. *Science* **274**, 982–985.

Kaname T. & Huxley C. (2001) Simple and efficient vectors for retrofitting BACs and PACs with mammalian neo^R and *EFGP* marker genes. *Gene* **266**, 147–153.

Kanno S. *et al.* (2000) Assembling of engineered IgG binding protein on gold surface for highly oriented antibody immobilization. *Journal of Biotechnology* **76**, 207–214.

Karas M. & Hillenkamp F. (1988) Laser desorption ionization of proteins with molecular masses exceeding 10 000 daltons. *Analytical Chemistry* **60**, 2299–2301.

Karlin S. (2001) Detecting anomalous gene clusters and pathogenicity islands in diverse bacterial genomes. *Trends in Microbiology* **9**, 335–343.

Kassua A. & Thornton J.M. (1999) Three-dimensional structure analysis of PROSITE patterns. *Journal of Molecular Biology* **286**, 1673–1691.

Kato-Maeda M. *et al.* (2001) Comparing genomes within the species *Mycobacterium tuberculosis*. *Genome Research* **11**, 547–554.

Kawasaki S. *et al.* (2001) Gene expression profiles during the initial phase of salt stress in rice. *Plant Cell* **13**, 899–905.

Kempin S.A., Liljegre S.J., Block L.M., Roundsley S.D., Yanofsky M.F. & Lam E. (1997) Targeted disruption in *Arabidopsis*. *Nature* **389**, 802–803.

Kim K.K., Hung L.W., Yokota H., Kim R. & Kim S.H. (1998) Crystal structures of eukaryotic translation initiation factor 5A from *Methanococcus jannaschii* at 1.8 A resolution. *Proceedings of the National Academy of Sciences, USA* **95**, 10419–10424.

Kim U-J. *et al.* (1996) Construction and characterization of a human bacterial artificial chromosome library. *Genomics* **34**, 213–218.

Kjemtrup S. *et al.* (1998) Gene silencing from plant DNA carried by a geminivirus. *Plant Journal* **14**, 91–100.

Klenow H. & Henningsen I. (1970) Selective elimination of the exonuclease activity of the deoxyribonucleic acid polymerase from *E.coli* B by limited proteolysis. *Proceedings of the National Academy of Sciences, USA* **65**, 168–175.

Klose J. (1975) Protein mapping by combined isoelectric focussing and electrophoresis of mouse tissues: a novel approach to testing for induced point mutations in mammals. *Humangenetik* **26**, 231–243.

Knight S.J. & Flint J. (2000) Perfect endings: a review of subtelomeric probes and their use in clinical diagnosis. *Journal of Medical Genetics* **37**, 401–409.

Ko M.S.H. (2001) Embryogenomics: developmental biology meets genomics. *Trends in Biotechnology* **19**, 511–518.

Ko M.S.H. *et al.* (2000) Large-scale cDNA analysis reveals phased gene expression patterns during preimplantation mouse development. *Development* **127**, 1737–1749.

Kohara Y., Akiyama K. & Isono K. (1987) The physical map of the whole *E.coli* chromosome: application of a new strategy for rapid analysis and sorting of a large genomic library. *Cell* **50**, 495–508.

Kolkman J.A. & Stemmer W.P.C. (2001) Directed evolution of proteins by exon shuffling. *Nature Biotechnology* **19**, 423–428.

Komiya T., Tanigawa Y. & Hirohashi S. (1997) A large-scale *in situ* hybridization system using an equalized cDNA library. *Analytical Biochemistry* **254**, 23–30.

Konfortov B.A. *et al.* (2000) A high-resolution HAPPY map of *Dictyostelium discoideum* chromosome 6. *Genome Research* **10**, 1737–1742.

Konieczny A. & Ausubel F.A. (1993) A procedure for mapping *Arabidopsis* mutations using co-dominant ecotype-specific PCR-based markers. *Plant Journal* **4**, 403–410.

Koob M. & Szbalski W. (1990) Cleaving yeast and *Escherichia coli* genomes at a single site. *Science* **1250**, 271–273.

Koonin E.V., Aravind L. & Kondrashov A.S. (2000) The impact of comparative genomics on our understanding of evolution. *Cell* **101**, 573–576.

Koppensteiner W.A., Lackner P., Wiederstein M. & Sippl M.J. (2000) Characterization of novel proteins based on known protein structures. *Journal of Molecular Biology* **296**, 1139–1152.

Koradi R., Billeter M., Engeli M., Güntert P. & Wüthrich K. (1998) Automated peak picking and peak integration in macromolecular NMR spectra using AUTOPSY. *Journal of Magnetic Resonance* **135**, 288–297.

Kornfeld K. (1997) Vulval development in *Caenorhabditis elegans*. *Trends in Genetics* **13**, 55–61.

Koshland D., Kent J.C. & Hartwell L.H. (1985) Genetic analysis of the mitotic transmission of minichromosomes. *Cell* **40**, 393–403.

Koski L.B., Morton R.A. & Golding G.B. (2001) Codon bias and base composition are poor indicators of horizontally transferred genes. *Molecular Biology and Evolution* **18**, 404–412.

Kouprina N., Eldarov M., Moyzis R., Resnick M. & Larionov V. (1994) A model system to assess the integrity of mammalian YACs during transformation and propagation in yeast. *Genomics* **21**, 7–17.

Kozal M. *et al.* (1996) Extensive polymorphisms observed in HIV-1 cladeB protease gene using high-density oligonucleotide arrays. *Nature Medicine* **2**, 753–759.

Kozlov G. *et al.* (2000) Rapid fold and structure determination of the archaeal translation elongation factor 1 from *Methanobacterium thermoautotrophicum*. *Journal of Biomolecular NMR* **17**, 187–194.

Krishnan B.R., Jamry I., Berg D.E., Berg C.M. & Chaplin D.D. (1995) Construction of a genomic DNA 'feature map' by sequencing from nested deletions: application to the HLA class 1 region. *Nucleic Acids Research* **23**, 117–122.

Krupp G., Bonatz G. & Parwaresch R. (2000) Telomerase, immortality and cancer. *Biotechnology Annual Review* **6**, 103–140.

Krutchinsky A.N., Zhang W.Z. & Chait B.T. (2000) Rapidly switchable matrix-assisted laser desorption/ionization and electrospray quadrupole–time-of-flight mass spectrometry for protein identification. *Journal of the American Society of Mass Spectrometry* **11**, 493–504.

Krysan P.J., Young J.C. & Sussman M.R. (1999) T-DNA as an insertional mutagen in *Arabidopsis*. *Plant Cell* **11**, 2283–2290.

Kumagai M.H., Donson J., Della-Cioppa G., Harvey D., Hanley K. & Grill L.K. (1995) Cytoplasmic inhibition of carotenoid biosynthesis with virus-derived RNA. *Proceedings of the National Academy of Sciences, USA* **92**, 1679–1683.

Kumar A. (1996) The adventures of the Ty1-*copia* group of retrotransposons in plants. *Trends in Genetics* **12**, 41–43.

Kunkel L.M. *et al.* (1986) Analysis of deletions in DAN from patients with Becker and Duchenne muscular dystrophy. *Nature* **322**, 73–77.

Kurata N. *et al.* (1994) A 300 kilobase interval genetic map of rice including 883 expressed sequences. *Nature Genetics* **8**, 365–372.

Kyrpides N., Overbeek R. & Ouzounis C. (1999) Universal protein families and the functional content of the last common ancestor. *Journal of Molecular Evolution* **49**, 413–423.

Laan M., Kallioniemi O-P., Hellsten E., Alitalo K., Peltonen L. & Palotie A. (1995) Mechanically stretched chromosomes as targets for high-resolution FISH mapping. *Genome Research* **5**, 13–20.

Lahaye T. & Bonas U. (2001) Molecular secrets of bacterial type III effector proteins. *Trends in Plant Science* **6**, 479–485.

Lahm H.W. & Langen H. (2000) Mass spectrometry: a tool for the identification of proteins separated by gels. *Electrophoresis* **21**, 2105–2114.

Lai C.S.L. *et al.* (2001) A forkhead-domain gene is mutated in a severe speech and language disorder. *Nature* **413**, 519–523.

Lai Z. *et al.* (1999) A shotgun optical map of the entire *Plasmodium falciparum* genome. *Nature Genetics* **23**, 309–313.

Lally E.T. *et al.* (1999) The interaction between RTX toxins and target cells. *Trends in Microbiology* **7**, 356–361.

Lamond A.I. & Mann M. (1997) Cell biology and the genome projects: a concerted strategy for characterizing multiprotein complexes by using mass spectrometry. *Trends in Cell Biology* **7**, 139–142.

Lamzin V.S. & Perrakis A. (2000) Current state of automated crystallographic data analysis. *Nature Structural Biology* **7**, 978–981.

Landegren U., Nilsson M. & Kwok P.Y. (1998) Reading bits of genetic information: methods for single-nucleotide polymorphism analysis. *Genome Research* **8**, 769–776.

Lander E.S. (1999) Array of hope. *Nature Genetics* **21** (Suppl), 3–4.

Lander E.S. & Botstein D. (1989) Mapping Mendelian factors underlying quantitative traits using RFLP linkage maps. *Genetics* **121**, 185–199.

Lander E.S. & Schork N.J. (1994) Genetic dissection of complex traits. *Science* **265**, 2037–2048.

Lander E.S. & Waterman M.S. (1998) Genomic mapping by fingerprinting random clones: a mathematical analysis. *Genomics* **4**, 231–239.

Larin Z., Monaco A.P. & Lehrach H. (1991) Yeast artificial chromosome libraries containing large inserts from mouse and human DNA. *Proceedings of the National Academy of Sciences, USA* **88**, 4123–4127.

Larionov V. *et al.* (1996) Specific cloning of human DNA as yeast artificial chromosomes by transformation-associated recombination. *Proceedings of the National Academy of Sciences, USA* **93**, 491–496.

Lashkari D.A. *et al.* (1997) Yeast microarrays for genome wide parallel genetic and gene expression analysis. *Proceedings of the National Academy of Sciences, USA* **94**, 13057–13062.

Laskowski R.A., MacArthur M.W., Moss D.S. & Thornton J.M. (1993) PROCHECK: a program to check the stereochemical quality of protein structures. *Journal of Applied Crystallography* **26**, 283–291.

Laskowski R.A., Luscombe N.M., Swindells M.B. & Thornton J.M. (1996) Protein clefts in molecular recognition and function. *Protein Science* **5**, 2438–2452.

Le Y. & Dobson M.J. (1997) Stabilization of yeast artificial chromosome clones in a *rad54–3* recombination-deficient host strain. *Nucleic Acids Research* **25**, 1248–1253.

Le Bourgeois P., Lautier M., Mara M. & Ritzenthaler P. (1992) New tools for the physical and genetic mapping of *Lactococcus* species. *Gene* **78**, 29–36.

Lee J-Y., Koi M., Stanbridge E.J., Oshimura M., Kumamoto A.T. & Feinberg A.P. (1994) Simple purification of human chromosomes to homogeneity using Muntjac hybrid cells. *Nature Genetics* **7**, 29–33.

Lee N. *et al.* (2001) A genomewide linkage-disequilibrium scan localizes the Saguenay-Lac-Saint-Jean cytochrome oxidase deficiency to 2p16. *American Journal of Human Genetics* **68**, 397–409.

Legrain P. & Selig L. (2000) Genome-wide protein interaction maps using two-hybrid systems. *FEBS Letters* **480**, 32–36.

Legrain P., Wojcik J. & Gauthier J.M. (2001) Protein–protein interaction maps: a lead towards cellular functions. *Trends in Genetics* **17**, 346–352.

Lemieux B., Aharon M. & Schena M. (1999) Overview of DNA chip technology. *Molecular Breeding* **4**, 277–289.

Leuking A. *et al.* (1999) Protein microarrays for gene expression and antibody screening. *Analytical Biochemistry* **270**, 103–111.

Lewin B. (1994) *Genes* V. Cell Press, Cambridge, MA/Oxford University Press, New York.

Lewis S., Ashburner M. & Reese M.G. (2000) Annotating eukaryote genomes. *Current Opinion in Structural Biology* **10**, 349–354.

Li Y., Mitaxov V. & Waksman G. (1999) Structure-based design of *Taq* DNA polymerases with improved properties of dideoxynucleotide incorporation. *Proceedings of the National Academy of Sciences, USA* **96**, 9491–9496.

Liao G., Rehm E.J. & Rubin G.M. (2000) Insertion site preferences of the P transposable element in *Drosophila melanogaster*. *Proceedings of the National Academy of Sciences, USA* **97**, 3347–3351.

Lim M. *et al.* (1997) A luminescent europium complex for the sensitive detection of proteins and nucleic acids immobilized on membrane supports. *Analytical Biochemistry* **245**, 184–195.

Limbach P.A., Crain P.F. & McCloskey J.A. (1995) Characterization of oligonucleotides and nucleic acids by mass spectrometry. *Current Opinion in Biotechnology* **6**, 96–102.

Lin J. *et al.* (1999a) Whole-genome shotgun optical mapping of *Deinococcus radiodurans*. *Science* **285**, 1558–1562.

Lin L.F., Posfai J., Roberts R.J. & Kong H. (2001) Comparative genomics of the restriction-modification systems in *Helicobacter pylori*. *Proceedings of the National Academy of Sciences, USA* **98**, 2740–2745.

Lin X. *et al.* (1999b) Sequence analysis of chromosome 2 of the plant *Arabidopsis thaliana*. *Nature* **402**, 761–768.

Lin Y., Ahn S., Murali N., Brey W., Bowers C.R. & Warren W.S. (2000) High-resolution, 1 GHz NMR in unstable magnetic fields. *Physical Review Letters* **85**, 3732–3735.

Lindsey K., Wei W., Clarke M.C., McArdle H.F., Rooke L.M. & Topping J.F. (1993) Tagging genomic sequences that direct transgene expression by activation of a promoter trap in plants. *Transgenic Research* **2**, 33–47.

Ling L.L., Ma N.S-F., Smith D.R., Miller D.D. & Moir D.T. (1993) Reduced occurrence of chimeric YACs in recombination-deficient hosts. *Nucleic Acids Research* **21**, 6045–6046.

Linial M. & Yona G. (2000) Methodologies for target selection in structural genomics. *Progress in Biophysics and Molecular Biology* **73**, 297–320.

Link A.J. *et al.* (1999) Direct analysis of protein complexes using mass spectrometry. *Nature Biotechnology* **17**, 676–682.

Lipshutz R.J., Fodor S.P., Gingeras T.R. & Lockhart D.J. (1999) High density synthetic oligonucleotide arrays. *Nature Genetics* **21**, 20–24.

Lisitsyn N., Lisitsyn N. & Wigler M. (1993) Cloning the difference between two complex genomes. *Science* **259**, 946–951.

Little P. (2001) The end of all human DNA maps. *Nature Genetics* **27**, 229–230.

Liu L.X. *et al.* (1999b) High-throughput isolation of *Caenorhabditis elegans* deletion mutants. *Genome Research* **9**, 859–867.

Liu S-L., Hessel A. & Sanderson K.E. (1993) Genomic mapping with I-*Cen*I an intron-encoded endonuclease specific for genes for ribosomal RNA, in *Salmonella* spp., *Escherichia coli* and other bacteria. *Proceedings of the National Academy of Sciences, USA* **90**, 6874–6878.

Liu Y-G. *et al.* (1999a) Complementation of plant mutants with large genomic DNA fragments by a transformation-competent artificial chromosome vector accelerates positional cloning. *Proceedings of the National Academy of Sciences, USA* **96**, 6535–6540.

Livak K.J., Marmaro J. & Todd J.A. (1995) Towards fully automated genome-wide polymorphism screening. *Nature Genetics* **9**, 341–342.

Livesey F.J., Furukawa T., Steffen M.A., Church G.M. & Cepko C.L. (2000) Microarray analysis of the transcriptional network controlled by the photoreceptor homeobox gene *Crx*. *Current Biology* **10**, 301–310.

Lockhart D.J. (1998) Mutant yeast on drugs. *Nature Medicine* **4**, 1235–1236.

Lockhart D.J. & Winzeler E.A. (2000) Genomics, gene expression and DNA arrays. *Nature* **405**, 827–836.

Lockhart D.J. *et al.* (1996) Expression monitoring by hybridization to high-density oligonucleotide arrays. *Nature Biotechnology* **14**, 1675–1680.

Louie A.Y. *et al.* (2000) *In vivo* visualization of gene expression using magnetic resonance imaging. *Nature Biotechnology* **18**, 321–325.

Lovett M. (1994) Fishing for complements: finding genes by direct selection. *Trends in Genetics* **10**, 352–357.

Lovett M., Kere J. & Hinton L.M. (1991) Direct selection: a method for the isolation of cDNAs encoded by large genomic regions. *Proceedings of the National Academy of Sciences, USA* **88**, 9628–9633.

Ludecke H.J., Senger G., Claussen U. & Horsthemke B. (1989) Cloning defined regions of the human genome by microdissection of banded chromosomes and enzymatic amplification. *Nature* **338**, 348–350.

Lukowitz W., Gillmor C.S. & Scheible W.R. (2000) Positional cloning in *Arabidopsis*. Why it feels good to have a genome initiative working for you. *Plant Physiology* **123**, 795–805.

Lund A.A., Blum P.H., Bhattramakki D. & Elthon T.E. (1998) Heat stress response of maize mitochondria. *Plant Physiology* **116**, 1097–1010.

Luning Prak E.T. & Kazazian H.H. (2000) Mobile elements and the human genome. *Nature Review Genetics* **1**, 134–144.

Lyamichev V. *et al.* (1999) Polymorphism identification and quantitative detection of genomic DNA by invasive cleavage of oligonucleotide probes. *Nature Biotechnology* **17**, 292–296.

Lyamichev V.I. *et al.* (2000) Experimental and theoretical analysis of the invasive signal amplification reaction. *Biochemistry* **39**, 9523–9532.

Lysov Y.P., Khorlin A.A., Khrapko K.R., Shick V.V., Florentiev V.L. & Mirzabekov A.D. (1988) DNA sequencing by hybridization with oligonucleotides: a novel method. *Proceedings of the National Academy of Sciences, USA* **303**, 1508–1511.

MacBeath G. & Schreiber S.L. (2000) Printing proteins as microarrays for high-throughput function determination. *Science* **289**, 1760–1763.

MacGregor P.F., Abate C. & Curran T. (1990) Direct cloning of leucine zipper proteins: Jun binds cooperatively to CRE with CRE-BP1. *Oncogene* **5**, 451–458.

Mackereth C.D., Arrowsmith C.H., Edwards A.M. & McIntosh L.P. (2000) Zinc-bundle structure of the essential RNA polymerase subunit RPB10 from *Methanobacterium thermoautotrophicum*. *Proceedings of the National Academy of Sciences, USA* **97**, 6316–6321.

Maeda I., Kohara Y., Yamamoto M. & Sugimoto A. (2001) Large-scale analysis of gene function in *Caenorhabditis elegans* by high-throughput RNAi. *Current Biology* **11**, 171–176.

Mahajan N.P., Linder K., Berry G., Gordon G.W., Heim R. & Herman B. (1998) *Bcl-2* and *bax* interactions in mitochondria probed with green fluorescent protein and fluorescence resonance energy transfer. *Nature Biotechnology* **16**, 547–552.

Maier E. *et al.* (1992) Complete coverage of the *Schizosaccharomyces pombe* genome in yeast artificial chromosomes. *Nature Genetics* **1**, 273–277.

Maier R.M., Neckermann K., Igloi G.L. & Kossel H. (1995) Complete sequence of the maize chloroplast genome: gene content, hotspots of divergence and fine tuning of genetic information by transcript editing. *Journal of Molecular Biology* **251**, 614–628.

Makalowski W., Zhang J. & Boguski M.S. (1996) Comparative analysis of 1196 orthologous mouse and human full-length mRNA and protein sequences. *Genome Research* **6**, 846–857.

Makarova K.S. *et al.* (2001) Genome of the extremely radiation-resistant bacterium *Deinococcus radiodurans* viewed from the perspective of comparative genomics. *Microbiology and Molecular Biology Reviews* **65**, 44–79.

Maleck K. *et al.* (2000) The transcriptome of *Arabidopsis thaliana* during systemic acquired resistance. *Nature Genetics* **26**, 403–410.

Malik V.S. & Saroha M.K. (1999) Marker gene controversy in transgenic plants. *Journal of Plant Biochemistry and Biotechnology* **8**, 1–13.

Malmqvist M. & Karlsson R. (1997) Biomolecular interaction analysis: affinity biosensor technologies for functional analysis of proteins. *Current Opinion in Chemical Biology* **1**, 378–383.

Mann K.L. & Huxley C. (2000) Investigation of *Schizosaccharomyces pombe* as a cloning host for human telomere and alphoid DNA. *Gene* **241**, 275–285.

Mann M. & Pandey A. (2001) Use of mass spectrometry-derived data to annotate nucleotide and protein sequence databases. *Trends in Biochemical Science* **26**, 54–61.

Mann M., Hojrup P. & Roepstorff P. (1993) Use of mass spectrometric molecular weight information to identify protein in sequence databases. *Biological Mass Spectrometry* **22**, 338–345.

Mar Alba M., Santibanez-Koref M.F. & Hancock J.M. (1999) Amino acid reiterations in yeast are overrepresented in particular classes of proteins and show evidence of a slippage-like mutational process. *Journal of Molecular Evolution* **49**, 789–797.

Marahrens Y. & Stillman B. (1992) A yeast chromosomal origin of DNA replication defined by multiple functional elements. *Science* **255**, 817–823.

Marcotte E.M., Pellegrini M., Ng H-L., Rice D.W., Yeates T.O. & Eisenberg D. (1999a) Detecting protein function and protein–protein interactions from genome sequences. *Science* **285**, 751–753.

Marcotte E.M., Pellegrini M., Thompson M.J., Yeates T.O. & Eisenberg D. (1999b) A combined algorithm for genome-wide prediction of protein function. *Nature* **402**, 83–86.

Marguiles E.H., Kardia S.L.R. & Innis J.W. (2001) A comparative molecular analysis of developing mouse forelimbs and hindlimbs using Serial Analysis of Gene Expression (SAGE). *Genome Research* **11**, 1686–1698.

Marienfeld J., Unseld M. & Brennicke A. (1999) The mitochondrial genome of *Arabidopsis* is composed of both native and immigrant information. *Trends in Plant Science* **4**, 495–502.

Marmur J., Rownd R. & Schildkraut C.L. (1963) Denaturation and renaturation of deoxyribonucleic acid. *Progress in Nucleic Acids Research* **1**, 231–300.

Marra M.A. *et al.* (1997) High throughput fingerprint analysis of large-insert clones. *Genome Research* **7**, 1072–1084.

Marshall E. & Pennisi E. (1998) Hubris and the human genome. *Science* **280**, 994–995.

Martin A.C. *et al.* (1998) Protein folds and functions. *Structure* **6**, 875–884.

Martinez F.D. *et al.* (1997) Association between genetic polymorphisms of the β_2-adrenoreceptor and response to albuterol in children with and without a history of wheezing. *Journal of Clinical Investigations* **100**, 3184–3188.

Martinez-Abarca F. & Toro N. (2000) Group II introns in the bacterial world. *Molecular Microbiology* **38**, 917–926.

Marton M.J. (1998) Drug target validation and identification of secondary drug target effects using DNA microarrays. *Nature Medicine* **4**, 1293–1301.

Martzen M.R. *et al.* (1999) A biochemical genomics approach for identifying genes by the activity of their products. *Science* **286**, 1153–1155.

Marx J. (2000) Medicine: DNA arrays reveal cancer in its many forms. *Science* **289**, 1670–1672.

Masood E. (1999) A consortium plans free SNP map of human genome. *Nature* **398**, 545–546.

Maxam A. & Gilbert W. (1977) A new method for sequencing DNA. *Proceedings of the National Academy of Sciences, USA* **74**, 560–564.

McClelland M., Jones R., Patel Y. & Nelson M. (1987) Restriction endonucleases for pulsed field mapping of bacterial genomes. *Nucleic Acids Research* **15**, 5985–6005.

McClelland M. *et al.* (2001) Complete genome sequence of *Salmonella enterica* serovar Typhimurium LT2. *Nature* **413**, 852–856.

McCraith S., Hotzam T., Moss B. & Fields S. (2000) Genome-wide analysis of vaccinia virus protein–protein interactions. *Proceedings of the National Academy of Sciences, USA* **97**, 4879–4884.

McCreath K.J., Howcroft J., Campbell K.H.S., Colman A., Schnieke A.E. & Kind A.J. (2000) Production of gene-targeted sheep by nuclear transfer from cultured somatic cells. *Nature* **405**, 1066–1069.

McInnes L.A. *et al.* (1996) A complete genome screen for genes predisposing to severe bipolar disorder in two Costa Rican pedigrees. *Proceedings of the National Academy of Sciences, USA* **93**, 13060–13065.

McKinney E.C. *et al.* (1995) Sequence based identification of T-DNA insertion mutations in *Arabidopsis*: actin mutants *act2-1* and *act4-1*. *Plant Journal* **8**, 613–622.

McKusick V.A. (1997) Genomics: structural and functional studies of genomics. *Genomics* **45**, 244–249.

Medek A., Olejniczak E.T., Meadows R.P. & Fesik S.W. (2000) An approach for high-throughput structure determination of proteins by NMR spectroscopy. *Journal of Biomolecular NMR* **18**, 229–238.

Meissner R., Chague V., Zhu Q., Emmanuel E., Elkind Y. & Levy A. (2000) A high throughput system for transposon tagging and promoter trapping in tomato. *Plant Journal* **22**, 265–274.

Meldrum D. (2000a) Automation for genomics. I. Preparation for sequencing. *Genome Research* **10**, 1081–1092.

Meldrum D. (2000b) Automation for genomics. II. Sequencers, microarrays, and future trends. *Genome Research* **10**, 1288–1303.

Mendoza L.G., McQuary P., Mongan A., Gangadharan R., Brignac S. & Eggers M. (1999) High-throughput microarray-based enzyme-linked immunosorbent assay (ELISA). *Biotechniques* **27**, 778–788.

Mewes H.W. *et al.* (2000) MIPS: a database for genomes and protein sequences. *Nucleic Acids Research* **28**, 37–40.

Miki R. *et al.* (2001) Delineating developmental and metabolic pathways *in vivo* by expression profiling using the RIKEN set of 18 816 full-length enriched mouse cDNA arrays. *Proceedings of the National Academy of Sciences, USA* **98**, 2199–2204.

Miki Y. *et al.* (1992) Disruption of the APC gene by a retro-transposal insertion of L1 sequence in a colon cancer. *Cancer Research* **52**, 643–645.

Millen R.S. *et al.* (2001) Many parallel losses of *infA* from chloroplast DNA during angiosperm evolution with multiple independent transfers to the nucleus. *Plant Cell* **13**, 645–658.

Mintz P.J., Patterson S.D., Neuwald A.F., Spahr C.S. & Spector D.L. (1999) Purification and biochemical characterization of interchromatin granule clusters. *EMBO Journal* **18**, 4308–4320.

Mirzabekov A. & Kolchinsky A. (2001) Emerging array-based technologies in proteomics. *Current Opinion in Chemical Biology* **6**, 70–75.

Mitas M. (1997) Trinucleotide repeats associated with human disease. *Nucleic Acids Research* **25**, 2245–2253.

Mittl P.R.E. & Grutter M.G. (2001) Structural genomics: opportunities and challenges. *Current Opinion in Chemical Biology* **5**, 402–408.

Mizukami T. *et al.* (1993) A 13 kb resolution cosmid map of the 14 Mb fission yeast genome by non random sequence-tagged site mapping. *Cell* **73**, 121–132.

Monaco A.P. & Larin Z. (1994) YACs, BACs, and MACs: artificial chromosomes as research tools. *Trends in Biotechnology* **12**, 280–286.

Monaco A.P., Neve R.L., Colletti-Feener C., Bertelson C.J., Kurnit D.M. & Kunkel L.M. (1986) Isolation of candidate cDNAs for portions of the Duchenne muscular dystrophy gene. *Nature* **323**, 646–650.

Monckton D.G. & Jeffreys A.J. (1993) DNA profiling. *Current Opinion in Biotechnology* **4**, 660–664.

Moody M.D., Van Arsdell S.W., Murphy K.P., Orencole S.F. & Burns C. (2001) Array-based ELISAs for high-throughput analysis of human cytokines. *Biotechniques* **31**, 186–194.

Mooney J.F., Hunt A.J., McIntosh J.R., Liberko C.A., Walba D.M. & Rogers C.T. (1996) Patterning of functional antibodies and other proteins by photolithography of silane monolayers. *Proceedings of the National Academy of Sciences, USA* **93**, 12287–12291.

Moons A., Gielen J., Vandekerckhove J., Van der Straeten D., Gheysen G. & Van Montagu M. (1997) An abscisic acid and salt stress responsive rice cDNA from a novel plant gene family. *Planta* **202**, 443–454.

Moore G., Devos K.M., Wang Z. & Gale M.D. (1995) Grasses, line up and form a circle. *Current Biology* **5**, 737–739.

Moore M.J. (1996) When the junk isn't junk. *Nature* **379**, 402–403.

Moran J.V., DeBerardinis R.J. & Kazazian H. (1999) Exon shuffling by L1 retrotransposition. *Science* **283**, 1530–1534.

Moult J. & Melamud E. (2000) From fold to function. *Current Opinion in Structural Biology* **10**, 384–389.

Moult J., Hubbard T., Fidelis K. & Pedersen J.T. (1999) Critical assessment of methods of protein structure prediction (CASP): round III. *Proteins* **S3**, 2–6.

Mozo T. *et al.* (1999) A complete BAC-based physical map of the *Arabidopsis thaliana* genome. *Nature Genetics* **22**, 265–270.

Mueller U. *et al.* (2001) Development of a technology for automation and miniaturisation of protein crystallisation. *Journal of Biotechnology* **85**, 7–14.

Muller U. (1999) Ten years of gene targeting: targeted mouse mutants, from vector design to phenotype analysis. *Mechanisms of Development* **82**, 3–21.

Mullins M.C., Hammerschmidt M., Haffter P. & Nusslein-Volhard C. (1994) Large-scale mutagenesis in the zebrafish: in search of genes controlling development in a vertebrate. *Current Biology* **4**, 189–201.

Murray A.W. & Szostak J.W. (1983) Construction of artificial chromosomes in yeast. *Nature* **305**, 189–193.

Murzin A. (2001) Progress in protein structure prediction. *Nature Structural Biology* **8**, 110–112.

Murzin A., Brenner S., Hubbard T. & Chothia C. (1995) SCOP – A structural classification of proteins database for the investigation of sequences and structures. *Journal of Molecular Biology* **247**, 536–540.

Mushegian A. (1999) The minimal genome concept. *Current Opinion in Genetics and Development* **9**, 709–714.

Mushegian A.R. & Koonin E.V. (1996) A minimal gene set for cellular life derived by comparison of complete bacterial genomes. *Proceedings of the National Academy of Sciences, USA* **93**, 10268–10273.

Myers E.W. *et al.* (2000) A whole-genome assembly of *Drosophila*. *Science* **287**, 2196–2204.

Neidhardt L. *et al.* (2000) Large-scale screen for genes controlling mammalian embryogenesis, using high-throughput gene expression analysis in mouse embryos. *Mechanisms of Development* **98**, 77–93.

Neilson L. *et al.* (2000) Molecular phenotype of the human oocyte by PCR–SAGE. *Genomics* **63**, 13–24.

Nilsson M., Krejci K., Koch J., Kwiatkowski M., Gustavsson P. & Landegren U. (1997) Padlock probes reveal single-nucleotide differences, parent of origin and *in situ* distribution of centromeric sequences in human chromosomes 13 and 21. *Nature Genetics* **16**, 252–255.

Nekrutenko A. & Li W-H. (2001) Transposable elements are found in a large number of human protein-coding genes. *Trends in Genetics* **17**, 619–621.

Nelson K.E. *et al.* (1999) Genome sequencing of *Thermotoga maritima*: evidence for lateral gene transfer between bacteria and archaea. *Nature* **399**, 323–329.

Nelson, R.W. *et al.* (2000) Biosensor chip mass spectrometry: a chip-based proteomics approach. *Electrophoresis* **21**, 1155–1163.

Neubauer G., Gottschalk A., Fabrizio P., Séraphin B., Lührmann R. & Mann M. (1997) Identification of the proteins of the yeast U1 small nuclear ribonucleoprotein complex by mass spectrometry. *Proceedings of the National Academy of Sciences, USA* **94**, 385–390.

Neubauer G. *et al.* (1998) Mass spectrometry and EST-database searching allows characterization of the multi-protein spliceosome complex. *Nature Genetics* **20**, 46–50.

Nguyen C. *et al.* (1995) Differential gene expression in the murine thymus assayed by quantitative hybridisation of arrayed cDNA clones. *Genomics* **29**, 207–216.

Nierman W.C. *et al.* (2001) Complete genome sequence of *Caulobacter crescentus*. *Proceedings of the National Academy of Sciences, USA* **98**, 4136–4141.

Nolan P.M. *et al.* (2000) A systematic, genome-wide, phenotype-driven mutagenesis programme for gene function studies in the mouse. *Nature Genetics* **25**, 440–443.

Noordewier M.O. & Warren P.V. (2001) Gene expression microarrays and the integration of biological knowledge. *Trends Biotechnology* **19**, 412–415.

Nordstrom T., Nourizad K., Ronaghi M. & Nyren P. (2000a) Methods enabling pyrosequencing on double-stranded DNA. *Analytical Biochemistry* **282**, 186–193.

Nordstrom T. *et al.* (2000b) Direct analysis of single-nucleotide polymorphism on double-stranded DNA by pyrosequencing. *Biotechnology and Applied Biochemistry* **31**, 107–112.

Norin M. & Sundstrom M. (2002) Structural proteomics: developments in structure-to-function predictions. *Trends in Biotechnology* **20**, 79–84.

Nussbaum R.L., McInnes R.R. & Willard H.F. (2001) *Genetics in Medicine*, 6th edn. W.B. Saunders, Philadelphia.

Nusslein-Volhard C., Wieschaus E. & Kluding H. (1984) Mutations affecting the pattern of the larval cuticle in *Drosophila melanogaster*. I. Zygotic loci on the second chromosome. *Wilheim Roux Archives of Developmental Biology* **193**, 267–282.

O'Farrell P.H. (1975) High-resolution two-dimensional electrophoresis of proteins. *Journal of Biological Chemistry* **250**, 4007–4021.

O'Farrell P.H. & Ivarie R.D. (1979) The glucocorticoid domain of response: measurement of pleiotropic cellular responses by two-dimensional gel electrophoresis. *Monographs in Endocrinology* **12**, 189–201.

Ohyama K. *et al.* (1986) Chloroplast gene organization deduced from complete squence of liverwort *Marchantia polymorpha* chloroplast DNA. *Nature* **322**, 572–574.

O'Kane C.J. & Gehring W.J. (1987) Detection *in situ* of genetic regulatory elements in *Drosophila*. *Proceedings of the National Academy of Sciences, USA* **84**, 9123–9127.

O'Kane C.J. & Moffat K.G. (1992) Selective cell ablation and genetic surgery. *Current Opinion in Genetic Development* **2**, 602–607.

Okamoto T., Suzuki T. & Yamamoto N. (2000) Microarray fabrication with covalent attachment of DNA using bubble jet technology. *Nature Biotechnology* **18**, 438–441.

Okubo K. *et al.* (1992) Large scale cDNA sequencing for analysis of quantitative and qualitative aspects of gene expression. *Nature Genetics* **2**, 173–179.

Old R.W. & Primrose S.B. (1994) *Principles of Gene Manipulation*, 5th edn. Blackwell Scientific Publications, Oxford.

Oliver S.G. (1996a) From DNA sequence to biological function. *Nature* **379**, 597–600.

Olson M.V. *et al.* (1986) A random-clone strategy for restriction mapping in yeast. *Proceedings of the National Academy of Sciences, USA* **83**, 7826–7830.

Orengo C., Michie A., Jones S., Jones D., Swindells M. & Thornton J. (1997) CATH – A hierarchic classification of protein domain structures. *Structure* **5**, 1093–1108.

Orengo C.A., Jones D.T. & Thornton J.M. (1994) Protein superfamilies and domain superfolds. *Nature* **372**, 631–634.

Osbourne B.I. & Baker B. (1995) Movers and shakers: maize transposons as tools for analyzing other plant genomes. *Current Opinion in Cell Biology* **7**, 406–413.

Osoegawa K. *et al.* (2000) Bacterial artificial chromosome libraries for mouse sequencing and functional analysis. *Genome Research* **10**, 116–128.

Osoegawa K. *et al.* (2001) A bacterial artificial chromosome library for sequencing the complete human genome. *Genome Research* **11**, 483–496.

Overbeek R., Fonstein M., D'Souza M., Pusch G.D. & Maltsev N. (1999) The use of gene clusters to infer functional coupling. *Proceedings of the National Academy of Sciences, USA* **96**, 2896–2901.

Palmer J.D. *et al.* (2000) Dynamic evolution of plant mitochondrial genomes: mobile genes and introns and highly variable mutation rates. *Proceedings of the National Academy of Sciences* **97**, 6960–6966.

Palwletz C.P. *et al.* (2001) Reverse phase protein microarrays which capture disease progression show activation of pro-survival pathways at the cancer invasion front. *Oncogene* **20**, 1981–1989.

Pandey A. & Mann M. (2000) Proteomics to study genes and genomes. *Nature* **405**, 837–846.

Pandey A., Podtelejnikov A.V., Blagoev B., Bustelo X.R., Mann M. & Lodish H.F. (2000) Analysis of receptor signaling pathways by mass spectrometry: identification of *vav-2* as a substrate of the epidermal and platelet-derived growth factor receptors. *Proceedings of the National Academy of Sciences, USA* **97**, 179–184.

Papp T., Palagyi Z., Ferenczy L. & Vagvolgyi C. (1999) The mitochondrial genome of *Mucor piriformis*. *FEMS Microbiology Letters* **171**, 67–72.

Pappin D.J.C., Horjup P. & Bleasby A.J. (1993) Rapid identification of proteins by peptide mass fingerprinting. *Current Biology* **3**, 327–332.

Pardue M.L., Danilevskaya O.N., Lowenhaupt K., Slot F. & Traverse K.L. (1996) *Drosophila* telomeres: new views on chromosome evolution. *Trends in Genetics* **12**, 48–52.

Parimoo S., Patanjali S.R., Shukle H., Chaplin D.D. & Weisman S.M. (1991) cDNA selection: efficient PCR approach for the selection of cDNAs encoded in large chromosomal DNA fragments. *Proceedings of the National Academy of Sciences, USA* **88**, 9623–9627.

Parinov S., Sevugan M., Ye D., Yang W.C., Kumaran M. & Sundaresan V. (1999) Analysis of flanking sequences from *Dissociation* insertion lines: a database for reverse genetics in *Arabidopsis*. *Plant Cell* **11**, 2263–2270.

Park J. *et al.* (1998) Sequence comparisons using multiple sequences detect three times as many remote homologues as pairwise methods. *Journal of Molecular Biology* **284**, 1201–1210.

Parkhill J. *et al.* (2001a) Complete genome sequence of a multiple drug resistant *Salmonella enterica* serovar Typhi CT18. *Nature* **413**, 848–852.

Parkhill J. *et al.* (2001b) Genome sequence of *Yersinia pestis*, the causative agent of plague. *Nature* **413**, 523–527.

Parra I. & Windle B. (1993) High resolution visual mapping of stretched DNA by fluorescent hybridization. *Nature Genetics* **5**, 17–21.

Paterson A.H., Lander E.S., Hewitt J.D., Peterson S., Lincoln S.E. & Tanksley S.D. (1988) Resolution of quantitative traits into Mendelian factors by using a complete linkage map of restriction fragment length polymorphisms. *Nature* **335**, 721–726.

Pearson W.R. & Lipman D.J. (1988) Improved tools for biological sequence comparison. *Proceedings of the National Academy of Sciences, USA* **85**, 2444–2448.

Pease A.C., Solas D., Sullivan E.J., Cronin M.T., Holmes C.P. & Fodor S.P.A. (1994) Light-generated oligonucleotide arrays for rapid DNA sequence analysis. *Proceedings of the National Academy of Science, USA* **91**, 5022–5026.

Pellegrini M., Marcotte E.M., Thompson M.J., Eisenberg D. & Yeates T.O. (1999) Assigning protein functions by comparative genome analysis: protein phylogenetic profiles. *Proceedings of the National Academy of Sciences, USA* **96**, 4285–4288.

Pelleschi S. *et al.* (1999) *Ivr2*, a candidate gene for a QTL of vacuolar invertase activity in maize leaves: gene-specific expression under water stress. *Plant Molecular Biology* **39**, 373–380.

Pelletier J. & Sidhu S. (2001) Mapping protein–protein interactions with combinatorial biology methods. *Current Opinion in Biotechnology* **12**, 340–347.

Perna N.T. *et al.* (2001) Genome sequence of enterohaemorrhagic *Escherichia coli* O157:H7. *Nature* **409**, 529–533.

Perou C.M. *et al.* (1999) Distinctive gene expression patterns in human mammary epithelial cells and breast cancers. *Proceedings of the National Academy of Sciences, USA* **96**, 9212–9217.

Perou C.M. *et al.* (2000) Molecular portraits of human breast tumours. *Nature* **406**, 747–752.

Perucho M., Hanahan D., Lipsich L. & Wigler M. (1980) Isolation of the chicken thymidine kinase gene by plasmid rescue. *Nature* **285**, 207–210.

Peters, D.G. *et al.* (1999) Comprehensive transcript analysis in small quantities of mRNA by SAGE-lite. *Nucleic Acids Research* **27**, e39.

Pflieger S., Lefebvre V., Caranta C., Blattes A., Goffinet B. & Palloix A. (1999) Disease resistance gene analogs as candidates for QTLs involved in pepper–pathogen interactions. *Genome* **42**, 1100–1110.

Phizicky E.M. & Fields S. (1995) Protein–protein interactions: methods for detection and analysis. *Microbiol Reviews* **59**, 94–123.

Picard P., Bourgoin-Greneche M. & Zivy M. (1997) Potential of two-dimensional electrophoresis in routine identification of closely related durum wheat lines. *Electrophoresis* **18**, 174–181.

Pierce J.C., Sauer B. & Sternberg N. (1992) A positive selection vector for cloning high molecular weight DNA by the bacteriophage P1 system: improved cloning efficiency. *Proceedings of the National Academy of Sciences, USA* **89**, 2056–2060.

Pietu G. *et al.* (1996) Novel gene transcripts preferentially expressed in human muscles revealed by quantitative hybridization of a high-density cDNA array. *Genome Research* **6**, 492–503.

Pih K.Y., Jang H.J., Kang S.G., Piao H.L. & Hwang I. (1997) Isolation of molecular markers for salt stress responses in *Arabidopsis thaliana*. *Molecular Cell* **7**, 567–571.

Pinkel D., Straume T. & Gray J.W. (1986) Cytogenetic analysis using quantitative, high-sensitivity, fluorescence hybridization. *Proceedings of the National Academy of Sciences, USA* **83**, 2934–2938.

Piper M.B., Bankier A.T. & Dear P.H. (1998) A HAPPY map of *Cryptosporidium parvum*. *Genome Research* **8**, 1299–1307.

Pizza M.G. *et al.* (2000) Whole genome sequencing to identify vaccine candidates against serogroup B meningococcus. *Science* **287**, 1816–1820.

Plano G.V., Day J.B. & Ferracci F. (2001) Type III export: new uses for an old pathway. *Molecular Microbiology* **40**, 284–293.

Pleissner K.-P., Oswald H. & Wegner S. (2001) Image analysis of two-dimensional gels. In: *Proteomics: From Protein Sequence to Function*, pp. 131–150. BIOS Scientific Publishers, Oxford.

Pollack D.J. (2001) *Ureaplasma urealyticum*: an opportunity for combinatorial genomics. *Trends in Microbiology* **9**, 169–175.

Polyak K., Xia Y., Zweier J.L., Kinzler K.W. & Vogelstein B. (1997) A model for p53 induced apoptosis. *Nature* **389**, 300–304.

Ponting C.P. *et al.* (2000) Evolution of domain families. *Advances in Protein Chemistry* **54**, 185–244.

Postlethwait J.H. *et al.* (1994) A genetic linkage map for the zebrafish. *Science* **264**, 699–704.

Primig M. *et al.* (2000) The core meiotic transcriptome in budding yeasts. *Nature Genetics* **26**, 415–423.

Primrose S.B., Twyman R.M. & Old R.W. (2001) *Principles of Gene Manipulation*, 6th edn. Blackwell Science, Oxford.

Qi X. *et al.* (2001) Development of simple sequence repeat markers from bacterial artificial chromosomes without subcloning. *Biotechniques* **31**, 355–362.

Rabilloud T., Adessi C., Giraudel A. & Lunardi J. (1997) Improvement of the solubilization of proteins in two-dimensional electrophoresis with immobilized pH gradients. *Electrophoresis* **18**, 307–316.

Raczniak G., Ibba M. & Soll D. (2001) Genomics-based identification of targets in pathogenic bacteria for potential therapeutic and diagnostic use. *Toxicology* **160**, 181–189.

Rain J.C. *et al.* (2001) The protein–protein interaction map of *Helicobacter pylori*. *Nature* **409**, 211–215.

Ramachandran S. & Sundaresan V. (2001) Transposons as tools for functional genomics. *Plant Physiology and Biochemistry* **39**, 243–252.

Ramakrishna R. & Srinivasan R. (1999) Gene identification in bacterial and organellar genomes using GeneScan. *Computers and Chemistry* **23**, 165–174.

Ranade K. *et al.* (2001) High-throughput genotyping with single nucleotide polymorphisms. *Genome Research* **11**, 1262–1268.

Ray P.N. *et al.* (1985) Cloning of the breakpoint of an X;21 translocation associated with Duchenne muscular dystrophy. *Nature* **318**, 672–675.

Raychaudhuri S., Sutphin P.D., Chang J.T. & Altman R.B. (2001) Basic microarray analysis: grouping and feature reduction. *Trends in Biotechnology* **19**, 189–193.

Read T.D. *et al.* (2000) Genome sequences of *Chlamydia trachomatis* MoPN and *Chlamydia pneumoniae* AR39. *Nucleic Acids Research* **28**, 1397–1406.

Reese M. *et al.* (2000) Genome annotation assessment in *Drosophila melanogaster*. *Genome Research* **10**, 483–501.

Reich D.E. *et al.* (2001) Linkage disequilibrium in the human genome. *Nature* **411**, 199–204.

Reiter R.S., Williams J.G.K., Feldmann K.A., Rafalsta A., Tingey S.V. & Scolnick P.A. (1992) Global and local genome mapping in *Arabidopsis thaliana* by using recombinant inbred lines and random amplified polymorphic DNAs. *Proceedings of the National Academy of Sciences, USA* **89**, 1477–1481.

Renau T.E. *et al.* (1999) Inhibitors of efflux pumps in *Pseudomonas aeruginosa* potentiate the activity of the fluoroquinolone antibacterial levofloxacin. *Medical Chemistry* **42**, 4928–4931.

Rey P., Pruvot G., Becuwe N., Eymery F., Rumeau D. & Peltier G. (1998) A novel thioredoxin-like protein located in the chloroplast is induced by water deficit in *Solanum tuberosum* L. plants. *Plant Journal* **13**, 97–101.

Reymond P., Weber H., Damond M. & Farmer E.E. (2000) Differential gene expression in response to mechanical wounding and insect feeding in *Arabidopsis*. *Plant Cell* **12**, 707–719.

Rhodes S.J. & Smith R.C. (1998) Using the power of developmental biology for drug discovery. *Drug Discovery Today* **3**, 361–369.

Richard G.F., Hennequin C., Thierry A. & Dujon B. (1999) Trinucleotide repeats and other microsatellites in yeast. *Research in Microbiology* **150**, 589–602.

Richmond C.S., Glasner J.D., Mau R., Jin H. & Blattner F.R. (1999) Genome-wide expression in *Escherichia coli* K-12. *Nucleic Acid Research* **27**, 3821–3835.

Richmond T. & Somerville S. (2000) Chasing the dream: plant EST microarrays. *Current Opinion in Plant Biology* **3**, 108–116.

Riechmann J.L. *et al.* (2000) *Arabidopsis* transcription factors: genome-wide comparative analysis among eukaryotes. *Science* **290**, 2105–2109.

Riek R., Pervushin K. & Wüthrich K. (2000) TROSY and CRINEPT: NMR with large molecular and supramolecular structures in solution. *Trends in Biochemical Science* **25**, 462–468.

Riethman H.C. *et al.* (1989) Cloning human telomeric DNA fragments into *Saccharomyces cerevisiae* using a yeast-artificial-chromosome vector. *Proceedings of the National Academy of Sciences, USA* **86**, 6240–6244.

Riethman H.C. *et al.* (2001) Integration of telomere sequences with the draft human genome sequence. *Nature* **409**, 948–951.

Rigaut G., Shevchenko A., Rutz B., Wilm M., Mann M. &

Seraphin B. (1999) A generic protein purification method for protein complex characterization and proteome exploration. *Nature Biotechnology* **17**, 1030–1032.

Rijkers T., Peetz A. & Ruther U. (1994) Insertional mutagenesis in transgenic mice. *Transgenic Research* **3**, 203–215.

Rinchik E.M., Carpenter D.A. & Selby P.B. (1990) A strategy for fine-structure functional analysis of a 6- to 11-centimorgan region of mouse chromosome 7 by high-efficiency mutagenesis. *Proceedings of the National Academy of Sciences, USA* **87**, 896–900.

Rine J. (1991) Gene overexpression studies of *Saccharomyces cerevisiae*. *Methods in Enzymology* **194**, 239–251.

Ringwald M., Eppig J.T., Kadin J.A. & Richardson J.E. (2000) GXD: a Gene Expression Database for the laboratory mouse: current status and recent enhancements. *Nucleic Acids Research* **28**, 115–119.

Rioux J.D. (2001) Genetic variation in the 5q31 cytokine gene cluster confers susceptibility to Crohn disease. *Nature Genetics* **29**, 223–228.

Rioux J.D. *et al.* (2000) Genomewide search in Canadian families with inflammatory bowel disease reveals two novel susceptibility loci. *American Journal of Human Genetics* **66**, 1863–1870.

Robinson K., Gilbert W. & Church G.M. (1994) Large scale bacterial gene discovery by similarity search. *Nature Genetics* **7**, 205–214.

Rogic S., Macksworth A.K. & Oulette F.B.F. (2001) Evaluation of gene-finding programs on mammalian sequences. *Genome Research* **11**, 817–832.

Ronaghi M. (2001) Pyrosequencing sheds light on DNA sequencing. *Genome Research* **11**, 3–11.

Ronaghi M. *et al.* (1996) Real-time DNA sequencing using detection of pyrophosphate release. *Analytical Biochemistry* **242**, 84–89.

Ronaghi M., Petersson B., Uhlen M. & Nyren P. (1998a) PCR-introduced loop structure as primer in DNA sequencing. *BioTechniques* **25**, 876–884.

Ronaghi M., Uhlen M. & Nyren P. (1998b) A sequencing method based on real-time pyrophosphate. *Science* **281**, 363–365.

Rong Y.S. & Golic K.G. (2000) Gene targeting by homologous recombination in *Drosophila*. *Science* **288**, 2103–2108.

Rorth P. (1996) A modular misexpression screen in *Drosophila* detecting tissue-specific phenotypes. *Proceedings of the National Academy of Sciences, USA* **93**, 12418–12422.

Rorth P. *et al.* (1998) Systematic gain-of-function genetics in *Drosophila*. *Development* **125**, 1049–1057.

Ross D.T. *et al.* (2000) Systematic variation in gene expression patterns in human cancer cell lines. *Nature Genetics* **24**, 227–235.

Ross-Macdonald P., Sheehan A., Roeder G.S. & Snyder M. (1997) A multipurpose transposon system for analyzing protein production, localization, and function in *Saccharomyces cerevisiae*. *Proceedings of the National Academy of Sciences, USA* **94**, 190–195.

Ross-MacDonald P. *et al.* (1999) Large-scale analysis of the yeast genome by transposon tagging and gene disruption. *Nature* **402**, 413–418.

Rout M.P., Aitchison J.D., Suprapto A., Hjertaas K., Zhao Y. & Chait B.T. (2000) The yeast nuclear pore complex: composition, architecture, and transport mechanism. *Journal of Cell Biology* **148**, 635–651.

Ruan Y., Gilmore J. & Conner T. (1998) Towards *Arabidopsis* genome analysis: monitoring expression profiles of 1400 genes using cDNA microarrays. *Plant Journal* **15**, 821–833.

Rubin G.M. *et al.* (2000) Comparative genomics of the eukaryotes. *Science* **287**, 2204–2215.

Rudnicki M.A., Braun T., Hinuma S. & Jaenisch R. (1992) Inactivation of MyoD in mice leads to up-regulation of the myogenic HLH gene *Myf-5* and results in apparently normal muscle development. *Cell* **71**, 383–390.

Rudnicki M.A., Schnegelsberg P.N., Stead R.H., Braun T., Arnold H.H. & Jaenisch R. (1993) *MyoD* or *Myf-5* is required for the formation of skeletal muscle. *Cell* **75**, 1351–1359.

Ruiz M.T., Voinnet O. & Baulcombe D.C. (1998) Initiation and maintenance of virus-induced gene silencing. *Plant Cell* **10**, 937–946.

Russell R.B., Saqi M.A., Bates P.A., Sayle R.A. & Sternberg M.J. (1998) Recognition of analogous and homologous protein folds: assessment of prediction success and associated alignment accuracy using empirical substitution matrices. *Protein Engineering* **11**, 1–9.

Russell W.L., Kelly P.R., Hunsicker P.R., Bangham J.W., Maddux S.C. & Phipps E.L. (1979) Specific-locus test shows ethylnitrosourea to be the most potent mutagen in the mouse. *Proceedings of the National Academy of Sciences, USA* **76**, 5918–5922.

Rychlewski L., Zhang B. & Godzik A. (1998) Fold and function predictions for *Mycoplasma genitalium* proteins. *Folding Design* **3**, 229–238. [URL: http://cape6.scripps.edu/leszek/genome/cgi–bin/genome.pl?mp]

Rychlewski L., Zhang B.H. & Godzik A. (1999) Functional insights from structural predictions: analysis of the *Escherichia coli* genome. *Protein Science* **8**, 614–624.

Ryo A. *et al.* (1999) Serial analysis of gene expression in HIV-1–infected T cell lines. *FEBS Letters* **462**, 182–186.

Saccone S. *et al.* (1999) Identification of the gene-richest bands in human prometaphase chromosomes. *Chromosome Research* **7**, 379–386.

Sachidanandam R. *et al.* (2001) A map of the human genome sequence variation containing 1.42 million single nucleotide polymorphisms. *Nature* **409**, 928–933.

Saitoh Y. & Laemmli U.K. (1994) Metaphase chromosome structure: bands arise from a differential folding path of the highly AT-rich scaffold. *Cell* **76**, 609–622.

Salzberg S.L., Pertea M., Delcher A.L., Gardner M.J. & Tettelin H. (1999) Interpolated Markov models for eukaryotic gene-finding. *Genomics* **59**, 24–31.

Salzberg S.L. *et al.* (2001) Microbial genes in the human genome: lateral transfer or gene loss? *Science* **292**, 1903–1906.

Sambrook J. & Russell D.W. (2001) *Molecular Cloning: A Laboratory Manual*, 3rd edn. CSHL Press, Cold Spring Harbor, NY.

Sanchez C. *et al.* (1999) Grasping at molecular interactions and genetic networks in *Drosophila melanogaster* using

FlyNets, an Internet database. *Nucleic Acids Research* **27**, 89–94.

Sanchez J.C. *et al.* (1997) Improved and simplified in-gel sample application using reswelling of dry immobilized pH gradients. *Electrophoresis* **18**, 324–327.

Sanchez R. & Sali A. (1998) Large-scale protein structure modeling of the *Saccharomyces cerevisiae* genome. *Proceedings of the National Academy of Sciences, USA* **95**, 13597–13602.

Sanger F. *et al.* (1977a) Nucleotide sequence of bacteriophage ΦX174DNA. *Nature* **265**, 687–695.

Sanger F., Nicklen S. & Coulson A.R. (1977b) DNA sequencing with chain terminating inhibitors. *Proceedings of the National Academy of Sciences, USA* **74**, 5463–5467.

Sasaki T. & Burr T. (2000) International Rice Genome Sequencing Project: the effort to completely sequence the rice genome. *Current Opinion in Plant Biology* **3**, 138–141.

Sassaman D.M. *et al.* (1997) Many human L1 elements are capable of retrotransposition. *Nature Genetics* **16**, 37–43.

Sato S. *et al.* (1999) Complete structure of the chloroplast genome of *Arabidopsis thaliana*. *DNA Research* **6**, 283–290.

Schaffer R., Landgraf J., Perez-Amador P. & Wisman E. (2000) Monitoring genome-wide expression in plants. *Current Opinion in Biotechnology* **11**, 162–167.

Scheele G.A. (1975) Two-dimensional gel analysis of soluble proteins: characterization of guinea pig exocrine pancreatic proteins. *Journal of Biological Chemistry* **250**, 5375–5385.

Schena M., Shalon D., Davis R.W. & Brown P.O. (1995) Quantitative monitoring of gene expression patterns with a complementary DNA microarray. *Science* **270**, 467–470.

Schena M. *et al.* (1996) Parallel human genome analysis: microarray-based expression monitoring of 1000 genes. *Proceedings of the National Academy of Sciences, USA* **93**, 10614–10619.

Schenk P.M. *et al.* (2000) Coordinated plant defense responses in *Arabidopsis* revealed by microarray analysis. *Proceedings of the National Academy of Sciences, USA* **97**, 11655–11660.

Scherf U. *et al.* (2000) A gene expression database for the molecular pharmacology of cancer. *Nature Genetics* **24**, 236–244.

Schisler N.J. & Palmer J.D. (2000) The IDB and IEDB: intron sequence and evolution databases. *Nucleic Acids Research* **28**, 181–184.

Schmid C.W. (1998) Does SINE evolution preclude Alu function? *Nucleic Acids Research* **26**, 4541–4550.

Schmitz-Linneweber C. *et al.* (2001) The plastid chromosome of spinach (*Spinacia oleracea*): complete nucleotide sequence and gene organization. *Plant Molecular Biology* **45**, 307–315.

Schmucker D. *et al.* (2000) *Drosophila* Dscam is an axon guidance receptor exhibiting extraordinary molecular diversity. *Cell* **101**, 671–684.

Schueler M.G. *et al.* (2001) Genomic and genetic definition of a functional human centromere. *Science* **294**, 109–115.

Schumacher A., Faust C. & Magnuson T. (1996) Positional cloning of a global regulator of anterior-posterior patterning in mice. *Nature* **383**, 250–253.

Schwartz D.C. & Cantor C.R. (1984) Separation of yeast chromosome-sized DNAs by pulsed field gradient gel electrophoresis. *Cell* **37**, 67–75.

Schwartz D.C., Li X., Hernandez L.I., Ramnarian S.P., Huff E.J. & Wang Y-K. (1993) Ordered restriction maps of *Sacharomyces cerevisiae* chromosomes constructed by optical mapping. *Science* **202**, 110–114.

Schwartz I. (2000) Microbial genomics: from sequence to function. *Emerging Infectious Diseases* **6**, 493–495.

Schweitzer B. & Kingsmore S.F. (2002) Measuring proteins on microarrays. *Current Opinion in Biotechnology* **13**, 14–19.

Schweitzer B. *et al.* (2000) Immunoassays with rolling circle DNA amplification: a versatile platform for ultrasensitive antigen detection. *Proceedings of the National Academy of Sciences, USA* **97**, 10113–10119.

Schwikowski B., Uetz P. & Fields S. (2000) A network of protein–protein interactions in yeast. *Nature Biotechnology* **18**, 1257–1261.

Seki M., Ito T., Shibata D. & Shinozaki K. (1999) Regional insertional mutagenesis of specific genes on the CIC5F11/CIC2B9 locus of *Arabidopsis thaliana* chromosome 5 using the Ac/Ds transposon in combination with the cDNA scanning method. *Plant Cell Physiology* **40**, 624–639.

Seki M. *et al.* (2001) Monitoring the expression pattern of 1300 *Arabidopsis* genes under drought and cold stresses by using a full-length cDNA microarray. *Plant Cell* **13**, 61–72.

Seoighe C. *et al.* (2000) Prevalence of small inversions in yeast gene order evolution. *Proceedings of the National Academy of Sciences, USA* **97**, 14433–14437.

Shabalina S.A. *et al.* (2001) Selective constraint in intergenic regions of human and mouse genomes. *Trends in Genetics* **17**, 373–376.

Shalon D., Smith S.J. & Brown P.O. (1996) A DNA microarray system for analysing complex DNA samples using two colour fluorescent prone hybridisation. *Genome Research* **6**, 639–645.

Shapiro L. & Harris T. (2000) Finding function through structural genomics. *Current Opinion in Biotechnology* **11**, 31–35.

Shatkay H., Edwards S., Wilbur W.J. & Boguski M. (2000) Genes, themes and microarrays: using information retrieval for large-scale gene analysis. *ISMB* **8**, 317–328.

Shedlovsky A., King T.R. & Dove W.F. (1988) Saturation germ line mutagenesis of the murine *t* region including a lethal allele at the *quaking* locus. *Proceedings of the National Academy of Sciences, USA* **85**, 180–184.

Shevchenko A., Loboda A., Shevchenko A., Ens W. & Standing K.G. (2000) MALDI quadrupole time-of-flight mass spectrometry: a powerful tool for proteomic research. *Analytical Chemistry* **72**, 2132–2141.

Shi H. *et al.* (1999) Template-imprinted nanosaturated surfaces for protein recognition. *Nature* **398**, 593–597.

Shibata K. *et al.* (2000) RIKEN integrated sequence analysis (RISA) system: 384-format sequencing pipeline with 384 multicapillary sequencer. *Genome Research* **10**, 1757–1771.

Shih H.M., Goldman O.S., DeMaggio A.J., Hollenberg S.M., Goodman R.H. & Joekstra M.F. (1996) A positive genetic selection for disrupting protein–protein interactions: identification of CREB mutations that prevent association with the coactivator CBP. *Proceedings of the National Academy of Sciences, USA* **93**, 13896–13901.

Shinozaki K. *et al.* (1986) The complete nucleotide sequence of the tobacco chloroplast genome: its gene organization and expression. *EMBO Journal* **5**, 2043–2049.

Shirasu K., Lahaye T., Tan M-W., Zhou F., Azevedo C. & Schulze-Lefert P. (1999) A novel class of eukaryotic zinc-binding proteins is required for disease resistance signaling in barley and development in *C. elegans. Cell* **99**, 355–366.

Shizuya H. *et al.* (1992) Cloning and stable maintenance of 300-kilobase-pair fragments of human DNA in *Escherichia coli* using and F-factor-based vector. *Proceedings of the National Academy of Sciences, USA* **89**, 8794–8797.

Shoemaker D.D., Lashkari D.A., Morris D., Mittman M. & Davis R.W. (1996) Quantitative phenotypic analysis of yeast deletion mutants using a highly parallel molecular barcoding strategy. *Nature Genetics* **14**, 450–456.

Shore D. (2001) Telomeric chromatin: replicating and wrapping up chromosome ends. *Current Opinion in Genetics and Development* **11**, 189–198.

Sidhu S.S., Lowman H.B., Cunningham B.C. & Wells J.A. (2000) Phage display for selection of novel binding peptides. *Methods in Enzymology* **328**, 333–363.

Simpson A.J.G. *et al.* (2000) The genome sequence of the plant pathogen *Xylella fastidiosa. Nature* **406**, 151–157.

Singer M. & Berg P. (1990) *Genes and Genomes.* Blackwell Scientific Publications, Oxford.

Singh-Gasson S. *et al.* (1999) Maskless fabrication of light-directed oligonucleotide microarrays using a digital micromirror array. *Nature Biotechnology* **17**, 974–978.

Siomi M.C. *et al.* (1998) Functional conservation of the transportin nuclear import pathway in divergent organisms. *Molecular Cell Biology* **18**, 4141–4148.

Skarnes W.C., Auerbach B.A. & Joyner A.L. (1992) A gene trap approach in mouse embryonic stem cells: the *lacZ* reporter is activated by splicing, reflects endogenous gene expression, and is mutagenic in mice. *Genes and Development* **6**, 903–918.

Skolnick J.F. & Kolinski J.S.A. (2000) Structural genomics and its importance for gene function analysis. *Nature Biotechnology* **18**, 283–287.

Smit A.F.A. (1999) Interspersed repeats and other mementos of transposable elements in mammalian genomes. *Current Opinion in Genetics and Development* **9**, 657–663.

Smith C.L., Econome J.G., Schutt A., Klco S. & Cantor C.R. (1987) A physical map of the *Escherichia coli* K12 genome. *Science* **236**, 1448–1453.

Smith G.P. (1985) Filamentous fusion phage: novel expression vectors that display cloned antigens on the virion surface. *Science* **228**, 1315–1317.

Smith N., Kilpatrick J.B. & Whitelam G.C. (2001) Superfluous transgene integration in plants. *Critical Reviews in Plant Science* **20**, 215–249.

Smith V., Botstein D. & Brown P.O. (1995) Genetic footprinting: a genomic strategy for determining a gene's function given its sequence. *Proceedings of the National Academy of Sciences, USA* **92**, 6479–6483.

Smith V., Chou K.N., Lashkari D., Botstein D. & Brown P.O.

(1996) Functional analysis of the genes of yeast chromosome V by genetic footprinting. *Science* **274**, 2069–2074.

Soderlund C. *et al.* (2000) Contigs built with fingerprints, markers and FPC V4.7. *Genome Research* **10**, 1772–1787.

Somerville C. & Somerville S. (1999) Plant functional genomics. *Science* **285**, 380–383.

Song W-Y. *et al.* (1995) A receptor kinase-like protein encoded by the rice disease resistance gene, Xa21. *Science* **270**, 1804–1806.

Southern E. (1996) DNA chips: analysing sequence by hybridization to oligonucleotides on a large scale. *Trends in Genetics* **12**, 110–115.

Southern E.M. (1988) Analyzing polynucleotide sequences. *International Patent Application PCT GB 89/01114.*

Southern E.M., Maskos U. & Elder J.K. (1992) Analyzing and comparing nucleic acid sequences by hybridization to arrays of oligonucleotides: evaluation using experimental models. *Genomics* **13**, 1008–1017.

Spellman P.T. *et al.* (1998) Comprehensive identification of cell cycle-regulated gene of the yeast *Saccharomyces cerevisiae* by microarray hybridization. *Molecular Biology of the Cell* **9**, 3273–3297. [http://cellcycle–www.stanford.edu]

Speulman E., Metz P.L., van Arkel G., te Lintel Hekkert B., Stiekema W.J. & Pereira A. (1999) A two-component enhancer-inhibitor transposon mutagenesis system for functional analysis of the *Arabidopsis* genome. *Plant Cell* **11**, 1853–1866.

Spradling A.C. & Rubin G.M. (1982) Transposition of cloned P elements into *Drosophila* germ line chromosomes. *Science* **218**, 341–347.

Spradling A.C., Stern D.M., Kiss I., Roote J., Laverty T. & Rubin G.M. (1995) Gene disruptions using P transposable elements: an integral component of the *Drosophila* genome project. *Proceedings of the National Academy of Sciences, USA* **92**, 10824–10830.

Spradling A.C. *et al.* (1999) The BDGP Gene Disruption Project: single P element insertions mutating 25% of vital *Drosophila* genes. *Genetics* **153**, 135–177.

Springer P.S. (2000) Gene traps: tools for plant development and genomics. *Plant Cell* **12**, 1007–1020.

Stanford W.L., Cohn J.B. & Cordes S.P. (2001) Gene-trap mutagenesis: past, present and beyond. *Nature Review Genetics* **2**, 756–768.

Staunton J.E. *et al.* (2001) Chemosensitivity prediction by transcriptional profiling. *Proceedings of the National Academy of Sciences, USA* **98**, 787–792.

Stebbins C.E. & Galan J.E. (2001) Structural mimicry in bacterial virulence. *Nature* **412**, 701–705.

Steemers F.J., Ferguson J.A. & Walt D.R. (2000) Screening unlabeled DNA targets with randomly ordered fiber-optic gene arrays. *Nature Biotechnology* **18**, 91–94.

Stein L. (2001) Genome annotation: from sequence to biology. *Nature Reviews, Genetics* **2**, 493–503.

Sternberg N. (1990) Bacteriophage P1 cloning system for the isolation, amplification and recovery of DNA fragments as large as 100 kilobase pairs. *Proceedings of the National Academy of Sciences, USA* **87**, 103–107.

Stevens R.C. (2000a) Design of high-throughput methods of protein production for structural biology. *Structure with Folding and Design* **8**, R177–R185.

Stevens R.C. (2000b) High-throughput protein crystallization. *Current Opinion in Structural Biology* **10**, 558–563.

Stover C.K. *et al.* (2000) Complete genome sequence of *Pseudomonas aeruginosa* PAO1, an opportunistic pathogen. *Nature* **406**, 959–964.

Streicher J., Donat M.A., Strauss B., Sporle R., Schughart K. & Muller G.B. (2000) Computer-based three-dimensional visualization of developmental gene expression. *Nature Genetics* **25**, 147–152.

Sutherland G.R. & Richards R.I. (1995) The molecular basis of fragile sites in human chromosomes. *Current Opinion in Genetics and Development* **5**, 323–327.

Suyama M. & Bork P. (2001) Evolution of prokaryotic gene order: genome rearrangements in closely related species. *Trends in Genetics* **17**, 10–13.

Suzuki H. *et al.* (2001) Protein–protein interaction panel using mouse full length cDNAs. *Genome Research* **11**, 1758–1765.

Swindells M., Orengo C., Jones D., Hutchinson E. & Thornton J. (1998) Contemporary approaches to protein structure classification. *Bioessays* **20**, 884–891.

Syed F. *et al.* (1999) CCR7 (EBI 1) receptor downregulation in asthma: differential gene expression in human CD4+ T lymphocyte. *Quarterly Journal of Medicine* **92**, 463–471.

Szustakowski J.D. & Weng Z. (2000) Protein structure alignment using a genetic algorithm. *Proteins* **38**, 428–440.

Szybalski W. (1997) RecA-mediated Achilles heel cleavage. *Current Opinion in Biotechnology* **8**, 75–81.

Tabor S. & Richardson C.C. (1995) A single residue in DNA polymerases of the *Escherichia coli* DNA polymerase I family is critical for distinguishing between deoxy- and dideoxyribonucleotides. *Proceedings of the National Academy of Sciences* **92**, 6339–6343.

Tao H., Bausch C., Richmond C., Blattner F.R. & Conway T. (1999) Functional genomics: expression analysis of *Escherichia coli* growing on minimal and rich media. *Journal of Bacteriology* **181**, 6425–6440.

Taton T.A., Mirkin C.A. & Letsinger R.L. (2000) Scanometric DNA array detection with nanoparticle probes. *Science* **289**, 1757–1760.

Tavernarakis N., Wang S.L., Dorovkov M., Ryazanov A. & Driscoll M. (2000) Heritable and inducible genetic interference by double-stranded RNA encoded by transgenes. *Nature Genetics* **24**, 180–183.

Teichmann S.A., Park J. & Chothia C. (1998) Structural assignments to the *Mycoplasma genitalium* proteins show extensive gene duplication and domain rearrangement. *Proceedings of the National Academy of Sciences, USA* **95**, 14658–14663. [URL: http://www.mrc–lmb.cam.ac.uk/genomes/MG_strucs.html]

Teichmann S.A., Chothia C. & Gerstein M. (1999) Advances in structural genomics. *Current Opinion in Structural Biology* **9**, 390–399.

Templin M.F., Stoll D., Schrenk M., Traub P.C., Vohringer C.F. & Joos T.O. (2002) Protein microarray technology. *Trends in Biotechnology* **20**, 160–166.

Terryn N., Rouze P. & Van Montagu M. (1999) Plant genomics. *FEBS Letters* **452**, 3–6.

Terwilliger T.C. (2000) Structural genomics in North America. *Nature Structural Biology* **7** (Suppl), 935–939.

Thiellement H. *et al.* (1999) Proteomics for genetical and physiological studies in plants. *Electrophoresis* **20**, 2013–2026.

Thierry A. & Dujon B. (1992) Nested chromosomal fragmentation in yeast using the meganuclease I-*Sce*I: a new method for physical mapping of eukaryotic genomes. *Nucleic Acids Research* **20**, 5625–5631.

Thompson A., Lucchini S. & Hinton J.C.D. (2001) It's easy to build your own microarrayer! *Trends in Microbiology* **9**, 154–156.

Thomson, G. (2001) Mapping of disease loci. *Pharmacogenomics* (eds Kalow, W. Meyer, U.A. & Tyndale, R.), pp. 337–361. Marcel Dekker, New York.

Tijssen P. (1993) *Hybridization with Nucleic Acid Probes. Part 1: Theory and Nucleic Acid Preparation.* Elsevier, Amsterdam.

Tingey S.V. & Del Tufo J.P. (1993) Genetic analysis with RAPD markers. *Plant Physiology* **101**, 349–352.

Tissier A. *et al.* (1999) Multiple independent defective *Suppressor–mutator* transposon insertions in *Arabidopsis*: a tool for functional genomics. *Plant Cell* **11**, 1841–52.

Tomkins J.P. *et al.* (2001) A marker-dense physical map of the *Bradyrhizobium japonicum* genome. *Genome Research* **11**, 1434–1440.

Tompa M. (2001) Identifying functional elements by comparative DNA sequence analysis. *Genome Research* **11**, 1143–1144.

Toth G., Gaspari Z. & Jurka J. (2000) Microsatellites in different eukaryotic genomes: survey and analysis. *Genome Research* **10**, 967–981.

Trask B. *et al.* (1992) Fluorescence *in situ* hybridization mapping of human chromosome 19: mapping and verification of cosmid contigs formed by random restriction fingerprinting. *Genomics* **14**, 162–167.

Trask B.J., Pinkel D. & Van den Engh G.J. (1989) The proximity of DNA sequences in interphase cell nuclei is correlated with genomic distance and permits ordering of cosmids spanning 250 kilobase pairs. *Genomics* **5**, 710–717.

Tucker C.L., Gera J.F. & Uetz P. (2001) Towards an understanding of complex protein networks. *Trends in Cell Biology* **11**, 102–106.

Twyman R.M. (2001) *Instant Notes Developmental Biology*, BIOS Scientific Publishers, Oxford UK.

Tyler-Smith C. & Willard H.F. (1993) Mammalian chromosome structure. *Current Opinion in Genetics and Development* **3**, 390–397.

Uberacher E. & Mural R. (1991) Locating protein-coding regions in human DNA sequences by a multiple sensor-neural network approach. *Proceedings of the National Academy of Sciences, USA* **88**, 11261–11265.

Uetz P. & Hughes R.E. (2000) Systematic and large-scale two-hybrid screens. *Current Opinion in Microbiology* **3**, 303–308.

Uetz P. *et al.* (2000) A comprehensive analysis of protein–protein interactions in *Saccharomyces cerevisiae*. *Nature* **403**, 623–627.

Umeda M. *et al.* (1994) Expressed sequence tags from cultured cells of rice (*Oryza sativa* L.) under stressed conditions: analysis of transcripts of genes engaged in ATP generating pathways. *Plant Molecular Biology* **25**, 469–478.

Vaguine A.A., Richelle J. & Wodak S.J. (1999) SFCHECK: a unified set of procedures for evaluating the quality of macromolecular structure–factor data and their agreement with the atomic model. *Acta Crystallographica* **55**, 191–205.

Valdes J.M., Tagle D.A. & Collins F.S. (1994) Island rescue PCR: a rapid and efficient method for isolating transcribed sequences from yeast artificial chromosomes. *Proceedings of the National Academy of Sciences, USA* **91**, 5377–5381.

Vasmatzis G. *et al.* (1998) Discovery of three genes specifically expressed in human prostate by expressed sequence tag database analysis. *Proceedings of the National Academy of Sciences, USA* **95**, 300–304.

Veculescu V.E., Zhang L., Vogelstein B. & Kinzler K.W. (1995) Serial analysis of gene expression. *Science* **270**, 484–487.

Veculescu V.E. *et al.* (1997) Characterization of the yeast transcriptome. *Cell* **88**, 243–251.

Veculescu V.E. *et al.* (1999) Analysis of human transcriptomes. *Nature Genetics* **23**, 387–388.

Venkatesh B., Gilligan P. & Brenner S. (2000) *Fugu*: a compact vertebrate reference genome. *FEBS Letters* **476**, 3–7.

Venter J.C., Smith H.O. & Hood L. (1996) A new strategy for genome sequencing. *Nature* **381**, 364–366.

Venter J.C. *et al.* (1998) Shotgun sequencing of the human genome. *Science* **280**, 1540–1542.

Venter J.C. *et al.* (2001) The sequence of the human genome. *Science* **291**, 1304–1351.

Vidal M., Brachmann R.K., Fattaey A., Harlow E. & Boeke J.D. (1996) Reverse two-hybrid and one-hybrid systems to detect dissociation of protein–protein and DNA–protein interactions. *Proceedings of the National Academy of Sciences, USA* **93**, 10315–10320.

Vivares C.P. & Metenier G. (2000) Towards the minimal eukaryotic parasitic genome. *Current Opinion in Microbiology* **3**, 463–467.

Vladutu C., McLaughlin J. & Phillips R.L. (1999) Fine mapping and characterization of linked quantitative trait loci involved in the transition of the maize apical meristem from vegetative to generative structures. *Genetics* **153**, 993–1007.

Volff J-N. & Altenbuchner J. (2000) A new beginning with new ends: linearisation of circular chromosomes during bacterial evolution. *FEMS Microbiology Letters* **186**, 143–150.

Voss H. *et al.* (1995) Efficient low redundancy large-scale sequencing at EMBL. *Journal of Biotechnology* **41**, 121–129.

Voytas D.F. (1996) Retroelements in genome organization. *Science* **274**, 737–738.

Walbot V. (2000) Saturation mutagenesis using maize transposons. *Current Opinion in Plant Biology* **3**, 103–107.

Walhout A.J. *et al.* (2000) Protein interaction mapping in *C. elegans* using proteins involved in vulval development. *Science* **287**, 116–122.

Walhout A.J.M. & Vidal M. (2001) Protein interaction maps for model organisms. *Nature Reviews of Molecular Cell Biology* **2**, 55–62.

Wallace A.C., Borkakoti N. & Thornton J.M. (1997) TESS: a geometric hashing algorithm for deriving 3D coordinate templates for searching structural databases: application to enzyme active sites. *Protein Science* **6**, 2308–2323.

Wallace B.A. & Janes R.W. (2001) Synchrotron radiation circular dichroism spectroscopy of proteins: secondary structure, fold recognition and structural genomics. *Current Opinion in Chemical Biology* **5**, 567–571.

Walsh M.A., Dementieva I., Evans G., Sanishvili R. & Joachimiak A. (1999) Taking MAD to the extreme: ultrafast protein structure determination. *Acta Crystallographica* **55**, 1168–1173.

Walter G. *et al.* (2000) Protein arrays for gene expression and molecular interaction screening. *Current Opinion in Microbiology* **3**, 298–302.

Walther D., Bartha G. & Morris M. (2001) Basecalling with LifeTrace. *Genome Research* **11**, 875–888.

Wang G-L., Holsten T.E., Song W-Y., Wang H-P. and Ronald P.C. (1995b) Construction of a rice bacterial artificial chromosome library and identification of clones linked to the Xa-21 disease resistance locus. *Plant Journal* **7**, 525–533.

Wang Y-K., Huff E.J. & Schwartz D.C. (1995a) Optical mapping of site-directed cleavages on single DNA molecules by the RecA-assisted restriction endonuclease technique. *Proceedings of the National Academy of Sciences, USA* **92**, 165–196.

Warren S.T. (1996) The expanding world of trinucleotide repeats. *Science* **271**, 1374–1375.

Washburn M.P., Wolters D. & Yates J.R. III (2001) Large-scale analysis of the yeast proteome by multidimensional protein identification technology. *Nature Biotechnology* **19**, 242–247.

Wasinger V.C. *et al.* (1995) Progress with gene product mapping of the Mollicutes: *Mycoplasma genitalium*. *Electrophoresis* **16**, 1090–1094.

Wassenaar T.M. & Gaastra W. (2001) Bacterial virulence: can we draw the line? *FEMS Microbiology Letters* **201**, 1–7.

Weber J.L. & Myers E.W. (1997) Human whole-genome shotgun sequencing. *Genome Research* **7**, 401–409.

Weigel D. *et al.* (2000) Activation tagging in *Arabidopsis*. *Plant Physiology* **122**, 1003–1013.

Weinberger S.R., Dalmasso E.A. & Fung E.T. (2001) Current achievements using ProteinChip Array technology. *Current Opinion in Chemical Biology* **6**, 86–91.

Whelan S., Lio P. & Goldman N. (2001) Molecular phylogenetics: state-of-the-art methods for looking into the past. *Trends in Genetics* **17**, 262–272.

White O. *et al.* (1999) Genome sequence of the radioresistant bacterium *Deinococcus radiodurans* R1. *Science* **286**, 1571–1577.

Wicks S.R. *et al.* (2001) Rapid gene mapping in *Caenorhabditis elegans* using a high density polymorphism map. *Nature Genetics* **28**, 160–164.

Wieschaus E., Nusslein-Volhard C. & Jurgens G. (1984) Mutations affecting the pattern of the larval cuticle in *Drosophila melanogaster*. III. Zygotic loci on the X-chromosome and fourth chromosome. *Wilhelm Roux Archives of Developmental Biology* **193**, 296–307.

Wigge P.A., Jensen O.N., Holmes S., Souès S., Mann M. & Kilmartin J.V. (1998) Analysis of the *Saccharomyces* spindle pole by matrix-assisted laser desorption/ionization (MALDI) mass spectrometry. *Journal of Cell Biology* **141**, 967–977.

Wiles M.V. *et al.* (2000) Establishment of a gene-trap sequence tag library to generate mutant mice from embryonic stem cells. *Nature Genetics* **24**, 13–14.

Wilkins Stevens P. *et al.* (2001) Analysis of single nucleotide polymorphisms with solid phase invasive cleavage reactions. *Nucleic Acids Research* **29**: E77.

Williams J.G.K., Kubelik A.R., Livak K.J., Rafalski J.A. & Tingey S.V. (1990) DNA polymorphisms amplified by arbitrary primers are useful as genetic markers. *Nucleic Acids Research* **18**, 6531–6535.

Williams S.M. & Robbins L.G. (1992) Molecular genetic analysis of *Drosophila* rDNA arrays. *Trends in Genetics* **8**, 335–340.

Wilson C.A., Kreychman J. & Gerstein M. (2000) Assessing annotation transfer for genomics: quantifying the relations between protein sequence, structure and function through traditional and probabilistic scores. *Journal of Molecular Biology* **297**, 233–249.

Wilson D.S. & Nock S. (2001) Functional protein microarrays. *Current Opinion in Chemical Biology* **6**, 81–85.

Wilson M. *et al.* (1999) Exploring drug-induced alterations in gene expression in *Mycobacterium tuberculosis* by microarray hybridization. *Proceedings of the National Academy of Sciences, USA* **96**, 12833–12838.

Winkler R.G., Frank M.R., Galbraith D.W., Feyereisen R. & Feldmann K.A. (1998) Systematic reverse genetics of transfer-DNA-tagged lines of *Arabidopsis*: isolation of mutations in the cytochrome P450 gene superfamily. *Plant Physiology* **118**, 743–749.

Winzeler E.A. *et al.* (1999) Functional analysis of the *S. cerevisiae* genome by gene deletion and parallel analysis. *Science* **285**, 901–906.

Wodicka L., Dong H., Mittmann M., Ho M-H. & Lockhart D.J. (1997) Genome-wide expression monitoring in *Saccharomyces cerevisiae*. *Nature Biotechnology* **15**, 1359–1367.

Wolf Y.I., Brenner S.E., Bash P.A. & Koonin E.V. (1999) Distribution of protein folds in the three superkingdoms of life. *Genome Research* 9, 17–26. [URL: ftp:// ftp.ncbi.nlm.nih.gov/pub/koonin/FOLDS/index.html]

Wolf Y.I., Grishin N.V. & Koonin E.V. (2000a) Estimating the number of protein folds and families from complete genome data. *Journal of Molecualr Biology* **299**, 897–905.

Wolf Y.I., Kondrashov A.S. & Koonin E.V. (2000b) Inter-kingdom gene fusions. *Genome Biology* **1**, research0013.1–13.13.

Wolf Y.I. *et al.* (2001) Genome alignment, evolution of prokaryotic genome organization, and prediction of gene function using genomic context. *Genome Research* **11**, 356–372.

Wood A.J. & Oliver M.J. (1999) Translational control in plant stress: the formation of messenger ribonucleoprotein particles (mRNPs) in response to desiccation of *Tortula ruralis* gametophytes. *Plant Journal* **18**, 359–370.

Wu N., Mo Y., Gao J. & Pai E.F. (2000) Electrostatic stress in catalysis: structure and mechanism of the enzyme orotidine monophosphate decarboxylase. *Proceedings of the National Academy of Sciences, USA* **97**, 2017–2022.

Wu R. & Taylor E. (1971) Nucleotide sequence analysis of DNA. II. Complete nucleotide sequence of the cohesive ends of bacteriophage λ DNA. *Journal Molecular Biology* **57**, 491–511.

Wurst W. *et al.* (1995) A large-scale gene-trap screen for insertional mutations in developmentally regulated genes in mice. *Genetics* **139**, 889–899.

Xenarios I. & Eisenberg D. (2001) Protein interaction databases. *Current Opinion in Biotechnology* **12**, 334–339.

Xenarios I., Rice D.W., Salwinski L., Baron M.K., Marcotte E.M. & Eisenberg D. (2000) DIP: the database of interacting proteins. *Nucleic Acids Research* **28**, 289–291.

Xenarios I. *et al.* (2001) DIP: the database of interacting proteins: update. *Nucleic Acids Research* **29**, 239–241.

Xiang C.C. & Chen Y. (2000) cDNA microarray technology and its applications. *Biotechnology Advances* **18**, 35–46.

Xu Y., Jablonsky M.J., Jackson P.L., Braun W. & Krishna N.R. (2001) Automatic assignment of NOESY cross peaks and determination of the protein structure of a new world scorpion neurotoxin using NOAH/DIAMOND. *Journal of Magnetic Resonance* **148**, 35–46.

Yamamoto K. & Sasaki T. (1997) Large-scale EST sequencing in rice. *Plant Molecular Biology* **35**, 135–144.

Yamamoto T., Lin H.X., Sasaki T. & Yano M. (2000) Identification of heading date quantitative trait locus *Hd6* and characterization of its epistatic interactions with *Hd2* in rice using advanced backcross progeny. *Genetics* **154**, 885–891.

Yano M. *et al.* (2000) *Hd1*, a major photoperiod sensitivity quantitative trait locus in rice, is closely related to the *Arabidopsis* flowering time gene *CONSTANS*. *Plant Cell* **12**, 2473–2484.

Yates J.R. III (2000) Mass spectrometry from genomics to proteomics. *Trends in Genetics* **16**, 5–8.

Yates J.R. III, Speicher S., Griffin P.R. & Hunkapiller T. (1993) Peptide mass maps: a highly informative approach to protein identification. *Analytical Biochemistry* **214**, 397–408.

Ye S.Q., Zhang L.Q., Zheng F., Virgil D. & Kwiterovich P.O. (2000) MiniSAGE: Gene expression profiling using serial analysis of gene expression from 1 μg total RNA. *Analytical Biochemistry* **287**, 144–152.

Yee A., Booth V., Dharamsi A., Engel A., Edwards A.M. & Arrowsmith C.H. (2000) Solution structure of the RNA polymerase subunit RPB5 from *Methanobacterium thermoautotrophicum*. *Proceedings of the National Academy of Sciences, USA* **97**, 6311–6315.

Yeh R.F., Lim L. & Burge C.B. (2001) Computational inference of homologous gene structures in the human genome. *Genome Research* **11**, 803–816.

Yershov K. *et al.* (1996) DNA analysis and diagnostics on oligonucleotide microchips. *Proceedings of the National Academy of Sciences, USA* **93**, 4913–4918.

Yokota H., Van Den Engh G., Hearst J.E., Sachs R.K. & Trask B.J. (1995) Evidence of the organization of chromatin in megabase pair-sized loops arranged along a random walk path in the human G0/G1 interphase nucleus. *Journal of Cell Biology* **130**, 1239–1249.

Yokota H. *et al.* (1997) A new method for straightening DNA molecules for optical restriction mapping. *Nucleic Acids Research* **25**, 1064–1070.

You Y. *et al.* (1997) Chromosomal deletion complexes in mice by radiation of embryonic stem cells. *Nature Genetics* **15**, 285–288.

Young R.A. (2000) Biomedical discovery with DNA arrays. *Cell* **102**, 9–15.

Zachariae W., Shin T.H., Galova M., Obermaier B. & Nasmyth K. (1996) Identification of subunits of the anaphase-promoting complex of *Saccharomyces cerevisiae*. *Science* **274**, 1201–1204.

Zambrowicz B.P., Friedrich G.A., Buxton E.C., Lilleberg S.L., Person C. & Sands A.T. (1998) Disruption and sequence identification of 2000 genes in mouse embryonic stem cells. *Nature* **392**, 608–611.

Zhang C. & DeLisi C. (1998) Estimating the number of protein folds. *Journal of Molecular Biology* **284**, 1301–1305.

Zhang L. *et al.* (1997) Gene expression profiles in normal and cancer cells. *Science* **276**, 1268–1272.

Zhang P. *et al.* (1994) An algorithm based on graph theory for the assembly of contigs in physical mapping of DNA. *CABIOS* **10**, 309–317.

Zhang X. & Smith T.F. (1997) The challenges of genome sequence annotation or 'the Devil is in the Details'. *Nature Biotechnology* **15**, 1222–1223.

Zhao N. *et al.* (1995) High-density cDNA filter analysis: a novel approach for large scale quantitative analysis of gene expression. *Gene* **156**, 207–213.

Zhao S. *et al.* (2000) Human BAC ends quality assessment and sequence analyses. *Genomics* **63**, 321–332.

Zhou H., Roy S., Schulman H. & Natan M.J. (2001) Solution and chip arrays in protein profiling. *Trends Biotechnol* **19** (Suppl., *A Trends Guide to Proteomics II*), S34–S39.

Zhu H. & Snyder M. (2001) Protein arrays and microarrays. *Current Opinion in Chemical Biology* **5**, 40–45.

Zhu H., Cong J.P., Mamtora G., Gingeras T. & Shenk T. (1998) Cellular gene expression altered by human cytomegalovirus: global monitoring with oligonucleotide arrays. *Proceedings of the National Academy of Sciences, USA* **95**, 14470–14475.

Zhu H. *et al.* (2000) Analysis of yeast protein kinases using protein chips. *Nature Genetics* **26**, 283–289.

Zhu H. *et al.* (2001) Global analysis of protein activities using proteome chips. *Science* **293**, 2101–2105.

Zhu T. *et al.* (2001) Toward elucidating the global gene expression patterns of developing *Arabidopsis*: parallel analysis of 8 300 genes by a high-density oligonucleotide probe array. *Plant Physiology and Biochemistry* **39**, 221–242.

Zoubak S., Clay O. & Bernardi G. (1996) The gene distribution of the human genome. *Gene* **174**, 95–102.

Zwaal R.R., Broeks A., Meurs J.V., Groenen J.T. & Plasterk R.H. (1993) Target-selected gene inactivation in *Caenorhabditis elegans* by using a frozen transposon insertion mutant bank. *Proceedings of the National Academy of Sciences, USA* **90**, 7431–7435.

Index